Microbiology Research Advances

nova
science publishers
www.novapublishers.com

Microbiology Research Advances

Endotoxins and their Importance
Arif Pandit, PhD (Editor)
R. S. Sethi, PhD (Editor)
2022. ISBN: 978-1-68507-839-3 (Softcover)
2022. ISBN: 978-1-68507-913-0 (eBook)

Understanding Antibiofilm Activity
Iram Liaqat, PhD (Editor)
2022. ISBN: 978-1-68507-927-7 (Hardcover)
2022. ISBN: 979-8-88697-005-0 (eBook)

Bacteriophages: Interaction, Diversity and Applications
Prasanth Manohar, PhD (Editor)
2022. ISBN: 978-1-68507-860-7 (Hardcover)
2022. ISBN: 978-1-68507-958-1 (eBook)

Endophytic Fungi: Biodiversity, Antimicrobial Activity and Ecological Implications
Qiang-Sheng Wu, PhD (Editor)
Ying-Ning Zou (Editor)
Yong-Jie Xu, PhD (Editor)
2021. ISBN: 978-1-68507-354-1 (Softcover)
2021. ISBN: 978-1-68507-442-5 (eBook)

Pathogenic Bacteria: Pathogenesis, Virulence Factors and Antibacterial Treatment Strategies
Keith D. Watts (Editor)
2021. ISBN: 978-1-68507-422-7 (Softcover)
2021. ISBN: 978-1-68507-424-1 (eBook)

More information about this series can be found at
https://novapublishers.com/product-category/series/microbiology-research-advances/

Pragya Tiwari
Editor

Endophytes: Types, Potential Uses and Mechanism(s) of Action

nova
science publishers
www.novapublishers.com

DOI: https://doi.org/10.52305/JETR3242

Library of Congress Cataloging-in-Publication Data

ISBN: 979-8-88697-045-6

Published by Nova Science Publishers, Inc. † New York

Contents

Foreword

Dr. Debdulal Banerjee, Professor
Department of Botany and Forestry,
Vidyasagar University, Midnapore- 721102
West Bengal, India
email: debu33@gmail.com

Endophytes are the unique entities of nature and are considered the hidden microorganisms that reside within the host plant either asymptomatically or in the form of a latent pathogen. They are involved in a symbiotic relationship with their host plant and protect the plant from biotic or abiotic stress situations. The symbiosis has led to horizontal gene sharing from plant to endophyte and this is the sole reason why endophytes are continuously explored for isolation of effective metabolomes. Production of taxol from the endophyte *Taxomyces andreanae* isolated from *Taxus baccata* is the first remarkable contribution of endophytes towards mankind. So, in one way the endangered/endemic plants are protected and on the other hand, fungal transcriptome can be manipulated for production enhancement or product modification-related opportunities. Endophyte biology deals with the diverse multi-domain application of a wide range of endophytic microbiota covering the species of fungi, bacteria, and actinobacteria. De Bary first coined the term endophyte and since then endophytes have been a marvelous arena of basic and applied scientific research. Their biotechnological importance, pharmacological acceptance, and agricultural utility totally depend on their unique diversity of bioactive metabolomes. In the 21st century, it is high time to re-visualize our hunt for therapeutic agents from sustainable novel biological sources and endophytes are the wisest option in that field. Endophytic products are generally non-toxic, environment friendly, and possess antimicrobial, antioxidative, anticancer, anti-tumor, anti-proliferative, cytotoxic actions, etc., and other broad-spectrum pharmaco-medical utilities. Endophytes or endophytic metabolomes both are effective in agricultural sectors too and are effective in managing cases of post-harvest disease in stocked agricultural products, controlling fungal or bacterial infections in standing crops, etc. As a whole, endophytes are the giants of modern agricultural or pharmaceutical research which addresses solutions to a wide range of problems by following the paths of the deep ecological movement and pampers the idea of sustainable agricultural practices. In recent times, especially for the past 15-20 years of worldwide research, scientists are expressing their eagerness toward the "endophytes" and these microorganisms are getting a lofty stature in the emerging topics of science and technology. High-end instruments are getting involved, unique methods are being generated, patents are being grabbed, national and international bodies are sanctioning funding, reputed publishing houses are expressing their interest, young and veteran minds are being engaged, and as a whole, a revolutionary shift is being noticed towards the domain of endophyte-related natural product research. So, the approach taken by the respected publishing authorities and the emerging scientist Dr. Pragya Tiwari is a very novel and well-appreciated one.

The book entitled *Endophytes: Types, Potential Uses and Mechanism of Action* contains the most emerging basic and applied topics in the world of endophyte biology. The readers will be enlightened with popular angles of present-day research along with next-generation concepts of dealing with the products and isolation of endophytic (both fungi and actinobacteria/ actinomycetes) microorganisms. The book is a balanced recipe of valuable topics related to the diversity of endophytes, their high potential utilities, and also mechanistic approaches towards the solution of modern-day problems. One will get a glimpse of a variety of subjects ranging from the very founding principles to high-demanding bombastic concepts in this arena. The authors have contributed with the most necessary topics covering the use of in-silico approach/ bioinformatics-related computational biology for enriching the database of endophyte biology research and finding solutions based on AI (artificial intelligence) system. Also, the book sheds light on the multi-omics tools used for increasing the utilities of agriculturally valuable endophytes. On the other hand, the methods of synthesis of bioactive nano-formulations and their vigorous utility are well documented here, along with a brief description of product/metabolite optimization and enhancement from the endophyte source. The concepts of environmental-agricultural sustainability are also covered and topics related to finding green solutions for plant growth promotion are also nurtured. Molecular mechanisms towards tolerance to hard environmental stresses along with some landmark cases of plant-associated endophytes are entailed here. The book is a package of endophyte biology serving all the dishes of this arena to cherish and will definitely act as a knowledge pool for readers of all ages and experiences. Researchers can benefit their scientific or industrial/ academic carrier through the contents and surely one will bend his/her mind toward endophyte biology after the diffusion of the concepts.

Preface

In recent times, plant-microbe associations have gained special attention from the scientific community and highlight a key area of research. An inexhaustible reservoir of natural products, biological species, and their associations have been widely investigated for new metabolites of natural origin and their multi-faceted significance in the environment and socio-economic application. Although, plants represent a key source of natural products, however, threatened habitats and overharvesting have contributed to declining natural resources. Presently, plant-microbe associations (particularly endophytes) have been extensively studied, showing a positive impact on plant growth promotion, nutrient cycling, and maintenance of ecological balance. Endophytes are ubiquitously present in diverse plant species, demonstrating a co-evolution with the plant host during the course of evolution. Furthermore, different classes of microbial species, including bacterial, fungi, and actinomycetes colonize the internal plant tissues, leading to a wide range of implications. It is now possible to perform the accurate identification of these microorganisms and learn about their interactions with the host and other microbes thanks to the progress of sophisticated tools and methods in modern biology. There has been considerable interest in screening the endophyte for bioactive compounds. A large number of compounds (secondary metabolites) have been extracted, isolated, and characterized from endophyte microbes. Several cultivation-independent tests, as well as fluorescence in situ hybridization-confocal laser scanning microscopy experiments, have recently established the presence of endophytes in host plants. Cultivation-based procedures rely on isolating and testing isolates, whereas cultivation-independent techniques look for differences in the overall endophytic community. Endophytes may be isolated from any microbial or plant growth medium, including agar, potato dextrose agar, and any nitrogen- or carbon-containing media. Isolation from surface-sterilized host plant tissue is the most common approach for detecting and counting endophytes.

The medicinal plants associated with endophytes have been extensively studied as a potential source of bioactive compounds, with pharmacological attributes and biotechnological applications. Scientific breakthroughs in whole genome sequencing and annotation of endophyte-associated gene/metabolic pathways provided key insights into functional mechanisms and their dynamics within plants. Studies have highlighted the socio-economic implications of fungal endophytes in the environment, agriculture, and medicine with considerable success. Although extensively studied as a "production platform" of novel pharmacological metabolites, the molecular mechanisms of plant-endophyte dynamics remain less understood/explored for their efficient utilization in drug discovery and other biotechnological applications. The book entitled Endophytes: Types, Potential Uses and Mechanism(s) of Action by Nova Publishers, USA aims to discuss and provide a comprehensive insight/knowledge about the multi-faceted applications of Endophytes in the socio-economic interface. This book includes 12 well-written and comprehensive chapters from authors working in different areas of biological sciences from South Africa, Taiwan, South Korea, and different regions of the Indian sub-continent. The multi-faceted expertise of distinguished contributors and their excellent chapters in both basic and applied areas of endophyte biology and biotechnology provide substantial knowledge on the emerging importance of plant-microbe interactions (endophytes) in the present era.

Chapter 1 provides an overview of endophytes and their emerging importance in the socio-economic context.

Chapter 2 discusses in detail the diverse molecular and biochemical approaches employed for the isolation and characterization of endophytes.

Chapter 3 provides insights into the bacterial and fungal endophytes colonizing medicinal plants and their bio-prospection for pharmacologically important bioactive metabolites.

Chapter 4 explores the role of multifaceted endophytes in enhancing plant growth and tolerance toward environmental stress, acting as a bio-resource to counter environmental challenges for the plants.

Chapter 5 provides some significant case studies on plant-associated endophytes and their dynamics with a specific discussion on endophytes from Phragmites australis.

Chapter 6 highlights a lesser-known group of endophytes, actinomycetes, and their increasing importance in endophyte biology and research.

Chapter 7 discusses how the yield improvement strategies in endophytes aim to enhance the production of high-value metabolites, a potential biotechnological application.

Chapter 8 provides an insight into the synthesis and applications of nanoparticles from bacterial and fungal endophytes, an emerging area in biological research.

Chapter 9 explores how different tools and resources in computational biology have substantially contributed to endophyte biology and research with significant success.

Chapter 10 discusses the recent insights and progress in applications of multi-omics approaches for the identification of endophytes in agriculture.

Chapter 11 extensively discusses the increased exploration and utilization of plant-associated endophytes (as bio formulations) for sustainable agriculture.

Chapter 12- discusses the emerging popularity and utilization of endophytes in multi-faceted socio-economic applications, which necessitate exploring the recent trends and prospects in endophyte biology and research.

The emerging importance of endophytes in multi-faceted applications, ranging from environmental bioremediation and biotic/abiotic stress tolerance to the production of high-value metabolites in pharmacological applications and from promoting plant growth and sustainable agriculture to the application of nanoparticles synthesized from endophytes forms the key component of the book. The book on endophytes would be of special relevance to the scientific community actively working on plant-microbe interactions, academicians, and students related/working in fields of biotechnology, microbiology, environmental science, and related areas.

Pragya Tiwari

Chapter 1

Endophytes and Their Emerging Importance in the Environment

R. O. Anyasi[1,*] and H. I. Atagana[2]
[1]Department of Environmental Sciences, University of South Africa, Johannesburg, South Africa.
[2]Institute for Nanotechnology and Water Sustainability, University of South Africa, Johannesburg, South Africa

Abstract

Endophytes are ubiquitous in internal tissues of plants. Endophytic relationships between plants and microorganisms have found their use in various biological applications among which is in the detoxification of several toxins facilitated by various microbial endophytes present in plants growing in contaminated soil. Plants in their natural state serve as hosts to endophytic microorganisms with which they form symbiotic associations. The benefits that the endophytes offer to the plants include amongst others enhancement of plants growth through the production of phytohormones; increase in the resistance of environmental stresses; production of important bioactive metabolites; help in the fixing of nitrogen in the plant's organelles; help in the metal translocation and accumulation in plants; assist in the production of enzymes involves in the degradation of organic contaminants. Therefore, recognizing these natural processes of the microbes will enable the understanding of effective mechanism for enhanced biological activities. This chapter surveys the progressive understanding of endophytes, its applications, their mechanism of operation as well as the type of interaction between them and the plants they inhabit.

*Corresponding Author's Email: ghnlwl@yahoo.com.

In: Endophytes
Editor: Pragya Tiwari
ISBN: 979-8-88697-045-6

Keywords: detoxifications, endophytes, microbes, phytoremediation, plant-microbes interactions, pollution

1. Introduction

Plants in nature are commonly in symbiotic relationship with endophytic organisms (Strobel and Daisy 2003). This group of highly diverse microbes greatly impacts plant communities by enhancing their adaptation and fitness in the environment through improvement of their abiotic and biotic stress tolerance, increasing biomass, decreasing water consumption and/or decreasing of fitness by altering the allocation of resources from soil to the plants (Rodriguez et al., 2009). It is important to note that although there has been increased interest and significant research reports in this field of study, the ecological significance of endophytes and their interactions with plants is yet to be fully realised.

Endophytes are microorganisms that inhabit plant tissues partly or in all of their lifecycle. They may be useful and or harmful to the plant host depending on the kind of interaction they have with their host plant (Strobel and Daisy 2003, Anyasi and Atagana, 2019). Although endophytes are only recently gaining recognition, most of the organisms studied for example *mycorrhizal* fungi and *rhizobia*, which are regarded as beneficial endophytes, and *Fusarium sp.*, which are also considered as endophytes are regarded as detrimental endophytes, because they are the causative agents of wilt disease in banana plants (Felde, 2011). Endophytes are, however, not directly detrimental to plants (Strobel and Daisy 2003). The focus of most studies on endophytes has been on exploring the physiology of mutualistic endophytes, as has been noted in some studies that reported the ability of endophytes to activate their host plant's existing defence mechanisms before or after pest and pathogens are supplied into the system (Strobel and Daisy 2003; Rodriguez et al., 2009; Sikora et al., 2010, Raymond Anyasi and Anyasi, 2020). Earlier research on endophytes focused on important properties such as antimicrobial, antioxidant, or antitumor properties (Palem et al., 2015; Pan et al., 2017; Wang et al., 2014) that were useful to the agricultural, medicinal, and pharmaceutical industry. However, recent studies have been diversified to other biotechnological functions of endophytes such as application in biosynthesis, biotransformation, biocatalysis and biodegradation of natural and xenobiotic compounds in the environment (Ting et al., 2021).

From the early definitions of symbiosis as 'the living together of dissimilar organisms' (De Bary, 1866), an array of symbiotic lifestyles have been defined based on fitness benefits or impacts to macroscopic hosts and microscopic symbionts (Hirosch and Murakanmi, 2011). Collectively, literature have reported that most, if not all plants in their natural ecosystems are symbiotic with microbial endophytes (Strobel and Daisy 2003; Petrini, 1992). These microbial symbionts can have profound effects on plant ecology, fitness, and evolution (Hardoim et al., 2015), shaping plant communities (Kozyrovska, 2013), and manifesting strong effects on the biological activities of the plants (Felde, 2011). Plants are associated with bacterial endophytes and mycorrhizal fungi for over 400 million years according to fossil record. This relationship starts the moment a land is colonized by plants, thus playing a long and important role in driving the evolution of life on land (Rodriguez et al., 2009; Krings et al., 2007; Garcia and Zimmermann, 2014). Unlike mycorrhizal fungi that colonize plant roots and grow into the rhizosphere, endophytes reside entirely within plant tissues and may grow within roots, stems and/or leaves, emerging to sporulate at plant or host-tissue senescence (Gazis et al., 2016; Xu et al., 2014; Suryanarayanan et al., 2009). These bio-activities were made known by the mechanisms that resulted in the evolution of endophytism and its genetic basis, although much understanding has not been garnered (Xu et al., 2014).

Endophytes are known to colonize ecological niches that are similar to that of phytopathogens, which present them as suitable biocontrol agents (Ryan et al., 2008). According to literature, endophytes possess the ability to control the pathogens of plant, insects and nematodes (Azevedo et al., 2000). Some can help to accelerate seedling emergence, promote plant establishment in adverse conditions and also enhance plant growth (Felde, 2011). Other classes of endophytes have been shown to be involved in synthesis of novel compounds and antifungal metabolites (Ryan et al., 2008). These biodiversity of strains of endophytes used for novel metabolites may assist in the identification of new drugs that could help in the effective treatment of diseases for humans, plants and animals (Strobel and Daisy 2003; Afzal et al., 2014; Aly et al., 2013). In addition to the production of novel chemicals, some other endophytes are implicated in their natural capacity for xenobiotic degradation, which may also act as vectors that introduces traits for degradation purposes (Stepniewska and Kuzniar, 2013). The benefits of endophytes are drawn from their exposition to diverse compounds in their interaction in plant-soil niche. At the moment, many investigations are underway as a way of improving the biocontrol capabilities of endophytes

(Andreolli et al., 2013; Doty, 2008; Fadiji and Babalola, 2020; Germaine et al., 2009; Khan and Doty, 2011; Ryan et al., 2008; Ting et al., 2021). This chapter aims to provide an insight to the study of endophytes and the level of their biological application as novel microorganisms.

2. Endophytism Explained

Dating back to the 18th century, series of debates have taken place on what endophytes are and how they can be described, based on the notion put forward by De Barry (1866). However, Petrini (1992) in his report, concluded that endophyte are those microorganisms that live in the plant organelles without causing diseases, which was widely accepted. Other schools of thought in trying to put meaning to the understanding of what an endophyte is, described the phenomenon as when an organism colonize the internal tissue of a plant during its life-cycle for mutualistic, commensalistic or sometimes parasitic relationship to the host (Backmann and Sikora, 2008). In other word, endophyte being an ecological concept is a plant-micro relationship or system which can thoroughly be explained by the understanding of the factors it entails. Endophytes exists in all plant, hence are described to be extensive in nature (Yu et al., 2010; Gao et al., 2015). However, literature reports estimate the existence of endophytes in microtubule of plant cells as well as in the intercellular spaces to several millionth species (Winther and Friedman, 2008). Various studies have provided information on different ways that endophytes could be isolated from variety of plant species as such the organisms are found distributed in several plant cells: roots, stems, leaves, fruits and seeds (Anyasi and Atagana, 2019). The roots of plant have the highest density of endophytes, which may reach 106 Colony Forming Units per gram (CFU/g) (Hardoim et al., 2008).

Clearly the type and numbers of the endophytes, as well as their rate of growth in the plant tissue is determined by the climatic and health condition of the plant (Petrini, 1992). Although, different techniques exist for identification of endophytes, the sources of endophytes are still under debate owing to the diversity of host's environmental conditions and the complexity of the relationship between them (Wen et al., 2004). This helps to form the two assumptions that propose the endogenous and exogenous system. Proponents of the endogenous system believes that endophytes evolve from the internal tissues of the plants - the mitochondrion and chloroplast of the plant cells, as they have similar genetic characters with the host. While the proponents of the exogenous system on the other hand are of the opinion that endophytes emanates from outside the plant tissue and enters the host through its surfaces (Zhang et al., 2012). With the help of biomarkers, the key tools used for studying ecological processes, external colonization of plants by endophytes has been established (Ryan et al., 2008).

3. Isolated Microbial Endophytes

Based on the organism's classification protocols, endophytes are classified into bacterial, fungi and actinomyces or algae (Tan and Zou, 2001). Following literature reports, the first endophytic fungus (*Taxomyces andreanae*) was first isolated by a research study in 1993 (Aly et al., 2013) using *Taxus brevifolia*, a medicinal plant (Winther and Friedman, 2008). Forty-two endophytic fungal strains isolated from *Nerium oleander L* were reported by Huang et al. (2007). Endophytic fungi abound in every ecosystem, but they seem to be more abundant in tropical ecosystems, because of their high biodiversity (Tenberge, 1999). *Ulvella leptochaete* (Figure 1), an alga was recently discovered growing as an endophyte in a range of seaweeds in India as an alga, establishing the abundance of these organisms in the environment. The ecological advantage of this alga to the host plant is that it enhances the host's disease resistance (Bast et al., 2012) by secreting anticancer metabolites used in many cancer treatments (Rodriguez et al., 2009). Bacterial endophytes are another class of endophytes with a broad range of taxa, which include *α, β*, and *γ-Proteobacteria, Firmicutes, Actinibacteria* etc. These classes of endophytes are reported to be intracellular in roots and shoot cells of a variety of plants and are found to gain entrance through the cells of the meristem (Zhang et al., 2012). Table 1 presented some of the endophytes that have been isolated from different plants and used in different studies.

According to the literature, different endophytic bacteria have been isolated from a list of over 100 selected crop plants (Anyasi and Atagana, 2019; Raymond Anyasi and Anyasi, 2020). The species includes about 55 genera of both gram-positive and gram-negative bacteria. The bacterial taxa isolated were mostly the *Pseudomonads* and the *Enterobacteriaceae* (Bast et al., 2012). Li et al. (2008) reported 98 non-symbiotic endophytic bacterial strains isolated from a study involving 150 root nodules of soybean. Colonization by endophytic bacteria was also found in plants growing in toxic conditions. This was exemplified by the isolation of bacterial endophytes of various species resistant to extreme pH conditions in the study of Kan and his group (Kan et al., 2007).

Table 1. List of endophytes isolated from different plants tissues

S/No	Endophytes	Plant	Plant tissue	Source
1	*Staphylococcus, Microbacterium, Pseudomonas, Curtobacterium, Bscillus, Arthrobacter, Paenibacillus, Leifsonia*	*Alyssum bertoloni*	Leaves, stems, roots	Barzanti et al. 2007
2	*Bacillus sp.*	*Alnus firma*	Roots	Shin et al. 2011
3	*Pseudomonas fluorescens, Microbacterium sp.*	*Brassica napus*	Roots	Sheng et al. 2008
4	*Enterobacter sp., Xanthomonodaceae, Pseudomonas sp., Pseudomonas fulva, Stenotrophomonas sp., Clostridium aminovalericum, Sanguibacter sp*	*Nicotiana tabacum*	Seed	Mastreta et al. 2009
5	*Microbacterium, Bacillus, Arthrobacter, Flavobacterium, Chryseobacterium, Agrobacterium, Sphingomonas, Pseudomonas, Serratia, Curtobacterium*	*Solanum nigrum*	Leaves, stems, leaves	Luo et al. 2011
6	*Phoma sp., Alternaria Sp, Peyronellaea sp., Steganosporium sp., Aspergillus sp.**	*Arabis hirsute, Acacia decurrens, Symplocos paniculata.*	Stems, leaves	Li et al. 2011
7	*Methylobacterium oryzae, Burkholderia sp*	*Rice*	NM	Madhaiyan et al. 2007
8	*Microsphaeropsis sp.**	*Solanum nigrum*	NM	Xiao et al. 2010
9	*Pseudomonas putida*	*Papulus trichocarpa*	Stem	Germaine et al. 2009
10	*Burkholderia cepacia*	*Zea mays*	NM	Wang et al. 2010
11	*Pseudomonas putida*	Poplar	NM	Weyens et al. 2010
12	*Ulvella Leptochaete***	Green algae: *Cladophora glomerata*, red algae: *Laurencia obtusa*	NM	Bast et al. 2014

NM: Not mentioned, *: Fungi, **: Algae.

Figure 1. Micrograph image of Green Microalgae (Ulvella Leptochaete) (Bast et al. 2012).

Streptomyces known to colonize both the underground and the above ground tissues of plant with greater incident at the underground tissues have been reported as the most isolated actinomyces (Olanrewaju and Babalola, 2019; Wang et al., 2009; Zappelini et al., 2018). Medicinal plants are important sources of this endophytic actinomyces as such plants are known to induce production of secondary metabolites, an important factor for biological activity in drug-resistant according to some studies (Castillo et al., 2002; Yu et al., 2010). Although, endophytes have been used in pollution control and management for a long time (Ryan et al., 2008), there seem to be relatively limited research in its exploitation in this area unlike its application in the medical sciences (Bezerra et al., 2015). The reason behind this is in the difficulty in understanding the mechanism of action of the endophytes in the environmental medium. This can only be unravelled through further practical applications (Stepniewska and Kuzniar, 2013).

4. Distribution of Endophytes in Plants

Endophytes are usually found in the intercellular and the vascular tissues within the plants (Ryan et al., 2008; Afzal et al., 2014; Glick and Stearns, 2011). The state of a plant's tissue is the determining factor for successful colonization by endophytes (Yrjälä et al., 2010). Li et al. (2012) introduced the concept of compartmentalization when they fail to obtain the same isolates in adjacent plant compartment, even when the plants were growing in the same soil type. The study discovered that only *Pseudomanas sp.* could be identified in all the parts of the plants (roots, stem and leaves tissues), while *Bacillus* and *Arthrobacter* were only found in the roots and stem tissues. *Pseudomonas* was reported as the most abundant genus that was recovered from the plants rhizosphere and this was attributed to the rhizodeposition of organic acids and aromatics such as phenols in the rhizosphere (Teghavi et al., 2011). The reason for the predominance of *Pseudomonas sp.* could be based on the ability of the bacteria to breakdown organic compounds which they use as a source of food and energy. This abundance of *Pseudominas sp.* is evident in Table 1.

There are other studies that reported a less compartmentalised isolates. For example, the study to find the incidences of microbes in a hybrid aspen plant, *Burkholderia*, *Methylobacterium* and other unidentified bacteria were cultured from every parts of the plant (Yrjälä et al., 2010). *Arthrobacter*, *Pseudomonas* and another *Microbacteriaceae* were found in the root and leaf, while *Microbacteriaceae*, *Sphingomonas*, *Aerolata* and *Sphingomonas auranica* which are airborne related bacteria were isolated from the stem and leaves (Li et al., 2012). *Sphingomonas* were more in number in the above-ground portion of the plant as reported in this study, establishing its ability to withstand ultraviolet radiation. This explains the condition of life of the organism in that particular environment. Other schools of thought have also provided information on plant microbes that are often pigmented to enhance their condition on leaf (Weyens et al., 2013; Kim et al., 2012). The growing shortfalls in compartmentalization of isolated endophytes in plants may have been due to the growth media used and perhaps the differences in plants physiology. This should however attract further studies to truly ascertain its cause (Stepniewska and Kuzniar, 2013).

There have been conflicts of reports on the frequency of colonization of endophytic fungi in trees. While most studies indicated highest colonization in the stem, others reported on the leaves (Castillo et al., 2002). However, Petrini et al. (1992) cleared the misperception reporting that different plant tissues and organs could represent distinct microhabitat. The factors could enable the determination of the general frequency of endophytes colonization, there include host-plant age, associated vegetation, elevation of the vegetation and exposure to the environment (Kusari et al., 2013). This explains that the existence of endophytes could be influenced by seasonal, environmental and host tissue's variations.

5. Isolation and Colonization

Methods of isolation of bacterial endophytes are constantly being reviewed (Ryan et al., 2008). The use of sodium hypochlorite in the disinfection of plant surfaces has been a common practice. But most of the residual sodium hypochlorite often affects the growth and induces mutagenesis as well as death of the microorganisms, thereby necessitating rinsing with a separate sodium thiosulphate to remove all the residual hypochlorite (Dutta et al., 2014; Tanaka et al., 2016).

From the literature, dilution-plate method of isolation is the widely used method because there is no complication in its operation. This method entails a stepwise process which must be followed thoroughly. The protocols to be followed include collection and washing of samples, followed by surface-sterilization with disinfectant and finally the incubation in petri dish with growth medium (Xu et al., 2012). If the sterilization time goes for too long, the plant cell would be broken so that no endophytes will be isolated (Gao et al., 2015). Surface-sterilizing compounds used include: $HgCl_2$, ethanol, H_2O_2, NaOCl and DMSO. It is usually important to do a confirmation test so as to establish the success of surface-sterilization. This is done by testing for growth of microorganisms from the last rinse of distilled water. The plant is considered to be clean when no bacterial growth is found. Depending on the type of endophytic microbe in question, it most often requires the addition of fungicides to prevent extraneous organisms like fungi from covering the colony (Xu et al., 2012; Sun et al., 2014). Endophyte isolated from one plant could equally be introduced successfully into the tissues of other plants by inoculation. This was successfully demonstrated by Lopez-Fernandez et al. (2015) and liu et al. (2014) in their study using bacteria endophytes. It was also done using endophytic *fungic Acremonium* of F. *arundinacea*/*L. perenne*, and *M. acuminata* to establish their resistance in plants (Xu et al., 2012). With the use of confocal electron microscope to observe an Excalibur plant: presence of *Streptomyces sp.* could indicate the presence of green fluorescent protein infected in the seed of the plant. This is an evidence of pure colonization of endophytes (Peng et al., 2013). Biomarker system of Autofluorescent protein (AFP), known as Green Fluorescent Protein (GFP) is currently being used to establish colonization of endophytes (Ryan et al., 2008).

6. Plant-Endophyte Relationship

Although, most organic pollutants are toxic to plants and microorganisms, several plants that are found growing in polluted soil tend to host a large number of bacteria that show ability to degrade such pollutant (Afzal et al.,

2014). Endophytes, however, are known to exhibit their symbiotic responsibilities to their hosts in most of the occasions by different physiological interactions between them and the host plants such as the supply of nutrients to the endophyte by the plant and the microbe in turn releases active metabolites as a result of their metabolic and physiological activities (Ryan et al., 2008; Zhang et al., 2012). The released active metabolites help improve plants root development and nutrient uptake (Li et al., 2008). The actions of the endophyte further confers protection to the plant in the form of reducing desiccation, preventing insect and parasitic fungal attacks (Dutta et al., 2014). These symbiotic activities enable endophytes to promote host plant growth and the supply of resistance mechanisms as explained in Figure 2. A wide range of relationships exist between endophytes and plants and according to reports, such relationship may change from mutualistic to parasitic depending of environmental and other prevailing conditions (Strobel and Daisy 2003; Johnston-Monje and Raizada 2011).

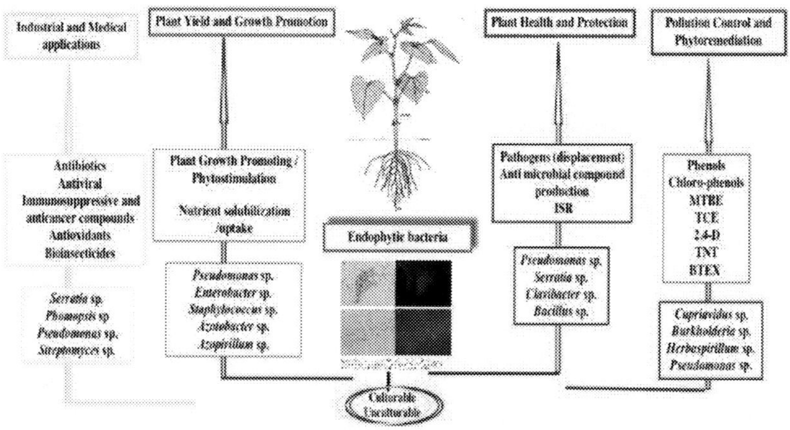

Figure 2. Schematic diagram of the different plant–bacterial endophyte interactions that have been studied and their applications (Ryan et al. 2008).

Nonetheless, altering the composition of any gene in the organism has the ability of changing the entire endophytic mechanism. This was demonstrated by *E. festucae*: a wild-type fungus found to grow in the intercellular spaces, which was reported not only to be responsible for the promotion of plant growth but also enhanced the resistance of plant in a contaminated environment (Tanaka et al., 2016). This study demonstrated the loss of apical dominance that eventually led to death of the plant the moment a plasmid of an enzyme was inserted into the coding region of the endophyte (Tanaka et al., 2016). It therefore means that inoculation of same endophytes into different hosts results into diverse activities. The study of Xu et al. (2012) provided an elucidation of this phenomenon explaining that *Colletotrichum magna* could cause a certain disease of the plant when the organism is inoculated in cucurbit plants; however, it grows asymptomatically in non-cucurbit species. Endophytes-plant relationship could either be in parasitism or mutualism. Symbiosis (mutualism), leads to the distinctions of their effects into bio-control, growth promotion and in bioremediation (Zabalgogeazcoa, 2008).

7. Biological Control by Endophytes

The literatures presents fungi, as the first endophytic organisms to be isolated, hence the notion that fungi are the most developed of the endophytes (Tenberge, 1999). These organisms have found their application in plants growth promotion and protection. Although, endophytes are known to be taxonomically diverse, their fungal classifications are broadly based on ecological categories or functional classes (Rodriguez et al., 2009). Further information is important in broadening the understanding of the biological activities that exists within the cells as a result of the interplay between plants and the associated orgasms in its tissues.

Development of agriculture and agricultural products has been saddled with problems of pests and diseases, hence it has experienced losses in its entire practice. However, the development of chemical pesticides was adjudged to provide solution to the menace of pests and pathogens, though it added in worsening the health of man and the environment (Afzal et al., 2014). The effect of pesticides to the emboldening soil and water has tremendously impacted the ecological status of the environment (Muponda, 2014). Certain bacterial endophytes have the ability to lessen harmful effects of some of these pests and its diseases (Ryan et al., 2008). For example, entomopathogenic microorganisms had been reported as an alternative means of controlling the effects of pests as to enable the control and reduction of the use of chemicals in agriculture (Dutta et al., 2014). The work of Elmi et al. (2000), proffered a reduction in the number of root-knot enthemopathogenic microbes (nematodes) found in tall fescue after inoculation with fungal endophytes for 35 days, without influencing the root dry weight. Other studies have also reported on the effects of plant colonised endophytic bacterial in preventing the infection of diseases (Sikora et al., 2010; Hirosch and Murakanmi, 2011).

This means that inoculation of entomopathogenic microbes in plants enhances plants microbial disease resistance (Elmi et al., 2000). Some other endophytic bacterial are implicated in the prevention of the effects of pathogen in plants (Gao et al., 2015). Figure 2 shows the various applications of endophyte. Although different symbiotic relationship occurs among endophytes and its host depending on the species involved, however, the two possible modes of action of endophytes according to the literature are the parasitic and/or mutualistic relationships with the hosts. It is from these two possible relationships that other actions are derived. This is explained in Figure 3, and it flows from there. For this reason, the interaction between endophytes and plants has been described as continuum (Fernandez et al., 2015).

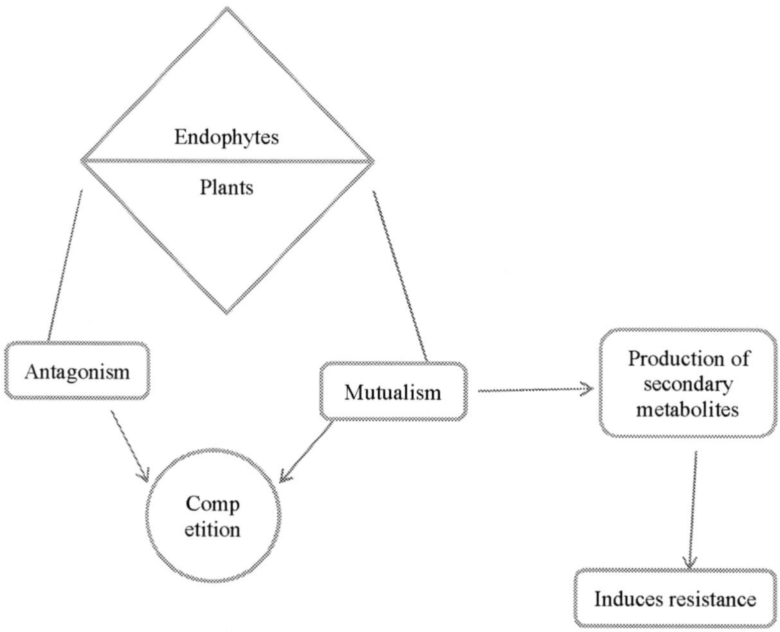

Figure 3. Schematic representation of Endophyte-Plant interaction and mode of action.

In parasitism, *Fusarium oxysporum* for example has been reported to infect the eggs of the banana weevil in the laboratory. This has been the mode of parasitic endophytes in attacking the pest or the causal agent of diseases (Xu et al., 2012). There are two factors that drive the endophytes towards its hosts, nutrients and infection sites. *Pseudomonas sp.*, for example as part of its biological activity produces siderophores that chelate iron molecule, depriving pathogens of this essential nutrient. This means that the organism will be competing for iron nutrient centres in the body of its host (Wang et al., 2009). The production of secondary metabolites is undoubtedly the major mode of action amongst endophytes. Other secondary metabolites have been identified in endophytes for example, isolated endophytes of *F. oxysporum* from tomatoes were found to cause juvenile mortality among nematodes by reducing the rate of hatching. *Colletotrichum sp.*, *Pseudomonas sp.* and *actinobacteria* are good candidates for the production of antibacterial, antifungal and antiviral metabolites (Zabalgogeazcoa, 2008; Sharma et al., 2021). Induced resistance, a phenomenon whereby the defence mechanism of plants could be activated is enhanced when it comes in contact with either biological or non-biological factors. This action usually results in reduced pest or pathogen damage (Tanaka et al., 2016; Fernandez et al., 2015). Before the inception of the molecular methods for identification of fungi in endophyte research, the real number of endophytic species in a sample has always been underestimated. The reason is because in this technique, obligate endophytes: mostly fungi with inhibited growth in the selected medium are usually not isolated. But with the introduction of rDNA sequencing the number of fungal species identified per host plant species increased substantially (Zabalgogeazcoa, 2008). This is because an important proportion of fungal isolates may be sterile in laboratory cultures. Genotypic identification methods have helped in the identification and distinction of sterile cultures (Gazis et al., 2016). Therefore, the introduction of modern biotechnological techniques enhanced the study of endophytes and its biocontrol activities.

8. Promotion of Growth by Endophytes

The ability of endophytes for plant growth promotion differs from that of the strains for biocontrol because they do not counteract the activities of pathogens rather they necessitates the increase in growth from the cycling of nutrients (Ryan et al., 2008). Biological activities of endophytes have been reported to enhance the synthesis of phytohormones that plays a part in root elongation and growth promotion in plants (Weyens et al., 2009a; Cosoveanu et al., 2021). Various metabolons are significant in the systemic activities of endophytes and such include indole acetic acid, gibberellin, indole acetonitrile, cytokinin etc. that are involved in plant growth promotion. There are other beneficial effects of endophytes on the growth of plants; they include adjustment of osmotic balance, regulation of stomata, root morphology modification, mineral uptake enhancement and alteration of metabolism (Weyens et al., 2009a). Hormones from endophytic fungi isolated from medicinal plants have been reported to enhance the growth of orchids (Zhang et al., 2000). Endophytes have also been tested in their seed germination enhancing abilities, and positive inferences have been derived with a report of reduction in disease symptoms that would have resulted from such practices (Bulgari et al., 2011). The plant growth-promoting bacterial endophytes and their effect are currently used in forest regeneration and phytoremediation activities (Khan and Doty, 2011). Endophytes enhance plant growth and adaptations through their growth-promoting activities (Stepniewska and Kuzniar, 2013) by using a series of mechanisms such as phosphate solubilisation, improvement of resistance to environmental stress of the host through enhancement of growth hormone production, promotion of siderophore production, increased synthesis of biological compounds such as vitamins in plants, aiding of osmotic regulation, stomatal adjustment, promotion of minerals' uptake, alteration of nitrogen accumulation as well as plant metabolism of compounds (Ryan et al., 2008).

9. Bioremediation

Plants growing in extreme environments exhibit environmental adaptation and such adaptation result in specialised physiological mechanism much of which help in strengthening the defence systems of plants through the expression of genes for such traits (Peng et al., 2013). Hence unique growth-promoting hormones are commonly found in endophytic strains growing in such environments (Stepniewska and Kuzniar, 2013). Such growth promoting abilities can be exploited in pollution management in contaminated soil by using plants colonized by an endophyte isolated from plants growing in a polluted environment. A demonstration of this principle was made when pyrene-degrading endophyte of an *Enterobacter* strain was isolated from a plant growing in soil contaminated by petroleum hydrocarbons, which on analysis was able to synthesize various metabolons with greater pyrene removal ability than the control experiment without the endophytes (Sheng et al., 2008). Furthermore, Germain et al. (2009) studied the degradation of herbicides with bacterial endophytes (*Pseudomonas sp.*); in this study, there was no sign of accumulation of the herbicide in the plant tissue and no sign of phytotoxicity was observed after the treatment, unlike the non-inoculated plant. Although some few studies have reported the potential of some fungi associated with grass to be effective in soil remediation of xenobiotic compounds, such action has not been demonstrated in the field (Kuldau and Bacon, 2008). Therefore, there is need for the creation of awareness about endophytic fungi present in plants to encourage further exploitation of these organisms in this field of subject (Mangunwardoyo and Gandjar, 2012). The importance of endophytic bioremediation range from the ability of the strain to reach larger population size because of reduced competition to their ability to degrade any absorbed xenobiotic in plants thereby reducing phytotoxic effects of contaminants (Khan and Doty, 2011).

There are two types of bioremediation, *in situ* and *ex situ* bioremediation based on whether the treatment is done on-site or out of site (Mcguinness and Dowling, 2009). *In situ* bioremediation which is remediation done at the site of contamination is also divided into different types based on the processes involved. Phytoremediation is a typical example of *in situ* bioremediation which uses plants to decontaminate soil or water using energy from sunlight (Doty, 2008). For some time now, there have been a focus on the exploitation of endophytes in resolving the shortcomings and constraints of phytoremediation (Germaine et al., 2009; Khandara et al., 2012; Minut et al., 2020; Ting et al., 2021). Endophytic bioaugmentation has proven to offer some benefits over traditional bioaugmentation where bacteria in the former case experiences less competition from the surrounding microbes and the plants in return provides the nutrient required to support growth and establishment (Khan and Doty, 2011).

10. Phytoremedation

Phytoremediation has been shown to be one of the most efficient and environmentally friendly means to restore soil to its original conditions when contaminated (Van Aken et al., 2010). The technology is generally accepted for the removal of environmental contaminants due to its minimal energy requirement and the beauty of the technique (Doty, 2008; Segura and Ramos, 2012; Yan et al., 2020; Nedjimi, 2021). However, heavy metals, persistent organic pollutants (POPs) and high molecular weight pesticides with their profound chemical properties, generally do not yield to conventional remediation practices and techniques (Mackinnon and Duncan, 2013). The limitation of phytoremediation is often as a result of the toxicity of these chemicals or their toxic end products in plants (Weyens et al., 2009b; Hur et al., 2011). This can however be ameliorated through endophyte-assisted phytoremediation either by means of natural colonization or by genetic manipulation of the plant as this has recently drawn attention (Doty, 2008; Wei et al., 2014, Yang et al., 2021). Various studies have reported on the impact of the dormancy in plants as a result of the toxicity of heavy chemicals. Sequel to this, modification of plants using plant-associated bacteria for a more potent environmental clean-up could become a remedy for such shortcoming (Van Aken et al., 2011; Eapen et al., 2007). Therefore, removal of toxic compounds from the soil can be enhanced by the use of endophyte-assisted phytoremediation; this is because endophytes assist their host plants through various means as elucidated in recent applications (Mcguinness and Dowling, 2009; Eissa and Almaroai, 2019; Nedjimi, 2021). Since all plant species are reported to have at least one endophytic species, the application of endophyte-assisted phytoremediation in treatment of contaminated environment is a feasible technology (Petrini, 1992; Khare et al., 2018).

Phytoremediation has been reported to amongst other advantages to have the ability of being used in combination with other techniques (phytostabilization, phytovolatilization, rhizofiltration etc), therefore the use of endophytes capable of degrading environmental contaminants in addition to the specific plants could offer an efficient, economic and sustainable remediation technology (Hur et al., 2011; Wei et al., 2014; Eapen et al., 2007; Liu et al., 2020). Literature also reported the ability for genetic transformation of the endophytic microbes (Transgenic preparation) for enhanced remediation activity (Sessitsch et al., 2013; Tang et al., 2014). This could be done by incorporating into plant endophyte those natural microbes' ability to conjugate with each other by means of movable DNA elements (vectors) between microbial populations (Khan and Doty, 2011). For example, by introducing bacterial genes pTOM-Bu61 involved in the metabolism of toluene and trichloroethylene (TCE), the tolerance and uptake of such compounds known to be phytotoxic to conventional plants in phytoremediation was improved (Doty, 2008).

11. Further Uses of Endophytes

Endophytes have some other applications, for example as decomposition agents of aging plant of salt marshes. Some other endophytes have demonstrated their ability to enhance osmoregulation in plant tissues, a boost to such plant's resistance to extreme condition, for example drought (Kuldau and Bacon, 2008). Most of these endophytes have been employed commercially in turf grasses to support their effort in resisting both biological and physical conditions. Some endophytes in their actions in plant growth promotion indirectly compete with pathogens and reduce their activity in the process (Bulgari et al., 2011). In genetic studies, endophytes serve as molecular tool used as a pool for gene carrier (Nagabhyru et al., 2013), for this reason it is possible to genetically engineer endophytes, which serve as protection for the host. This was first successful demonstrated by Fahey (1997) as exogenous gene expression. In the recent past, greater attention has been focused on the use of group of microorganisms in the phytoremediation of various environmental contaminants. In the soil bacteria gene of *Burkholderia cepacia* L.S.2.4 genetically modified by the introduction of a pTOM toluene degrading plasmid of *B. cepacia* G4, a natural endophyte of yellow lupine, was used for phytoremediation of toluene (Stepniewska and Kuzniar, 2013; Andreolli et al., 2013). This genetic modification increased the degradation of toluene by the recombinant strain. Germaine et al. (2009) also described inoculation of pea plant (*Pisum sativum*) with a genetically modified bacterial endophytes that naturally possessed the ability to degrade 2,4-dichlorophenoxyacetic acid. According to the results obtained through this transformation process, degradation capacity of the compound from the soil by plants inoculated with *Pseudomonas putida* was improved on.

In other subsequent studies, there was an improvement of toxic metal phytoremediation from the environment using engineered endophytes to ensure better clean-up of the environment (Stepniewska and Kuzniar, 2013). Such study provided the potential of microbe-plant relationship towards aiding the remediation of polluted sites, which could be useful in the development of efficient environmental contaminant removal systems. Hence, to further understand these interactions between plants and microorganisms a field trial will be necessary to establish the efficacy of the process.

Studies involving the resistance of host by endophytes are lacking, and the reasons behind it may not be far-fetched. One of the reasons could be due to the complex relationship existing between endophytes and hosts and the other may be the difficulty in understanding the mechanisms of life and conditions of endophytes to enable replication of their life processes. However, according to Xu et al. (2012), one may be able to build on such understanding using the following points:

1. Consider that the plant growth promotion mechanism, the pathogen and the endophytes ability to cause resistance to insects' attack might be related to the toxicant which they synthesis in the process. These

toxicants mostly inhibit the activities of such pathogens (Schulz and Boyle, 2006; Findlay et al., 1997). Therefore, endophytes growth promotion and pathogen resistance ability is relative to the rate at which chemicals are produces in its biocontrol activity.

2. Also the secondary metabolites produced by endophytes could be responsible for its potential resistance to various unfavourable conditions, which enhance the competitiveness of their host, as reported in the literature (Mousa and Raizada, 2013; Wang et al., 2015). However, the demonstration of this claim as put forward in the literature has not helped in resolving the disagreement that is growing on this subject. While the literature has provided the direct relationship between pathogens and the mechanism of endophytes actions, some researchers have different opinions, for instance, the inhibition of gall-forming insects provides the endophyte's host with resistance against such insect (Azevedo et al., 2000). The above phenomenon has generated many discussions in the recent past regarding the effect of endophytes in harnessing disease control in insects (Schmidt et al., 2014; Lacava and Azevedo, 2013). The study of Zabalgogeazcoa (2008) explained the mechanisms of operation of the endophytes and their interactions with plant pathogen. The study indicated that some of the endophytes are specialists in their infections while others are generalist (Zabalgogeazcoa, 2008). However, in resistance induction to disease facilitation, Adame-Alvarez et al. (2014) reported that the order of arrival of endophytes in plants tissues, affects endophytes-pathogen interactions which could lead to facilitation (positive inducement of disease pathogen), negatively enhance resistance by host or causing no effect (Adame-Alvarez et al., 2014). These require further studies for more confirmation.

Discussions on Future Prospects

The importance of endophytes in managing as well as exploiting the natural environment cannot be over emphasized (Figure 4). However, there have been limitations to the complete understanding of these organisms. Such limitation includes the method of isolation of endophyte, which relies mostly on culture-dependent approaches, while a large number of microorganisms cannot be cultured in by using this approach (Ryan et al., 2008). This therefore presents non culture-dependent approaches as possible avenues for improving the studies on endophytes. Exploitation of plant endophyte interaction can also play a key role in promoting sustainable agriculture (Ryan et al., 2008). The inference concerning the insignificance of the amount of secondary metabolites synthesized by endophytes to the host, disproof's the antagonistic effect between endophytes and pathogens. To accept or not accept these claims require evidence in current and future environmental studies. Therefore, in order for an improvement in the prospects of endophyte-assisted biological applications most especially in the field, intensive future studies are expected to demonstrate a better understanding of the organisms and their relationship with the hosts. Thus, the mechanism used by endophytes in its relationship with the host plant to accomplish the various advantages that it confers on the host plant is an important topic for future research.

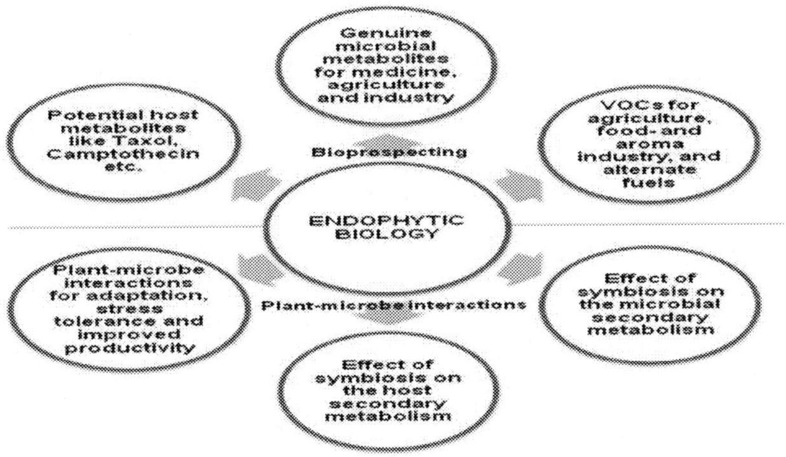

Figure 4. Current interests in the study of endophytic biology (Wani et al. 2015).

Author's Contribution

Raymond O Anyasi performed the literature search and writing of the chapter. Harrison I Atagana contributed immensely in editing and preparation of the final chapter.

Conflicts of Interest

The authors declare no conflict of interest.

Acknowledgments

The financial contribution by the National Research Foundation (NRF) of South Africa is immensely acknowledged.

References

Adame-Alvarez, R.-M. Mendiola-Soto, J. and Heil, M., (2014) Order of arrival shifts endophyte– pathogen interactions in bean from resistance induction to disease facilitation. *FEMS Microbiological Letters,* vol. 355, pp. 100-107.

Afzal, M. Khan, Q. M. and Sessitsch, A. 2014. Endophytic bacteria: prospects and applications for the phytoremediation of organic pollutants. *Chemosphere,* vol. 117, pp. 232-242.

Aly, A. H. Debbab, A. and Proksch, P., 2013. Fungal endophytes-secret producers of bioactive plant metabolites. *Die Pharmazie* [*The Pharmacy*] *International Journal of Pharmaceutical Science,* vol. 68, pp. 499-505.

Andreolli, M. Lampis, S. Poli, M. Gullner, G. Biro, B. and Vallini, B., 2013. Endophytic *Burkholderia fungorum* DBT1 can improve phytoremediation efficiency of polycyclic aromatic hydrocarbons. *Chemosphere,* vol. 92, pp. 688-694.

Anyasi, R. O. and Atagana, H. I. 2019. Endophyte: understanding the microbes and its applications. *Pakistan Journal of Biological Sciences,* vol. 22(4), pp. 154-167.

Azevedo, J. L. Maccheroni, J. Jr, Pereira, O. and Ara, W. L., 2000. Endophytic microorganisms: a review on insect control and recent advances on tropical plants. *Electronic Journal of Biotechnology,* vol. 3, pp. 40-65.

Backman, P. A. and Sikora, R. A., 2008. Endophytes: an emerging tool for biological control. *Biological Control,* vol. 46, pp. 1-3.

Bast, F. Bhushan, S. and John, A. A., 2012. DNA barcoding of a new record of epi-endophyte green algae *Ulvella leptochaete* (Ulvellaceae, Chlorophyta) in India. *Journal of Bioscience* 39:711-716.

Bezerra, J. D. P. Nascimento, C. C. F. Barbosa, R. N. Da Silva, D. C. V. Svedese, V. M. Silva-Nogueira, E. B. Gomes, E. S. Paiva, L. M. and Souza-Motta, C. M., 2015. Endophytic fungi from medicinal plant Bauhinia forficate: Diversity and biotechnological potential. *Brazilian Journal Microbiology,* vol. 46, pp. 49-57.

Bulgari, D. Casati, P. Crepaldi, P. Daffonchio, D. Quaglino, F. Brusetti, L. and Bianco, P. A. 2011. Restructuring of endophytic bacterial communities in grapevine yellows-diseased and recovered *Vitis vinifera* L. plants. *Applied Environment and Microbiology* 77:5018-5022.

Castillo, U. F. Strobel, G. A. Ford, E. J. Hess, W. M. Porter, H. Jensen, J. P. Albert, H. Robinson, R. Condron, M. A. Teplow, D. B. Stevens, D. and Yaver, D. 2002. Munumbicins, wide-spectrum antibiotics produced by *Streptomyces* NRRL 30562, endophytic on *Kennedia nigriscans*. *Microbiology,* vol. 148, pp. 2675-2685.

Cosoveanu, A., Chowdhary, K., Cabrera, R., Sharma, S. 2021. Role of Phytohormones-Producing Fungal Endophytes in Plant–Microbial Interactions Under Stress. In: Patil, R. H., Maheshwari, V. L. (eds.) *Endophytes.* Springer, Singapore. https://doi.org/10.1007/978-981-15-9371-0_10.

De Bary, A., 1866. Morphologie und Physiologie der Pilze, Flechten, und Myxomyceten. *Hofmeister's Handbook of physiological Botany,* Leipzig, Germany: Vol 2.

Doty, S. L., 2008. Enhancing phytoremediation through the use of transgenics and endophytes. *New Phytology,* vol. 179, pp. 318-333.

Dutta, D. Puzari, K. C. Gogoi, R. and Dutta, P. 2014. Endophytes: Exploitation as a tool in plant protection. *Brazilian Archives of Biology and Technology,* vol. 57, pp. 621-629.

Eapen, S. Singh, S. and D'Souza, S. F., 2007. Advances in development of transgenic plants for remediation of xenobiotic pollutants. *Biotechnological Advances,* vol. 25, pp. 442-451.

Eissa, M. A. and Almaroai, Y. A. 2019. Phytoremediation capacity of some forage plants grown on metals-contaminated soil. *Soil and Sediment Contamination: An International Journal.* Vol. 28 (6) 569-581.

Elmi, A. A. West, C. P. Robbins, R. T. and Kirkpatrick, T. L. 2000. Endophyte effects on reproduction of a root-knot nematode (Meloidogyne marylandi) and osmotic adjustment in tall fescue. *Grass Forage Science,* vol. 55, pp. 166-172.

Fadiji, A. E. and Babalola, O. O. 2020. Elucidating mechanisms of endophytes used in plant protection and other bioactivities with multifunctional prospects. *Frontiers of Bioengineering and Biotechnology,* vol 8, pp 467.

Felde, A. Z., 2011. Endophytes: novel weapons in the IPM arsenal. CGIAR SP-IPM 2011. [online] Available from: http://spipm.cgiar.org; Accessed 20/08/2015.

Findlay, J. A. Buthelezi, S. Li, G. Seveck, M. and Miller, J. D., 1997. Insect toxins from an endophytic fungus from Wintergreen. *Journal of Natural Products,* vol. 60, pp. 1214-1215.

Gao, Y. Liu, Q. Zang, P. Li, X. Ji, Q. He, Z. Zhao, Y. Yang, H. Zhao, X. and Zhang, L., 2015. An endophytic bacterium isolated from Panax ginseng C. A. Meyer enhances growth, reduces morbidity, and stimulates ginsenoside biosynthesis. *Phytochemistry Letters,* vol. 11, pp. 132-138.

Garcia, K. and Zimmermann, S. D., 2014. The role of mycorrhizal associations in plants potassium nutrition. *Frontier of Plant Science,* vol. 5, pp. 1-9.

Gazis, R. Kuo, A. Riley, R. Labutti, K. Lipzen, A. Lin, J. Amirebrahimi, M. Hesse, C. N., Spatafora, J. W. Henrissat, B. Hainaut, M. Grigoriev, I. V. and Hibbett, D. S., 2016. The genome of *xylona heveae* provides a window into fungal endophytism. *Fungal Biology,* vol. 120, pp. 26-42.

Germaine, K. Liu, X. Cabellos, G. Hogan, J. Ryan, D. and Dowling, D. N., 2009. Bacterial endophyte-enhanced phyto-remediation of the organochlorine herbicide 2,4-dichlorophenoxyacetic acid. *FEMS Microbiology and Ecology,* vol. 57, pp. 302-310.

Glick, B. R. and Stearns, J. C. 2011. Making phytoremediation work better: maximizing a plant's growth potential in the midst of adversity. *International Journal of Phytoremediation,* vol. 13, pp. 4-16.

Hardoim, P. R. Van Overbeek, L. S. and Elsas, J. D. V., 2008. Properties of bacterial endophytes and their proposed role in plant growth. *Trends in Microbiology,* vol. 16, pp. 463-471.

Hardoim, P. R. Van Overbeek, L. S. Berg, G. Pirttila, A. M. Compant. S. Campisano, A. Doring, M. and Sessitsch, A. 2015. The hidden world within plants: Ecology and evolutionary considerations for defining functionality of microbial endophytes. *Microbial and Molecular Biology Review,* vol. 79, pp. 293-320.

Hirosch, E. and Murakanmi, A., 2011. Microscopic anatomy and pigment characterization of coral-encrusting black sponge with cyanobacterial symbiont, *Terpio hoshinota.* Zoological Science, vol 28, pp. 199-205.

Huang, W. Y. Cai, Y. Z. Xing, J. Corke, H. and Sun, M., 2007 A potential antioxidant resource: endophytic fungi isolated from traditional Chinese medicinal plants. *Economic Botany,* vol. 61, pp. 14-30.

Hur, M. Kim, Y. Song, H. R. Kim, J. M. Choi, Y. I. and Yi, H., 2011. Effect of genetically modified poplars on soil microbial communities during the phytoremediation of waste mine tailings. *Applied Environment and Microbiology,* vol. 77, pp. 7611-7619.

Kan, F. L. Chen, Z. Y. Wang, E. T. Tian, C. F. Sui, X. H., and Chen, W. X., 2007. Characterization of symbiotic and endophytic bacteria isolated from root nodules of herbaceous legumes grown in Qinghai–Tibet Plateau and in other zones of China. *Archives of Microbiology,* vol. 188, pp. 103-115.

Khan, Z. and Doty, S., 2011. Endophyte-assisted phytoremediation. *Current Topics in Plant Biology,* vol 12, pp. 97-105.

Khandare, R. V., Rane, N. R., Waghmode, T. R., and Govindwar, S. P. 2012. Bacterial assisted phytoremediation for enhanced degradation of highly sulfonated diazo reactive dye. *Environmental Science and Pollution Research,* 19:1709–1718.

Khare, E., Jitendra Mishra, J. and Naveen Kumar Arora, N. K. 2018. Multifaceted Interactions Between Endophytes and Plant: Developments and Prospects, *Frontier in Microbiology,* vol. 9, article 2732. doi.org/10.3389/fmicb.2018.02732.

Kim, T.-U. Cho, S.-H. Han, J.-H. Shin. Y.-M. Lee, H. B. and Kim, S. B., 2012. Diversity and physiological properties of root endophytic Actinobacteria in native herbaceous plants of Korea. *Journal of Microbiolology,* vol 50, pp 50-57.

Kozyrovska, N. O., 2013. Crosstalk between endophytes and a plant host within information processing networks. *Biopolymer Cell,* vol. 29, pp. 234-243.

Krings, M. Taylor, T. N. Hass, H. Kerp, H. Dotzler, N. Hermsen, E. J., 2007. Fungal endophytes in a 400-million-yr-old land plant: infection pathways, spatial distribution, and host responses. *New Phytology,* vol. 174, pp. 648-657.

Kuldau, G. and Bacon, C. 2008. Clavicipitaceous endophytes: Their ability to enhance resistance of grasses to multiple stresses. *Biological Control,* vol. 46, pp. 57-71.

Kusari, S. Pandey, S. P. and Spiteller, M., 2013. Untapped mutualistic paradigms linking host plant and endophytic fungal production of similar bioactive secondary metabolites. *Phytochemistry,* vol. 91, pp. 81-87.

Lacava, P. T. and Azevedo, J. L., 2013. Biological control of insect-pest and diseases by endophytes. Springer, India. 1-10.

Li, H.-Y. Wei, D.-Q. Shen, M. and Zhou, Z.-P., 2012. Endophytes and their role in phytoremediation. *Fungal diversities,* vol. 54, pp. 11-18.

Li, J. Zhao, G. Z. Chen, H. H. Qin, S. Xu, L. H. Jiang, C. L. 2008. *Rhodococcus cercidiphylli* sp. nov., a new endophytic actinobacterium isolated from a *Cercidiphyllum japonicum* leaf. *Systemic and Applied Microbiology,* vol. 31, pp. 108-113.

Liu, J. Liu, S. Sun, K. Sheng, Y. Gu, Y. and Gao, Y., 2014. Colonization on root surface by a phenanthrene-degrading endophytic bacterium and its application for reducing plant phenanthrene contamination. *Plos One,* vol. 9, pp. 1-8.

Liu, S., Yang, B., Liang, Y., Xiao, Y. and Fang, J. 2020. Prospect of phytoremediation combined with other approaches for remediation of heavy metal-polluted soils. *Environmental Science and Pollution Research,* vol. 27, pp 16069–16085. https://doi.org/10.1007/s11356-020-08282-6

Lopez-Fernandez, S. Compant, S. Vrhovsek, U. Bianchedi, P. Z., Sessitsch, A., Pertot, L. and Compisano, A. 2015. Grapevine colonization by endophytic bacteria shifts secondary metabolism and suggests activation of defence pathways. *Plant and soils,* vol. 2639, pp. 1-21.

Mackinnon, G. and Duncan, H. J., 2013. Phytotoxicity of branched cyclohexanes found in the volatile fraction of diesel fuel on germination of selected grass species. *Chemosphere,* vol. 90, pp. 952-957.

Lòpez-Fernàndez, S. Sonego, P. Moretto, M. Pancher, M. Engelen, K. Pertot, I. and Campisano, A. 2015. Whole-genome comparative analysis of virulence genes unveils similarities and differences between endophytes and other symbiotic bacteria. *Frontier of Microbiology,* vol. 6, pp. 1-15.

Mangunwardoyo, W. and Gandjar, S.I., 2012. Frequency of endophytic fungi isolated from Dendrobium crumenatum (Pigeon orchid) and antimicrobial activity. *Biodiversity,* vol. 13, pp. 34-39.

Mcguinness, M. and Dowling, D. 2009. Plant-associated bacterial degradation of toxic organic compound in soil. *International Journal of Environmental Research and Public Health,* vol. 6, pp. 2226-2247.

Minut, M., Rosca, M., Cozma, P., Diaconu, M. and Gavrilescu, M. 2020. The Beneficial Role of Some Microorganism in Soil Phytoremediation and Mitigation of Health Risk. *The 8th IEEE International Conference on E-Health and Bioengineering - EHB 2020.* Grigore T. Popa University of Medicine and Pharmacy, Web Conference, Romania, October 29-30, 2020.

Mousa, W. K. and Raizada, M. N., 2013. Endophytes of anti-microbial secondary metabolites produced by fungal endophytes: An interdisciplinary perspective. *Frontier of Microbiology,* vol. 4, pp. 1-18.

Muponda, L., 2014. An investigation into the anthelmintic properties of seed extracts from endophyte containing pasture grasses. Masters Thesis at Lincoln University. NZ. Pp. 42-48.

Nagabhyru, P. Dinkins, R. D. Wood, C. L. Bacon, C. W. and Schardl, C. L. 2013. Tall fescue endophyte effects on tolerance to water-deficit stress. *BMC Plant Biology,* vol. 13, pp. 1-17.

Nedjimi, B. 2021. Phytoremediation: a sustainable environmental technology for heavy metals decontamination. *Soil Nutrient and Applied Sciences,* vol 3, pp 286.

Peng, A. P. Liu, J. Gao, Y. Z. and Chen, Z. Y. 2013. Distribution of endophytic bacteria in Alopecurus aequalis Sobol and Oxalis corniculata L. from soils contaminated by polycyclic aromatic hydrocarbons. *Plos One,* vol 8, pp. 1-10.

Petrini, O. 1992. Fungal endophytes of tree leaves. In: Andrews J. A., Hirano S. S. (eds.). *Microbial ecology of leaves*, Springer, New York. USA: pp. 1-10.

Raymond Anyasi, J. O. and Anyasi, R. O. 2020. Use of Bacterial Endophyte as a Control for White Aphid's Infestation in Tobacco Plant. *Sumerianz Journal of Agriculture and Veterinary,* vol. 2(10), pp. 88-95.

Rodriguez, R. J. White, J. F. Jr, Anold, A. E. and Redman, R. S., 2009. Fungal endophytes: diversity and functional roles. *New Phytolology,* vol. 189, pp. 314-330.

Ryan, R. P. Germaine, K. Franks, A. Ryan, D. J., Dowling, D. N., 2008. Bacterial endophytes: recent developments and applications. *FEMS Microbiology Letters,* vol. 278, pp. 1-9.

Schmidt, R. Köberl, M. Mostafa, A. Ramadan, E. M. Monschein, M. Jensen, K. B. Bauer, R. and Berg, G., 2014. Effects of bacterial inoculants on the indigenous microbiome and secondary metabolites of chamomile plants. *Frontier of Microbiology,* vol. 5, pp. 1-11.

Schulz, B. and Boyle, C., 2006. What are endophytes? In: *Microbial Root Endophytes.* Springer, Heidelberg, Germany. Pp. 1-10.

Segura, A. and Ramos, J. L., 2012. Plant–bacteria interactions in the removal of pollutants. *Current Opinion in Biotechnology,* vol. 24, pp. 467-473.

Sessitsch, A. Kuffner, M. Kidd, P. Vangronsveld, J. Wenzel, W. Fallmann, K. and Puschenreiter, M., 2013. The role of plant-associated bacteria in the mobilization and phytoextraction of trace elements in contaminated soils. *Soil Biology and Biochemistry,* vol. 60, pp. 182-194.

Sharma, H., Rai, A. K., Dahiya, D., Chettri, R., Nigam, P. S. 2021. Exploring endophytes for *in vitro* synthesis of bioactive compounds similar to metabolites produced *in vivo* by host plants. *AIMS Microbiology,* vol 7(2), pp 175-199.

Sheng, X. F. Xia, J. J. Jiang, C. Y. He, L. Y. and Qian, M., 2008. Characterization of heavy metal-resistant endophytic bacteria from rape (*Brassica napus*) roots and their potential in promoting the growth and lead accumulation of rape. *Environmental Pollution.* Vol.156, pp. 1164-1170.

Sikora, R. A. Zum Felde, A. Mendoza, A. Menjiver, R. and Pocasangre, L., 2010. In planta suppressiveness to nematodes and long-term root health stability through biological enhancement-do we need a cocktail? *Acta Horticulture,* vol. 879, pp. 553-560.

Stepniewska, Z. and Kuzniar, A. 2013. Endophytic microorganisms-promising applications in bioremediation of greenhouse gases. *Applied Microbiology and Biotechnology,* vol. 97, pp. 9589-9596.

Strobel, G. and Daisy, B., 2003. Bioprospecting for microbial endophytes and their natural products. *Microbiology and Molecular Biology Research,* vol. 67, pg. 491-502.

Sun, K. Liu, J. Jin, L. and Gao, Y. 2014. Utilizing pyrene-degrading endophytic bacteria to reduce the risk of plant pyrene contamination. *Plant and Soil,* vol. 374, pp. 251-262.

Suryanarayanan, T. S. Thirunavukkarasu, N. Govindarajulu, M. B. Sasse, F. Jansen, R. and Murali, T. S., (2009) Fungal endophytes and bioprospecting. *Fungal Biology Review,* vol. 23, pp. 9-19.

Taghavi, S. Weyens, N. Vangronsveld, J. Lelie, D. Pirttilä, A. M. and Frank, A. C., 2011. Improved phytoremediation of organic contaminants through engineering of bacterial endophytes of trees. *Endophytes of Forest Trees.* Springer, Netherlands. Pp. 1-10.

Tan, R. X. and Zou, W. X., 2001. Endophytes: a rich source of functional metabolites. *National Product Reports,* vol. 18, pp. 448-459.

Tanaka, A. Christensen, M. J., Takemoto D., Park P., Scott B. (2006) Reactive oxygen species play a role in regulating a fungus-perennial ryegrass mutualistic interaction. *Plant Cell* 18:1052-1066.

Tang, X. J. Hashmi, M. Z. Lond, D. Chen, L. Khan, M. L. and Shen, C., 2014. Influence of heavy metals and PCBs pollution on the enzyme activity and microbial community of paddy soils around an E-waste recycling workshop. *International Journal of Environmental Research and Public Health,* vol. 11, pp. 3118-3131.

Tenberge, K. B., 1999. *Biology and life strategy of Ergot fungi.* Institut fur Botanik. Munster-Gemany, pp. 5-11.

Ting A. S. Y., Chaverri P. and Edrada-Ebel R. A. 2021. Editorial: Endophytes and Their Biotechnological Applications. *Front. Bioeng. Biotechnol.* 9:795174. doi: 10.3389/fbioe.2021.795174

Van Aken, B. Correa, P. A. and Schnoor, J. L., 2010. Phytoremediation of polychlorinated biphenyls: New trends and promises. *Environmental Science and Technology,* vol. 44, pp. 27-67.

Van Aken, B. Tehrani, R. and Schnoor, J. 2011. Endophyte-assisted phytoremediation of explosives in poplar trees by Methylobacterium populi BJ001T. In: Pirttila, A. M., Frank, A. C. (Eds.), *Endophytes of Forest Trees.* Springer, Netherlands. pp. 1-10.

Wang, L. Zhang, W. M. Pan, Q. L. Li, H. H. Tao, M. H. and Gao, X. X., 2009. Isolation and molecular identification of endophytic fungi from *Aquilaria sinensis*. *Jounal of Fungal Research,* vol. 7, pp. 37-42.

Wang, Z. Ma, Z. Wang, L. Tang, C. Hu, Z. Chou, G. and Li, W. 2015. Active anti-acetylcholinesterase component of secondary metabolites produced by the endophytic fungi of Huperzia serrata. *Electronics Journal of Biotechnology,* vol. 18, pp. 399-405.

Wani, Z. A. Ashraf, N. Mohiuddin, T. and Riyaz-Ul-Hassan, S., 2015. Plant-endophyte symbiosis, an ecological perspective. *Applied Journal of Microbiology and Biotechnology,* vol. 99, pp.2955-2965.

Wei, J. Liu, X. Wang, Q. Wang, C. Chen, X. and Li, H., 2014. Effect of rhizodeposition on pyrene bioaccessibility and microbial structure in pyrene and pyrene–lead polluted soil. *Chemosphere,* vol. 97, pp. 92-97.

Wen, S.-H. Xu-Jian, H. Yang-Song, L. Liu-Guang, G. and Dai, S., 2004. Plant uptake of aldicarb from contaminated soil and its enhanced degradation in the rhizosphere. *Chemosphere,* vol. 54, pp. 569-574.

Weyens, N. Schellingen, K. Beckers, B. Janssen, J. Ceulemans, R. Lelie, D. Taghavi, S. Carleer, R. and Vangronsveld, J., 2013. Potential of willow and its genetically engineered associated bacteria to remediate mixed Cd and toluene contamination. *Journal of Soils and Sediments,* vol. 13, pp. 176-188.

Weyens, N. Van Der Lelie, D. Artois, T. Smeets, K. Taghavi, S. Newman, L. and Vangronsveld. J. (2009b) Bioaugmentation with engineered endophytic bacteria improves contaminant fate in phytoremediation. *Environmental Science and Technology,* vol. 43, pp. 9413-9418.

Weyens, N. Van Der Lelie, D. Taghavi, S. Newman, L. and Vangronsveld, J. 2009a Exploiting plant–microbe partnerships to improve biomass production and remediation. *Trends in Biotechnology,* vol. 27, pp. 591-598.

Winther, J. L. and Friedman, W. E., 2008. Arbuscular mycorrhizal associations in Lycopodiaceae. *New Phytolology,* vol. 177, pp. 790-801.

Xu, X. H. Su, Z. Z. Wang, C. Kubicek, C. P. Feng, X. X. Mao, L. J. Wang, J. Y. Chen, C. Lin, F. C. and Zhang, C. L., 2014. The rice endophyte Harpophora oryzae genome reveals evolution from pathogen to mutualistic endophytes. *Science Reports,* vol. 4, pp. 1-9.

Xu, Z. Gao, D. Sung, X. and Xu, Y., 2012. A review of endophyte and its use and function. *Advance Biomedical Engineering,* vol. 8, pp. 124-130.

Yan, A., Wang, Y., Tan S. N., Yusof, M. L. M., Ghosh, S. and Chen, Zhong. 2020. Phytoremediation: A Promising Approach for Revegetation of Heavy Metal-Polluted Land. *Frontiers in Plant Science,* vol. 11, pp 359.

Yang, K. M., Poolpak, T., Pokethitiyook, P., Kruatrachue, M., & Saengwilai, P. (2021). Responses of oil degrader enzyme activities, metabolism and degradation kinetics to bean root exudates during rhizoremediation of crude oil contaminated soil. *International Journal of Phytoremediation,* vol. 21(7), pp.1-9.

Yrjälä K., Mancano G., Fortelius C., Åkerman M. L., Sipilä T. P. (2010) The incidence of Burkholderia in epiphytic and endophytic bacterial cenoses in hybrid aspen grown on sandy peat. *Boreal Environmental Research,* vol 15, pp 81-96.

Yu, H. Zhang, L. Li, L. Zheng, C. Gao, L. Li, W. Sun, P. and Qin, L. 2010. Recent development and future prospects of antimicrobial metabolites produced by endophytes. *Microbiology Research,* vol. 65, pp. 437-449.

Zabalgogeazcoa, I., 2008. Fungal endophytes and their interaction with plant pathogens. *Spanish Journal of Agricultural Research,* vol. 6, pp. 138-146.

Zhang, J., Wang, C. and Guo, S. 2021. Studies on the plant hormones produced by 5 species of endophytic fungi isolated from medicinal plants (Orchidacea). *Tree Physiology.* vol. 22, pp 431-434.

Zhang, X. X. George, A. Bailey, M. J. Rainey, P. B., 2012. The histidine utilization (hut) genes of Pseudomonas fluorescens SBW25 are active on plant surfaces, but are not required for competitive colonization of sugar beet seedlings. *Microbiology,* vol. 152, pp. 1867-1875.

Chapter 2

Molecular and Biochemical Methods for Isolation and Characterization of Endophytes

Mangalam Bajpai[1], Jyotirmayee Das[1] and Pragya Tiwari[2,*]
[1]Global Product Compliance (GPC) Regulatory India Private Limited, 3rd floor, Samved Sankul, Civil Lines, Nagpur, Maharashtra, India
[2]Department of Biotechnology, Yeungnam University, Gyeongsan, Gyeongbuk, Korea

Abstract

Endophytic microorganisms are plant partners who have a mutually beneficial connection with the host plant. Despite the fact that these host plants are associated with microbial partners since the beginning, however, endophytes were identified in recent times. Because they produce a variety of compounds and interact with other microorganisms, both pathogenic and non-pathogenic, endophytes' beneficial effects are becoming more important due to the possibility of obtaining novel medicinally important compounds as well as their role in increasing crop productivity. The accurate identification of these microorganisms and their interactions with the host and other microbes is now feasible, thanks to the progress of sophisticated tools and methods in modern biology. Several scientific methods including biochemical and molecular methods are employed to isolate and characterize bacterial and fungal endophytes from diverse plants, including cultivation and non-cultivation-based methods. The chapter provides an overview of the multi-faceted importance of plant-associated endophytes and the isolation approaches

---

*Corresponding Author's Email: pragyatiwari@ynu.ac.kr.

In: Endophytes
Editor: Pragya Tiwari
ISBN: 979-8-88697-045-6

(molecular and biochemical) for further characterization and validation of endophytes.

Keywords: *Burkholderia cepacia*, cultivation methods, endophytes, growth medium, indole-3-acetic acid (IAA), Luria-Bertani (LB) medium, surface sterilization

1. Introduction

Endophytes are the microorganisms, which colonize the living plant tissue without causing any immediate, overt, negative effect on the plant. Anton de Bary first introduced the term epiphytes for fungi that live on the surface of their host and endophytes for those living inside the plant tissue (De Bary 1866). Later, this term was expanded to fungi and bacteria, including actinomycetes, which spend the whole or part of their life cycle colonizing inter- or intracellularly, inside the healthy living tissues of the host, typically causing no apparent symptom of the disease. The word endophyte is derived from two Greek words, 'endon' means within and 'phyton' means plant.

Endophytes were mentioned for the first time in the 19th century (Azevedo 1998) and have been defined in many ways. Wilson (1995) defined endophytes as the fungi that live internally and remain asymptomatic for at least part of their life cycle. He also described the symptomless nature of endophyte colonization in plant tissue and the symbiotic and mutualistic relationship between the endophytes and their host. Sometimes, they can also be aggressive saprophytes or opportunistic pathogens. Endophytes are plant-associated prokaryotes that form associations with their host plants by colonizing the internal tissues, which has made them valuable for agriculture as a tool for improving crop performance (Azevedo et al. 2000; Tiwari and Bae 2020a). There have been several studies on the plant endophyte relationship, especially for grasses such as tall fescue, where it has been shown that endophytes produce toxins that discourage insects and other grazing animals (Bultman and Murphy 2000). Persian darnel, an annual grass, considered a troublesome weed by wheat farmers, is extensively studied for endophytes biology. It has been discovered that grasses with high colonization with endophytes are often resistant to attack by certain insects (Clay 1988; Bills 1996).

There has been considerable interest in screening the endophytes for bioactive compounds (Kumar and Hyde 2004). A large number of compounds (secondary metabolites) have been extracted, isolated, and characterized from endophytic microbes (Tan and Zou 2001; Strobel et al. 2004). With an overview of endophyte biology and its occurrence, the chapter discusses the isolation approaches (molecular and biochemical) for further characterization and validation of endophytes.

2. The Plant–Endophyte Interface

In their native biological settings, microorganisms are seldom seen in axenic forms. Fungal endophytes are no exception; they generally live and work in a niche of micro-and macroorganisms that are rich and diverse (Kharwar et al. 2011; Kusari et al. 2012). Under multiple biotic (pathogens and feeders) and abiotic (environmental variables) selection pressures, these niches contain a plethora of varied crosstalk among fungal endophytes, bacterial endophytes, endophytic viruses, and related host plants. Such multiplexed interactions are definitely required for these and similar creatures to carry out ecological operations (Phelan et al. 2011). An analogy between such a complex and fascinating plant–endophyte cost-benefit relationship and the concept of an ecological marketplace, in which a large number of fair-trade partners, such as symbionts or mutualists, engage in cooperative and social interactions with cheaters, antagonists, and pathogens (West et al. 2006; Werner et al. 2014). The lack of precise knowledge of fungal endophytes' interactions with related species, is particularly intriguing because fungal endophytes produce a wide range of natural products when cultivated outside of their normal hosts or ecological niches *in vitro*. *In vitro* fermentation conditions, on the other hand, differ from the plant apoplast environment (e.g., medium type and composition, aeration, pO_2, pCO_2, pH, temperature, agitation, sampling, and harvest points) (Sattelmacher 2001). Nonetheless, during the colonization of endophytes inside living host plants, secondary metabolites generated by *in vitro* grown fungal endophytes are documented to alter host plant physiology, defense, and tolerance to biotic and abiotic stimuli. The amount to which fungal endophytes produce natural products inside the host, if at all, and the significance of these secondary metabolites in helping plants achieve fitness benefits are gaining momentum.

3. Plant Associated Endophytes

The vast array of root/shoot dynamic habitats known as the rhizosphere (in soil) and phyllosphere (in plants) play a role in plant-associated microbial interactions (plant aerial parts). Microbial communities interact with each other and with plants throughout in order to promote plant adaptability and tolerance to environmental restrictions (Tiwari et al. 2020b; Tiwari et al. 2021a; Tiwari et al. 2021b). Beneficial “endophytic interactions” occur in the rhizosphere and phyllosphere environments, where a broad collection of bacterial and fungal communities infiltrate plant inter-and intracellular spaces and constitute several types of plant-associated microbial interactions (Tiwari et al. 2021a; Tiwari et al. 2021b). The structural composition of endophytic microbial communities has been studied in relation to a few bacterial families and fungal species. The identities of the vast majority of microbial species in the plant endophytic microbiome, as well as their ecological activities, are unknown. A number of abiotic stresses can occur alone or in combination in the soil environment, generating both general and particularly negative impacts on plant growth and development. Abiotic stresses are known to affect the rhizosphere soil as well as plant-associated beneficial microbial interactions and hence crop production in a negative way, in addition to the direct detrimental impact on plant development. The term “induced systemic tolerance” (IST) was coined to describe several ways through which endophytic bacterial, fungal, or microbe-mediated systemic tolerance to abiotic stresses operates in the plants. As a result, modern agronomic practices must be redefined in light of our current understanding of the importance of plant-associated beneficial microbial communities (rhizosphere, symbiotic, and endophyte interactions) for plant productivity and health in abiotic stress environments

4. Bacterial Endophytes

Unlike soil fungus, bacteria in the rhizosphere known as rhizobacteria respond effectively to root exudates in terms of nutrient acquisition, and their metabolic activities tend to control plant growth and development in a favorable way. Species of the genera *Pseudomonas*, *Bacillus*, *Enterobacter*, *Serratia*, *Arthrobacter*, *Azotobacter*, *Azospirillum* (Gray and Smith 2005), and *Streptomyces* (Tokala et al. 2002; Dimkpa et al. 2009) have been discovered to have favorable impacts on diverse plants at the root zone. Many of the rhizobacteria that colonize the root environment and surface (exo-root) also penetrate into the root cortex (endo-root), where they exhibit plant growth-promoting qualities as endophytes (Kloepper and Beauchamp 1992; Sessitsch et al. 2004). Bacterial endophytes from various agricultural plants include *Bacillus*, *Pseudomonas*, *Enterobacter*, *Klebsiella*, *Serratia*, and *Streptomyces* sp. Bacterial endophytes belonged to *Proteobacteria* (54 percent), *Actinobacteria* (20 percent), *Firmicutes* (16 percent), and *Bacteroidetes* (16 percent) in a sequence-based analysis. Based on their meta-analysis, they also reported the occurrence of endophytic Archaea as endophytes. Another research using 16S-rRNA pyrosequencing found that *Proteobacteria* accounted for 90% of the bacterial endophytes populations in tomato leaves, followed by *Actinobacteria* (1.5 percent), *Planctomycetes* (1.4 percent), *Verrucomicrobia* (1.1 percent), and *Acidobacteria* (0.5 percent) (Romero et al. 2014). Non-rhizobial root nodule endophytes isolated from native and foreign legume plants in Flanders (Belgium) displayed a wide range of taxa from the groups- *Alphaproteobacteria, Betaproteobacteria, Gammaproteobacteria*, *Actinobacteria*, *Firmibacteria*, *Flavobacteria*, and *Sphingobacteria* (De Meyer et al. 2015). Endophytes such as *Proteobacteria* (alpha, beta, gamma), *Firmicutes*, and *Actinobacteria* were found in *Cucumis melo* fruits using a cultivation-independent method called fluorescence *in situ* hybridization (Glassner et al. 2015).

5. Fungal Endophytes

Endophytes include fungi that participate in protective mutualisms, despite the fact that some create epiphytic structures. However, the word 'endophyte' is also used to refer to any fungus, bacterium, or another microbe that colonizes live plants without exhibiting harmful symptoms and has no visible exterior evidence of infection. Rodriguez et al. (2009) classified fungal endophytes into four groups based on the host range, tissues colonized, *in planta* colonization, biodiversity, mechanism of transmission, and the nature of the advantages provided to the host. The ability for vertical endophyte transmission through seed and a constrained host range, both of which are characteristics of class 1 endophytes (Rodriguez et al. 2009), and are diagnostic in characterizing fungal endophytes that participate in defensive mutualisms. Vertically transmitted hereditary symbionts are fully reliant on host reproduction for their own proliferation and reproductive success. When a vertically transmitted symbiont benefits the host, it indirectly promotes the symbiont's own fitness. Any symbiont that is vertically transmitted and is a disadvantage to its host would become extinct because non-infected conspecifics would outcompete it (Ewald 1987). While the horizontally transmitted symbionts, on the other hand, take advantage of the host's (Darwin 1859) capacity to live and spread the infection to non-infected organisms and are typically more virulent since the symbiont's reproduction is not entirely dependent on the host's reproductive success (Ferdy and Godelle 2005; Tiwari and Bae 2020a). Clavicipitaceous endophytes with vertical transmission are only found in a few Poaceae and Convolvulaceae species and have limited host ranges (Clay and Schardl 2002; Steiner et al. 2011). A particular fungal taxon is mostly linked to a certain plant host species (Schardl et al. 2011). Many horizontally transmitted endophytes, on the other hand, belong to fungal genera that are common in the environment and are frequently connected with several host taxa (Rodriguez 2009).

6. Beneficial Effects of Bacterial Endophytes on the Host Plant

Microorganisms can stimulate plant growth by fixing nitrogen (Sevilla et al. 2001; Hurek et al. 2002; Iniguez et al. 2004), production of phytohormones, biocontrol of phytopathogens in the root zone (by producing antifungal or antibacterial agents, siderophore production, nutrient competition, and inducing systematic acquired host resistance, or immunity, or increasing mineral availability (Sessitsch et al. 2002; Benhizia et al. 2004; Tiwari et al. 2021c; Tiwari et al. 2021d). The understanding of the mechanisms that promote plant development will aid in the selection of species and situations that benefit plants more. Volatile chemicals generated by bacteria, such as 2-3 butanediol and acetoin, represent a newly identified method for plant growth stimulation (Ryu et al. 2003). It is fascinating to see whether volatiles might be produced within plants. Endophytes create adenine ribosides, which promote pine tissue development and reduce browning (Pirttilä et al. 2004). Red clover bacterial endophyte appear to be responsible for the allelopathic effects reported with these plants over maize, resulting in decreased plant emergence and height (Christiansen et al. 1992; Seghers et al. 2004). Clover bacterial endophyte replicated the maize-damaging effects.

In the near future, genetically altered endophytes with biological control capability in agricultural crops highlight attractive alternatives. To manage insect pests, the endophytes *Herbaspirillum serope dicae* and *Clavibacter xylii* were genetically engineered to generate and excrete the -endotoxin of *Bacillus thuringiensis* (Turner et al. 1991; Downing et al. 2000). Endophytic populations of bacteria that degrade refractory chemicals are more common than the rhizosphere of plants in polluted locations (Siciliano et al. 2001), suggesting that endophytes have a role in metabolizing these substances. Endophytic *Burkholderia cepacia* strains with better phytoremediation and toluene tolerance were developed (Barac et al. 2004). There is an increasing interest in genetically engineered endophytes (Andreote et al. 2004). The advantages and obstacles to using bioengineered endophytes have been clearly discussed (Van et al. 2004; Newman and Reynolds 2005).

Table 1. Types of media used for the isolation of Bacterial endophytes (Suman et al. 2016)

S. No.	Name of Media	Composition of the medium
1.	Nutrient agar	5 gm peptone; 5 gm NaCl; 3 gm beef extract; 20 gm agar
2.	T3 agar	3 gm tryptone; 2 gm tryptose; 1.5 gm yeast extract; 0.005 gm $MnCl_2$; 0.05 gm NaH_2PO_4; 20 gm agar
3.	Tryptic soy agar	17 gm tryptone; 3 gm soya meal; 2.5 gm $C_6H_{12}O_6$; 5 gm NaCl; 2.5 gm K_2HPO_4; 20 gm agar
4.	King's B agar	2 gm protease peptone; 1.5 gm K_2HPO_4; 1.5 gm $MgSO_4.7H_2O$; 10 mL glycerol; 20 gm agar
5.	Jensen's agar	20 gm Sucrose;1 gm K_2HPO_4; 0.5 gm Mg_2SO_4; 0.5 gm NaCl; 0.001 gm Na_2MoO_4; 0.01 gm $FeSO_4$; 2 gm $CaCO_3$; 20 gm agar
6.	R2A agar	0.5 gm proteose peptone; 0.5 gm amino acids; 0.5 gm yeast extract; 0.5 gm dextrose; 0.5 gm soluble starch; 0.3 gm K_2HPO_4; 0.05 gm $MgSO_4.7H_2O$; 0.3 gm sodium pyruvate; 20 gm agar
7.	Ammonium minerals salt	0.70 gm K_2HPO_4; 0.54 gm KH_2PO_4; 1.00 gm $MgSO_4$. $7H_2O$; 0.20 gm $CaCl_2.2H_2O$; 4.00 mg $FeSO_4.7H_2O$; 0.50 gm NH_4Cl; 100 μg $ZnSO_4.7H_2O$; 30 μg $MnCl_2.4H_2O$; 300 μg H_3BO_3; 10 μg $CuCl_2.2H_2O$; 200 μg $CoCl_2.6H_2O$; 20 μg $NiCl_2.6H_2O$; 60 μg $Na_2MoO_4.2H_2O$; 20 gm agar
8.	Luria-Bertani Media	10 gm casein acid hydrolysate; 5 gm yeast extract; 10 gm NaCl; 20 gm agar
9.	Modified Dobereiner medium	10 gm sucrose; 5 gm malic acid; 0.2 gm $K_2HPO_4.H_2O$; 0.4 gm $KH_2PO_4.H_2O$; 0.1 gm NaCl; 0.01 gm $FeCl_3$; 0.002 gm Na_2MoO_4; 0.2 gm $MgSO_4.7H_2O$; 0.02 gm $CaCl_2.H_2O$; 20 gm agar
10.	Yeast extract mannitol agar	1 gm yeast extract; 10 gm mannitol 0.5 gm $K_2HPO_4.H_2O$; 0.002 gm $MgSO_4.7H_2O$ 0.1 gm NaCl; 20 gm agar

7. Isolation and Identification of Endophytes from Different Sources

Many endophytes that live inside plants are typically regarded as symbionts, with a distinct and treasured association with the plants with which they develop (Mitter et al. 2013; Berg et al. 2014). Several cultivation-independent tests as well as fluorescence in situ hybridization-confocal laser scanning microscopy experiments have recently established the presence of endophytes (Mitter et al. 2013; Berg et al. 2014). Cultivation-based procedures rely on isolating and testing isolates, whereas cultivation-independent techniques look for differences in the overall endophytic community (Menpara and Chanda 2013). Endophytes may be isolated from any microbial or plant growth medium, including agar, potato dextrose agar, and any nitrogen- or carbon-containing media (Manias et al. 2020; Yadav et al. 2021). Isolation from surface-sterilized host plant tissue is the most common approach for detecting and counting endophytes. The genotype of the plant, growth stage, physiological status, type of plant tissue, environmental condition of the soil in which it is grown, sampling season, surface sterility, selective media and culture conditions, as well as various agricultural practices (Gaiero et al. 2013; Golinska et al. 2015) are the main factors that may regulate endophyte colonization within a plant or microbial species. Ecological awareness of the role of endophytes in nature can also provide the best clues for targeting a particular type of endophyte bioactivity with the greatest potential for bioprospecting (Strobel and Daisy 2003).

8. Bioactive Compounds from Endophytes

Endophytes have been shown to produce a variety of bioactive metabolites in a single plant or microbe, making them a good source of pharmaceuticals for treating a variety of illnesses and having potential uses in agriculture, medicine, food, and cosmetics (Strobel and Daisy 2003; Jalgaonwala et al. 2011; Godstime et al. 2014; Shukla et al. 2014). Alkaloids, benzopyranones, chinones, flavonoids, phenolic acids, quinones, steroids, saponins, tannins, terpenoids, tetralones, xanthones, and many more secondary metabolites were classified into distinct functional categories (Schulz et al. 2002; Strobel and Daisy 2003; Joseph and Priya 2011; Pimentel et al. 2011; Godstime et al. 2014; Shukla et al. 2014). The season of sample collection, the meteorological condition, and geographical location influence the extraction of metabolites from endophytes (Jalgaonwala et al. 2011). However, extraction from plants and other natural sources has become more possible, efficient, and easy, thanks to a novel synthetic approach developed in recent years (Hussain et al. 2012; Tiwari et al. 2021e; Tiwari et al. 2022a). The evolution of host microorganisms, which may have incorporated genetic information from higher plants, allowing them to better adapt to the host plant and perform some functions, such as protection from pathogens, insects, and grazing animals, has been directly linked to the production of bioactive substances by endophytes (Strobel and Daisy 2003). One of the most promising anticancer drugs identified or synthesized so far is taxol (paclitaxel), a complex diterpene alkaloid generated by the endophyte *Metarhizium anisopliae* found in the bark of the *Taxus* tree (Strobel and Daisy 2003; Zhang et al. 2009; Visalakchi and Muthumary 2010). *Nothapodytes foetida* produces camptothecin, which has cytotoxic and antifungal effects (Han and Rahman 2012; Shukla et al. 2014). Huperzine-A (HupA) is a cholinesterase inhibitor that comes from the *Huperzia serrata* (Nair and Padmavathy 2014). Lignans isolated from endophytic *Podophyllum hexandrum*, such as cathartics, emetics, and cholagogue, have been found to serve as anticancer agents (Konuklugil 1995; Tiwari et al. 2022b). *P. emodi* resins, such as etoposide and teniposide, have potent anticancer properties (Konuklugil 1995). Bactericidal properties have been discovered in compounds such as oxacillin, ampicillin, catechin, gallic acid, and cephalexin (Akiyama et al. 2001). Terpenoids have anticancer, antibacterial, antiviral, and gastrointestinal stimulant properties (Strobel and Daisy 2003; Godstime et al. 2014; Shukla et al. 2014). *Cytonaema* sp., a fungal endophyte, produces triterpenoid helvolic acid, which has potent antibacterial properties (Kumar et al. 2014).

9. Scientific Methods for Screening and Isolation of Endophytes

Extraction from a natural source, microbial synthesis via fermentation, and microbial transformation are all methods for obtaining bioactive chemicals. Extraction from natural sources has several drawbacks, such as a reliance on seasonal, climatic, and political factors, as well as the possibility of environmental issues, necessitating novel techniques to access such substances (Bicas et al. 2009). As a result, biotechnology approaches using various microbes appear to be attractive options for creating an endless, cost-effective, and renewable supply of high-value bioactive chemicals and fragrance compounds (Dhole and Shelat 2022). The biotransformation process has a wide range of applications (Borges et al. 2009; Bakhtiyarifar et al. 2021), for example, it has been widely used to create volatile chemicals (Krings et al. 2006; Bicas et al. 2008; Bicas et al. 2009). These volatile compounds have antimicrobial (vanillin, essential oil constituents), antifungal and antiviral (some alkanolides), antioxidant (eugenol, vanillin), somatic fat reducing (nootkatone), blood pressure regulating (2-[E]-hexenal), anti-inflammatory properties (1, 8-cineole), and other desirable properties (Berger 2009; Dubey et al. 2021).

Although there is still a substantial need for synthetic products in the medicine sector owing to cost and time constraints, bioactive natural chemicals generated by endophytes have shown promise in terms of safety and human health problems (Strobel et al. 2004; Dubey et al. 2021). Human health issues such as the emergence of antibiotic resistance in human pathogenic bacteria, fungal infections, and life-threatening viruses call for novel therapeutic agents to treat illnesses in humans, plants, and animals that are currently untreated (Strobel and Daisy 2003; Strobel 2003; Borges et al. 2009).

10. Molecular Methods

10.1. Sample Collection and Preparation

Endophytes have been isolated from a variety of plants, including mosses and woody trees like citrus (Gardner et al. 1982; Strobel et al. 2004; Zhikai et al. 2017). Endophytic microorganisms may be found in all of the plant's organs, but the roots have the largest number of endophytes (Lodewyckx et al. 2002). Another element to consider while choosing plants is the plant's overall health. Collecting samples from a healthy plant with no evident symptoms lowers the risk of isolating pathogens with endophytes (Kusari et al. 2015). Within 24 hours of collection from the target plant, the samples are processed. The plant material, on the other hand, is kept at 4°C in tight plastic bags to maintain the tissue freshness and prevent microbe death. The surface sterilizing step is required after the samples have been collected to guarantee that the isolated microorganisms are endophytes rather than epiphytes (Verma et al. 2009). In nutshell, surface sterilization begins with a 5–10 time sterile water rinse, followed by a wash of the plant tissue in 70%–75% ethanol for several minutes, and then a final rinse with distilled water under a laminar hood (Shah et al. 2022). For complete sterilization, extra phases of washing with sodium hypochlorite (NaOCl) or sodium thiosulfate ($Na_2S_2O_3$) might be added (Qin et al. 2009). Some of the spent water can be utilized as a negative control after the final wash of the plant material; effective sterilization is confirmed by the lack of microbial growth on the agar plate seeded with this water sample (Shi et al. 2013). For the isolation of the required microorganism, the sterile plant materials are sliced into small pieces and then transferred to the growing media. Figure 1. Sample preparation and endophyte isolation procedure. The plant sample is gathered, disinfected completely using water and ethanol, then fragmented into little pieces. Plant tissue is plated on a growth medium that is suited for them. Bacteria/fungi were added to the mixture and cultured until colonies appeared. Colonies should be sub-cultured in fresh plates until pure isolates may be obtained.

10.2. Isolation and Identification of Endophytic Microorganisms

Bacterial and fungal endophytes have been isolated from a wide range of plant tissues. According to sequences deposited in the NCBI nucleotide database, about 70% of the endophytes now isolated belong to the fungus kingdom, while the remaining 30% are bacteria. Both bacteria and fungi go through the same procedures, from collecting samples through surface sterilization and fragmentation of plant tissue, as detailed in the previous section (Surendirakumar et al. 2022). Later methods, such as growth medium selection, antibiotics, incubation duration, and molecular marker, differ between bacteria and fungus. Figure 2. Endophyte identification and characterization techniques. The diagram depicts the most popular methods for identifying and characterizing a pure endophytic isolate.

10.3. Biochemical Methods

Endophytic microbes promote plant development by altering phytohormone levels, either by boosting auxin, cytokines, and gibberellin synthesis or by reducing ethylene levels (Santoyo et al. 2016). Endophytic species ranging from bacteria (*Streptomyces*, *Enterobacter*, and *Bacillus*) to fungus (*Fusarium* and *Colletotrichum*) have been identified to be involved in this process

(Laguerre et al. 2001; Sánchez-Cruz et al. 2019). Calorimetric tests are used to assess the number of phytohormones in the body.

11. Auxin Indole-3-Acetic Acid

A typical technique for characterizing an endophytic microbe is the quantification of the auxin indole-3-acetic acid (IAA). Various endophytes, including *Herbaspirillum* and *Pseudomonas* species, have been identified using this technique. Depending on the investigating species, the isolates are cultivated in a liquid medium in the dark for a few days. The medium is supplied with or without tryptophan to elucidate the reliance of IAA production on the tryptophan pathways. The cells are centrifuged after incubation, and Salkowski's reagent is added to the extracted supernatant, followed by a brief (30-minute) incubation in the dark.

12. 1-Aminocyclopropane-1-Carboxylic Acid

The production of the growth-inhibiting hormone ethylene requires 1-aminocyclopropane-1-carboxylic acid (ACC). Some genera with enhanced ACC deaminase activity include *Neorhizobium*, *Arthrobacter*, and *Ensifer*. The enzyme ACC deaminase is found in PGP bacteria, and it cleaves ACC into -ketobutyrate and ammonia (Shah et al. 2018). Under stress conditions, plant growth-promoting (PGP) bacteria can degrade ACC molecules secreted from plant cells, reducing the availability of ACC and, as a result, the generation of ethylene. For PGP bacteria isolates, assays to measure ACC deaminase activity are often utilized. A colorimetric test that detects -ketobutyrate, the result of the ACC deaminase process, using the 2,4-dinitrophenylhydrazine reagent (Penrose and Glick 2003). Another way to check for ACC deaminase activity in isolates is to incubate them on a minimal medium (Dworkin and Foster-DF salt media) and see if they can grow on ACC as the sole source of nitrogen. Active ACC deaminase is shown by the presence of growth on DF agar plates (Sun et al. 2008).

Conclusion

Endophytes are a category of bacteria that can synthesize bioactive substances that can be utilized to fight a variety of infections. These have shown to be reliable sources of bioactive and chemically unique molecules, and have been beneficial in the development of new drugs. Biotransformation processes offer a wide range of applications, including the synthesis of antibacterial (vanillin, essential oils), antifungal and antiviral (alkaloids), antioxidant (eugenol), anti-inflammatory (cineole), and other bioactive molecules. To draw the attention of the research community to this emerging field and possible exploitation of the available sources for therapeutic uses in various fields such as medicine, pharmaceuticals, food, and cosmetics, it is critical to review and highlight previous successes, ongoing research, and recent developments in research associated with endophytic microorganisms. Understanding the morphological, physiological, and molecular processes unique to bacterially (rhizo- and bacterial endophytes) mediated resistance to biotic pressures has progressed faster than tolerance to abiotic challenges. As a result, additional research into the processes prompting IST by rhizobacteria/endophytes, including fungi and other fungal endophytes, will be critical in decoding signaling cascades and those leading to increased abiotic stress tolerance. Fungal endophytes use a newly identified biological phenomenon termed habitat-adapted symbiosis to allow their host plants to thrive in high-stress environments.

Acknowledgements

The authors acknowledge their respective institutions for encouragement and support.

Conflict of Interests

No conflict of interest was declared with any individual or organization.

References

Akiyama, Hisanori, Kazuyasu Fujii, Osamu Yamasaki, Takashi Oono, and Keiji Iwatsuki. "Antibacterial action of several tannins against *Staphylococcus aureus*." *Journal of antimicrobial chemotherapy* 48, no. 4 (2001): 487-491.

Azevedo, JL. "Microorganisms as endofiticos". In: Melo IS and Azavedo JL, editors. Ecologia Microbiana, Brazil: Editora Embrap, Jaguariuna, Sao Paulo, (1998) 117-137.

Azevedo, João Lúcio, Walter Maccheroni Jr, José Odair Pereira, and Welington Luiz De Araújo. "Endophytic microorganisms: a review on insect control and recent advances on tropical plants." *Electronic Journal of Biotechnology* 3, no. 1 (2000): 15-16.

Bakhtiyarifar, Marzieh, Naeimeh Enayatizamir, and Khosro Mehdi Khanlou. "Biochemical and molecular investigation of non-rhizobial endophytic bacteria as potential biofertilizers." *Archives of Microbiology* 203, no. 2 (2021): 513-521.

Barac, Tanja, Safiyh Taghavi, Brigitte Borremans, Ann Provoost, Licy Oeyen, Jan V. Colpaert, Jaco Vangronsveld, and Daniel van der Lelie. "Engineered endophytic bacteria improve phytoremediation of water-soluble, volatile, organic pollutants." *Nature biotechnology* 22, no. 5 (2004): 583-588.

Baucom, Deana L., Marie Romero, Robert Belfon, and Rebecca Creamer. "Two new species of *Undifilum*, fungal endophytes of *Astragalus* (locoweeds) in the United States." *Botany* 90, no. 9 (2012): 866-875.

Benhizia, Yacine, Hayet Benhizia, Ammar Benguedouar, Rosella Muresu, Alessio Giacomini, and Andrea Squartini. "Gammaproteobacteria can nodulate legumes of the genus *Hedysarum*." *Systematic and Applied Microbiology* 27, no. 4 (2004): 462-468.

Berg, Gabriele, Martin Grube, Michael Schloter, and Kornelia Smalla. "Unraveling the plant microbiome: looking back and future perspectives." *Frontiers in microbiology* 5 (2014): 148.

Berger, Ralf G. "Biotechnology of flavors-the next generation." *Biotechnology letters* 31, no. 11 (2009): 1651-1659.

Bicas, Juliano Lemos, Ana Paula Dionisio, and Glaucia Maria Pastore. "Bio-oxidation of terpenes: an approach for the flavor industry." *Chemical reviews* 109, no. 9 (2009): 4518-4531.

Bicas, Juliano Lemos, Francisco Fábio Cavalcante Barros, Roger Wagner, Helena Teixeira Godoy, and Gláucia Maria Pastore. "Optimization of R-(+)-α-terpineol production by the biotransformation of R-(+)-limonene." *Journal of Industrial Microbiology and Biotechnology* 35, no. 9 (2008): 1061-1070.

Bills, G. F. "Isolation and analysis of endophytic fungal communities from woody plants." In: Redlin SC, Cams LM, editors. *Endophytic Fungi in Grasses and Woody Plants; Systemic Ecology and Evolution.* St Paul; ABS Press, (1996) pp. 31-65.

Borges, Keyller Bastos, Warley de Souza Borges, Rosa Durán-Patrón, Mônica Tallarico Pupo, Pierina Sueli Bonato, and Isidro González Collado. "Stereoselective biotransformations using fungi as biocatalysts." *Tetrahedron: Asymmetry* 20, no. 4 (2009): 385-397.

Bultman, Thomas L., and John Charles Murphy. "Do fungal endophytes mediate wound-induced resistance?" *Microbial endophytes* (2000): 421-455.

Christiansen-Weniger, C., A. F. Groneman, and J. A. Van Veen. "Associative N2 fixation and root exudation of organic acids from wheat cultivars of different aluminum tolerance." *Plant and Soil* 139, no. 2 (1992): 167-174.

Clay, Keith, and Christopher Schardl. "Evolutionary origins and ecological consequences of endophyte symbiosis with grasses." *The American naturalist* 160, no. S4 (2002): S99-S127.

Clay, Keith, and Jenny Holah. "Fungal endophyte symbiosis and plant diversity in successional fields." *Science* 285, no. 5434 (1999): 1742-1744.

Clay, Keith, Susan Marks, and Gregory P. Cheplick. "Effects of insect herbivory and fungal endophyte infection on competitive interactions among grasses." *Ecology* 74, no. 6 (1993): 1767-1777.

Clay, Keith, Tad N. Hardy, and Abner M. Hammond. "Fungal endophytes of grasses and their effects on an insect herbivore." *Oecologia* 66, no. 1 (1985): 1-5.

Clay, Keith. "Fungal endophytes of grasses: a defensive mutualism between plants and fungi." *Ecology* 69, no. 1 (1988): 10-16.

Colegate, Steven M., Peter R. Dorling, and Clive R. Huxtable. "A spectroscopic investigation of swainsonine: An α-mannosidase inhibitor isolated from *Swainsona canescens*." *Australian Journal of Chemistry* 32, no. 10 (1979): 2257-2264.

Darwin, Charles. The origin of species by means of natural selection. Pub One Info, 1859.

De Bary, A. "Morphology and Physiology der Pilze, Flechten and Myxomyceten." *Holfmeister's Handbook of Physiological Botany*, Germany: (1866) Leipzig; Vol. 2.

De Meyer, Sofie E., Kristof De Beuf, Bram Vekeman, and Anne Willems. "A large diversity of non-rhizobial endophytes found in legume root nodules in Flanders (Belgium)." *Soil Biology and Biochemistry* 83 (2015): 1-11.

Dhole, Archana, and Harsha Shelat. "Non-Rhizobial Endophytes associated with nodules of *Vigna radiata* L. and their combined activity with *Rhizobium* sp." *Current Microbiology* 79, no. 4 (2022): 1-13.

Dimkpa, C. O., D. Merten, A. Svatoš, G. Büchel, and E. Kothe. "Siderophores mediate reduced and increased uptake of cadmium by *Streptomyces tendae F4* and sunflower (*Helianthus annuus*), respectively." *Journal of Applied Microbiology* 107, no. 5 (2009): 1687-1696.

Downing, Katrina J., Graeme Leslie, and Jennifer A. Thomson. "Biocontrol of the sugarcane borer *Eldana saccharina* by expression of the *Bacillus thuringiensis cry1Ac7* and *Serratia marcescens chiA* genes in sugarcane-associated bacteria." *Applied and Environmental Microbiology* 66, no. 7 (2000): 2804-2810.

Dubey, Anamika, Diksha Saiyam, Ashwani Kumar, Abeer Hashem, Elsayed Fathi Abd_Allah, and Mohammed Latif Khan. "Bacterial root endophytes: characterization of their competence and plant growth promotion in soybean (*Glycine max (L.) Merr.*) under drought stress." *International Journal of Environmental Research and Public Health* 18, no. 3 (2021): 931.

Ewald, Paul W. "Transmission modes and evolution of the parasitism-mutualism continuum." *Annals of the New York Academy of Sciences* 503 (1987): 295-306.

Ferdy, Jean-Baptiste, and Bernard Godelle. "Diversification of transmission modes and the evolution of mutualism." *The American Naturalist* 166, no. 5 (2005): 613-627.

Fernando, Dini, Marcelo-Jose Mortatti, Andre-Oliveira de Souza, Maccheroni Walter, Lucio Joao, and Luiz Welington. "Impact of genetically modified *Enterobacter cloacae* on the indigenous endophytic community of *Citrus sinensis* seedlings." *Journal of Microbiology* 42, no. 3 (2004): 169-173.

Filipov, N. M., F. N. Thompson, N. S. Hill, D. L. Dawe, J. A. Stuedemann, J. C. Price, and C. K. Smith. "Vaccination against ergot alkaloids and the effect of endophyte-infected fescue seed-based diets on rabbits." *Journal of animal science* 76, no. 9 (1998): 2456-2463.

Finkes, Laura K., Alan B. Cady, Juliana C. Mulroy, Keith Clay, and Jennifer A. Rudgers. "Plant–fungus mutualism affects spider composition in successional fields." *Ecology Letters* 9, no. 3 (2006): 347-356.

Fuentes-Ramírez, Luis E., Jesús Caballero-Mellado, Jorge Sepúlveda, and Esperanza Martínez-Romero. "Colonization of sugarcane by *Acetobacter diazotrophicus* is inhibited by high N-fertilization." *FEMS Microbiology Ecology* 29, no. 2 (1999): 117-128.

Gadberry, M. S., T. M. Denard, D. E. Spiers, and E. L. Piper. "Effects of feeding ergovaline on lamb performance in a heat stress environment." *Journal of animal science* 81, no. 6 (2003): 1538-1545.

Gaiero, Jonathan R., Crystal A. McCall, Karen A. Thompson, Nicola J. Day, Anna S. Best, and Kari E. Dunfield. "Inside the root microbiome: bacterial root endophytes and plant growth promotion." *American journal of botany* 100, no. 9 (2013): 1738-1750.

Gallagher, R. T., A. G. Campbell, A. D. Hawkes, P. T. Holland, D. A. McGaveston, E. A. Pansier, and I. C. Harvey. "Ryegrass staggers: the presence of *lolitrem* neurotoxins in perennial ryegrass seed." *New Zealand veterinary journal* 30, no. 11 (1982): 183-184.

Gallagher, R. T., E. P. White, and P. H. Mortimer. "Ryegrass staggers: isolation of potent neurotoxins lolitrem A and lolitrem B from staggers-producing pastures." *New Zealand veterinary journal* 29, no. 10 (1981): 189-190.

Gallagher, Rex T., Allan D. Hawkes, Pieter S. Steyn, and Robert Vleggaar. "Tremorgenic neurotoxins from perennial ryegrass causing ryegrass staggers disorder of livestock: structure elucidation of lolitrem B." *Journal of the Chemical Society, chemical communications* 9 (1984): 614-616.

Gardner, John M., Albert W. Feldman, and Robert M. Zablotowicz. "Identity and behavior of xylem-residing bacteria in rough lemon roots of Florida citrus trees." *Applied and Environmental Microbiology* 43, no. 6 (1982): 1335-1342.

Glassner, Hanoch, Einat Zchori-Fein, Stéphane Compant, Angela Sessitsch, Nurit Katzir, Vitaly Portnoy, and Sima Yaron. "Characterization of endophytic bacteria from cucurbit fruits with potential benefits to agriculture in melons (*Cucumis melo* L.)." *FEMS Microbiology Ecology* 91, no. 7 (2015).

Golinska, Patrycja, Magdalena Wypij, Gauravi Agarkar, Dnyaneshwar Rathod, Hanna Dahm, and Mahendra Rai. "Endophytic actinobacteria of medicinal plants: diversity and bioactivity." *Antonie Van Leeuwenhoek* 108, no. 2 (2015): 267-289.

Gray, E. J., and D. L. Smith. "Intracellular and extracellular PGPR: commonalities and distinctions in the plant–bacterium signaling processes." *Soil biology and biochemistry* 37, no. 3 (2005): 395-412.

Guo, Zhikai, Cuijuan Gai, Caihong Cai, Liangliang Chen, Shoubai Liu, Yanbo Zeng, Jingzhe Yuan, Wenli Mei, and Haofu Dai. "Metabolites with insecticidal activity from *Aspergillus fumigatus* JRJ111048 isolated from mangrove plant *Acrostichum specioum* endemic to Hainan Island." *Marine drugs* 15, no. 12 (2017): 381.

Hallmann, J., R. Rodrıguez-Kábana, and J. W. Kloepper. "Chitin-mediated changes in bacterial communities of the soil, rhizosphere and within roots of cotton in relation to nematode control." *Soil Biology and Biochemistry* 31, no. 4 (1999): 551-560.

Hartley, W. J. "Some observations on the pathology of *Swainsona spp.* poisoning in farm livestock in eastern Australia." *Acta neuropathologica* 18, no. 4 (1971): 342-355.

Hill, N. S., Frederick Nimrod Thompson, D. L. Dawe, and John Alfred Stuedemann. "Antibody binding of circulating ergot alkaloids in cattle grazing tall fescue." *Am. J. Vet. Res* 55 (1994): 419-424.

Huang, Y. Q., Zhang, E. Y. & Pan, W. F. Current status of locoweed toxicity. *Shandong Science* (2003) 16, 34–39.

Hurek, Thomas, Linda L. Handley, Barbara Reinhold-Hurek, and Yves Piché. "Azoarcus grass endophytes contribute fixed nitrogen to the plant in an unculturable state." *Molecular Plant-Microbe Interactions* 15, no. 3 (2002): 233-242.

Hussain, Md Sarfaraj, Sheeba Fareed, Md Saba Ansari, Akhlaquer Rahman, Iffat Zareen Ahmad, and Mohd Saeed. "Current approaches toward production of secondary plant metabolites." *Journal of pharmacy & bioallied sciences* 4, no. 1 (2012): 10.

Iniguez, A. Leonardo, Yuemei Dong, and Eric W. Triplett. "Nitrogen fixation in wheat provided by *Klebsiella pneumoniae 342*." *Molecular Plant-Microbe Interactions* 17, no. 10 (2004): 1078-1085.

Jalgaonwala, Ruby Erach, Bhavna Vishwas Mohite, and Raghunath Totaram Mahajan. "A review: natural products from plant-associated endophytic fungi." *J Microbiol Biotechnol Res* 1, no. 2 (2011): 21-32.

James, Lynn F., Kent R. Van Kampen, and William J. Hartley. "Comparative pathology of *Astragalus* (locoweed) and *Swainsona* poisoning in sheep." *Pathologia veterinaria* 7, no. 2 (1970): 116-124.

Johnson, M. C., D. L. Dahlman, M. R. Siegel, L. P. Bush, G. C. M. Latch, D. A. Potter, and D. R. Varney. "Insect feeding deterrents in endophyte-infected tall fescue." *Applied and Environmental Microbiology* 49, no. 3 (1985): 568-571.

Jones, T. C., Hunt, R. D. and King, N. W. "*Veterinary Pathology*", Williams and Wilkins, Baltimore (1997) pp. 1–1392.

Joseph, Baby, and R. Mini Priya. "Bioactive compounds from endophytes and their potential in pharmaceutical effect: a review." *Am J Biochem Mol Biol* 1, no. 3 (2011): 291-309.

Kharwar, Ravindra N., Ashish Mishra, Surendra K. Gond, Andrea Stierle, and Donald Stierle. "Anticancer compounds derived from fungal endophytes: their importance and future challenges." *Natural product reports* 28, no. 7 (2011): 1208-1228.

Kloepper, Joseph W., and Chantal J. Beauchamp. "A review of issues related to measuring colonization of plant roots by bacteria." *Canadian journal of microbiology* 38, no. 12 (1992): 1219-1232.

Konuklugil, Belma. "The importance of Aryltetralin (*Podophyllum*) lignans and their distribution in the plant kingdom." *Journal of Faculty of Pharmacy of Ankara University* 24, no. 2 (1995): 109-125.

Krings, Ulrich, Björn Hardebusch, Dieter Albert, Ralf G. Berger, Mario Maróstica, and Glaucia M. Pastore. "Odor-active alcohols from the fungal transformation of α-farnesene." *Journal of Agricultural and Food Chemistry* 54, no. 24 (2006): 9079-9084.

Kumar, D. Siva Sundara, and Kevin D. Hyde. "Biodiversity and tissue-recurrence of endophytic fungi in *Tripterygium wilfordii*." *Fungal diversity* (2004).

Kumar, Suneel, Ravindra Prasad Aharwal, Harshita Shukla, R. C. Rajak, and Sardul Singh Sandhu. "Endophytic fungi: as a source of antimicrobials bioactive compounds." *World J Pharm Pharm Sci* 3, no. 2 (2014): 1179-1197.

Kusari, Parijat, Souvik Kusari, Michael Spiteller, and Oliver Kayser. "Implications of endophyte-plant crosstalk in light of quorum responses for plant biotechnology." *Applied microbiology and biotechnology* 99, no. 13 (2015): 5383-5390.

Kusari, Souvik, Christian Hertweck, and Michael Spiteller. "Chemical ecology of endophytic fungi: origins of secondary metabolites." *Chemistry & biology* 19, no. 7 (2012): 792-798

Laguerre, Gisèle, Sarah M. Nour, Valérie Macheret, Juan Sanjuan, Pascal Drouin, and Noëlle Amarger. "Classification of rhizobia based on nodC and nifH gene analysis reveals a close phylogenetic relationship among *Phaseolus vulgaris* symbionts. The GenBank accession numbers for the sequences reported in this paper are AF217261 through AF217272 for nodC and AF218126, AF275670 and AF275671 for nifH." *Microbiology* 147, no. 4 (2001): 981-993.

Lodewyckx, Cindy, Jaco Vangronsveld, Fiona Porteous, Edward RB Moore, Safieh Taghavi, Max Mezgeay, and Daniel van der Lelie. "Endophytic bacteria and their potential applications." *Critical reviews in plant sciences* 21, no. 6 (2002): 583-606.

Lorenz, Nicole, Thomas Haarmann, Sylvie Pažoutová, Manfred Jung, and Paul Tudzynski. "The ergot alkaloid gene cluster: functional analyses and evolutionary aspects." *Phytochemistry* 70, no. 15-16 (2009): 1822-1832.

Manias, Dimitrios, Akanksha Verma, and Dharmendra Kumar Soni. "Isolation and characterization of endophytes: Biochemical and molecular approach." In *Microbial Endophytes*, Woodhead Publishing, (2020): pp. 1-14.

Matthews, Jeffrey W., and Keith Clay. "Influence of fungal endophyte infection on plant-soil feedback and community interactions." *Ecology* 82, no. 2 (2001): 500-509.

Menpara, Disha, and Sumitra Chanda. "Endophytic bacteria-unexplored reservoir of antimicrobials for combating microbial pathogens." *Microbial pathogens and strategies for combating them: Science, technology, and education* (2013): 1095-1103.

Mitter, Birgit, Alexandra Petric, Maria W. Shin, Patrick SG Chain, Lena Hauberg-Lotte, Barbara Reinhold-Hurek, Jerzy Nowak, and Angela Sessitsch. "Comparative genome analysis of *Burkholderia phytofirmans PsJN* reveals a wide spectrum of endophytic lifestyles based on interaction strategies with host plants." *Frontiers in plant science* 4 (2013): 120.

Molyneux, Russell J., and Lynn F. James. "Loco intoxication: indolizidine alkaloids of spotted locoweed (*Astragalus lentiginosus*)." *Science* 216, no. 4542 (1982): 190-191.

Nair, Dhanya N., and S. Padmavathy. "Impact of endophytic microorganisms on plants, environment, and humans." *The Scientific World Journal* 2014 (2014). doi: 10.1155/2014/250693

Newman, Lee A., and Charles M. Reynolds. "Bacteria and phytoremediation: new uses for endophytic bacteria in plants." *Trends in Biotechnology* 23, no. 1 (2005): 6-8.

Oldrup, Erik, Jennifer McLain-Romero, Anna Padilla, Andrew Moya, Dale Gardner, and Rebecca Creamer. "Localization of endophytic *Undifilum* fungi in locoweed seed and influence of environmental parameters on a locoweed *in vitro* culture system." *Botany* 88, no. 5 (2010): 512-521.

Omojate Godstime, C., O. Enwa Felix, O. Jewo Augustina, and O. Eze Christopher. "Mechanisms of antimicrobial actions of phytochemicals against enteric pathogens–a review." *J Pharm Chem Biol Sci* 2, no. 2 (2014): 77-85.

Panaccione, D. G. "Ergot alkaloids". *The Mycota, Vol. X, Industrial Applications*, 2nd edn (ed. M. Hofrichter), pp. 195–214. Springer-Verlag, Berlin-Heidelburg. (2010)

Panter, K. E., L. F. James, B. L. Stegelmeier, M. H. Ralphs, and J. A. Pfister. "Locoweeds: effects on reproduction in livestock." *Journal of Natural Toxins* 8, no. 1 (1999): 53-62.

Parish, J. A., M. A. McCann, R. H. Watson, C. S. Hoveland, L. L. Hawkins, N. S. Hill, and J. H. Bouton. "Use of non-ergot alkaloid-producing endophytes for alleviating tall fescue toxicosis in sheep." *Journal of animal science* 81, no. 5 (2003): 1316-1322.

Parish, J. A., M. A. McCann, R. H. Watson, N. N. Paiva, C. S. Hoveland, A. H. Parks, B. L. Upchurch, N. S. Hill, and J. H. Bouton. "Use of non-ergot alkaloid-producing endophytes for alleviating tall fescue toxicosis in stocker cattle." *Journal of animal science* 81, no. 11 (2003b): 2856-2868.

Penrose, Donna M., and Bernard R. Glick. "Methods for isolating and characterizing ACC deaminase-containing plant growth-promoting rhizobacteria." *Physiologia plantarum* 118, no. 1 (2003): 10-15.

Pfister, J. A., B. L. Stegelmeier, D. R. Gardner, and L. F. James. "Grazing of spotted locoweed (*Astragalus lentiginosus*) by cattle and horses in Arizona." *Journal of animal science* 81, no. 9 (2003): 2285-2293.

Phelan, Vanessa V., Wei-Ting Liu, Kit Pogliano, and Pieter C. Dorrestein. "Microbial metabolic exchange-the chemotype-to-phenotype link." *Nature chemical biology* 8, no. 1 (2012): 26-35.

Pimentel, Mariana Recco, Gustavo Molina, Ana Paula Dionísio, Mário Roberto Maróstica Junior, and Gláucia Maria Pastore. "The use of endophytes to obtain bioactive compounds and their application in biotransformation process." *Biotechnology research international* 2011 (2011).doi: 10.4061/2011/576286.

Pirttilä, Anna Maria, Päivi Joensuu, Helmut Pospiech, Jorma Jalonen, and Anja Hohtola. "Bud endophytes of Scots pine produce adenine derivatives and other compounds that affect morphology and mitigate browning of callus cultures." *Physiologia Plantarum* 121, no. 2 (2004): 305-312.

Powell, Richard G., Ronald D. Plattner, Shelly G. Yates, Keith Clay, and Adrian Leuchtmann. "Ergobalansine, a new ergot-type peptide alkaloid isolated from *Cenchrus echinatus* (sandbur grass) infected with *Balansia obtecta*, and produced in liquid cultures of *B. obtecta* and *Balansia cyperi*." *Journal of natural products* 53, no. 5 (1990): 1272-1279.

Pryor, B. M., R. Creamer, R. A. Shoemaker, J. McLain-Romero, and S. Hambleton. "Undifilum, a new genus for endophytic *Embellisia oxytropis* and parasitic *Helminthosporium bornmuelleri* on legumes." *Botany* 87, no. 2 (2009): 178-194.

Qin, Sheng, Jie Li, Hua-Hong Chen, Guo-Zhen Zhao, Wen-Yong Zhu, Cheng-Lin Jiang, Li-Hua Xu, and Wen-Jun Li. "Isolation, diversity, and antimicrobial activity of rare actinobacteria from medicinal plants of tropical rain forests in Xishuangbanna, China." *Applied and environmental microbiology* 75, no. 19 (2009): 6176-6186.

Ralphs, M. H., Daniel Cook, D. R. Gardner, and D. S. Grum. "Transmission of the locoweed endophyte to the next generation of plants." *Fungal Ecology* 4, no. 4 (2011): 251-255.

Rodriguez, R. J., J. F. White Jr, Anne E. Arnold, and A. RS and Redman. "Fungal endophytes: diversity and functional roles." *New phytologist* 182, no. 2 (2009): 314-330.

Romero, Fernando M., María Marina, and Fernando L. Pieckenstain. "The communities of tomato (*Solanum lycopersicum* L.) leaf endophytic bacteria, analyzed by 16S-ribosomal RNA gene pyrosequencing." *FEMS microbiology letters* 351, no. 2 (2014): 187-194.

Rosenblueth, Mónica, and Esperanza Martínez-Romero. "*Rhizobium etli* maize populations and their competitiveness for root colonization." *Archives of microbiology* 181, no. 5 (2004): 337-344.

Rowan, Daryl D. "Lolitrems, peramine and paxilline: mycotoxins of the ryegrass/endophyte interaction." *Agriculture, Ecosystems & Environment* 44, no. 1-4 (1993): 103-122.

Rowan, Daryl D., Jenny J. Dymock, and Margaret A. Brimble. "Effect of fungal metabolite peramine and analogs on feeding and development of Argentine stem weevil (*Listronotus bonariensis*)." *Journal of chemical ecology* 16, no. 5 (1990): 1683-1695.

Rowan, Daryl D., Martin B. Hunt, and David L. Gaynor. "Peramine, a novel insect feeding deterrent from ryegrass infected with the endophyte *Acremonium loliae*." *Journal of the Chemical Society, chemical communications* 12 (1986): 935-936.

Rudgers, Jennifer A., Jennifer M. Koslow, and Keith Clay. "Endophytic fungi alter relationships between diversity and ecosystem properties." *Ecology Letters* 7, no. 1 (2004): 42-51.

Rudgers, Jennifer A., Michelle E. Afkhami, Megan A. Rúa, Andrew J. Davitt, Samantha Hammer, and Valérie M. Huguet. "A fungus among us: broad patterns of endophyte distribution in the grasses." *Ecology* 90, no. 6 (2009): 1531-1539.

Ryu, Choong-Min, Mohamed A. Farag, Chia-Hui Hu, Munagala S. Reddy, Han-Xun Wei, Paul W. Paré, and Joseph W. Kloepper. "Bacterial volatiles promote growth in *Arabidopsis*." *Proceedings of the National Academy of Sciences* 100, no. 8 (2003): 4927-4932.

Saikia, Sanjay, Matthew J. Nicholson, Carolyn Young, Emily J. Parker, and Barry Scott. "The genetic basis for indole-diterpene chemical diversity in filamentous fungi." *Mycological Research* 112, no. 2 (2008): 184-199.

Sánchez-Cruz, Ricardo, Irán Tpia Vázquez, Ramón Alberto Batista-García, Erick Williams Méndez-Santiago, María del Rayo Sánchez-Carbente, Alfonso Leija, Verónica Lira-Ruan, Gerogina Hernández, Arnoldo Wong-Villarreal, and Jorge Luis Folch-Mallol. "Isolation and characterization of endophytes from nodules of *Mimosa pudica* with biotechnological potential." *Microbiological research* 218 (2019): 76-86.

Santoyo, Gustavo, Gabriel Moreno-Hagelsieb, Ma del Carmen Orozco-Mosqueda, and Bernard R. Glick. "Plant growth-promoting bacterial endophytes." *Microbiological research* 183 (2016): 92-99.

Sattelmacher, Burkhard. "The apoplast and its significance for plant mineral nutrition." *New Phytologist* 149, no. 2 (2001): 167-192.

Schardl, Christopher L. "The *Epichloae*, symbionts of the grass subfamily Poöideae." *Annals of the Missouri Botanical Garden* (2010): 646-665.

Schardl, Christopher L., Carolyn A. Young, Jerome R. Faulkner, Simona Florea, and Juan Pan. "Chemotypic diversity of *Epichloae*, fungal symbionts of grasses." *Fungal Ecology* 5, no. 3 (2012): 331-344.

Schardl, Christopher L., Robert B. Grossman, Padmaja Nagabhyru, Jerome R. Faulkner, and Uma P. Mallik. "Loline alkaloids: currencies of mutualism." *Phytochemistry* 68, no. 7 (2007): 980-996.

Schulz, Barbara, Christine Boyle, Siegfried Draeger, Anne-Katrin Römmert, and Karsten Krohn. "Endophytic fungi: a source of novel biologically active secondary metabolites." *Mycological research* 106, no. 9 (2002): 996-1004.

Seghers, Dave, Lieven Wittebolle, Eva M. Top, Willy Verstraete, and Steven D. Siciliano. "Impact of agricultural practices on the *Zea mays* L. Endophytic community." *Applied and Environmental Microbiology* 70, no. 3 (2004): 1475-1482.

Sessitsch, A., J. G. Howieson, X. Perret, H. Antoun, and E. Martinez-Romero. "Advances in *Rhizobium* research." *Critical Reviews in Plant Sciences* 21, no. 4 (2002): 323-378.

Sessitsch, Angela, Birgit Reiter, and Gabriele Berg. "Endophytic bacterial communities of field-grown potato plants and their plant-growth-promoting and antagonistic abilities." *Canadian Journal of microbiology* 50, no. 4 (2004): 239-249.

Sevilla, Myrna, Robert H. Burris, Nirmala Gunapala, and Christina Kennedy. "Comparison of benefit to sugarcane plant growth and 15N2 incorporation following inoculation of sterile plants with *Acetobacter diazotrophicus* wild-type and nif mutant strains." *Molecular plant-microbe interactions* 14, no. 3 (2001): 358-366.

Shah, Dawood, Mohammad Sayyar Khan, Shahkaar Aziz, Haidar Ali, and Lorenzo Pecoraro. "Molecular and biochemical characterization, antimicrobial activity, stress tolerance, and plant growth-promoting effect of Endophytic bacteria isolated from Wheat varieties." *Microorganisms* 10, no. 1 (2021): 21.

Shah, Sujit, Roshani Shrestha, Sabitri Maharjan, Marc-Andre Selosse, and Bijaya Pant. "Isolation and characterization of plant growth-promoting endophytic fungi from the roots of *Dendrobium moniliforme*." *Plants* 8, no. 1 (2018): 5.

Shi, YingWu, Xuebing Zhang, Kai Lou, and Michael Adams. "Isolation, characterization, and insecticidal activity of an endophyte of drunken horse grass, *Achnatherum inebrians*." *Journal of Insect Science* 13, no. 1 (2013).

Shukla, S. T., P. V. Habbu, V. H. Kulkarni, K. S. Jagadish, A. R. Pandey, and V. N. Sutariya. "Endophytic microbes: a novel source for biologically/pharmacologically active secondary metabolites." *Asian J Pharmacol Toxicol* 2, no. 3 (2014): 1-6.

Siciliano, Steven D., Nathalie Fortin, Anca Mihoc, Gesine Wisse, Suzanne Labelle, Danielle Beaumier, Danielle Ouellette et al. "Selection of specific endophytic bacterial genotypes by plants in response to soil contamination." *Applied and environmental microbiology* 67, no. 6 (2001): 2469-2475.

Sonaimuthu, Visalakchi, and Muthumary Johnpaul. "Taxol (anticancer drug) producing endophytic fungi: an overview." *International Journal of Pharma and Bio Sciences* 1, no. 3 (2010).

Steiner, Ulrike, Sarah Leibner, Christopher Lewis Schardl, Adrian Leuchtmann, and Eckhard Leistner. "*Periglandula*, a new fungal genus within the Clavicipitaceae and its association with Convolvulaceae." *Mycologia* 103, no. 5 (2011): 1133-1145.

Strobel, Gary A. "Endophytes as sources of bioactive products." *Microbes and infection* 5, no. 6 (2003): 535-544.

Strobel, Gary and Bryn Daisy. "Bioprospecting for microbial endophytes and their natural products." *Microbiology and molecular biology reviews* 67, no. 4 (2003): 491-502.

Strobel, Gary, Bryn Daisy, Uvidelio Castillo, and James Harper. "Natural products from endophytic microorganisms." *Journal of Natural products* 67, no. 2 (2004): 257-268.

Strobel, Gary, Bryn Daisy, Uvidelio Castillo, and James Harper. "Natural products from endophytic microorganisms." *Journal of Natural products* 67, no. 2 (2004): 257-268.

Sturz, A. V., and B. R. Christie. "Endophytic bacterial systems governing red clover growth and development." *Annals of Applied Biology* 126, no. 2 (1995): 285-290.

Suman, Archana, Ajar Nath Yadav, and Priyanka Verma. "Endophytic microbes in crops: diversity and beneficial impact for sustainable agriculture." In *Microbial inoculants in sustainable agricultural productivity*, pp. 117-143. Springer, New Delhi, 2016.

Sun, Lei, Fubin Qiu, Xiaoxia Zhang, Xin Dai, Xiuzhu Dong, and Wei Song. "Endophytic bacterial diversity in rice (*Oryza sativa* L.) roots estimated by 16S rDNA sequence analysis." *Microbial ecology* 55, no. 3 (2008): 415-424.

Surendirakumar, Kannaiah, Radha Raman Pandey, Thangavelu Muthukumar, Anbazhagan Sathiyaseelan, Surbala Loushambam, and Amit Seth. "Characterization and biological activities of melanin pigment from root endophytic fungus, *Phoma* sp. RDSE17." *Archives of Microbiology* 204, no. 3 (2022): 1-15.

Tan, Ren Xiang, and Wen Xin Zou. "Endophytes: a rich source of functional metabolites." *Natural product reports* 18, no. 4 (2001): 448-459.

Tanaka, Aiko, Brian A. Tapper, Alison Popay, Emily J. Parker, and Barry Scott. "A symbiosis expressed non-ribosomal peptide synthetase from a mutualistic fungal endophyte of perennial ryegrass confers protection to the symbiote from insect herbivory." *Molecular microbiology* 57, no. 4 (2005): 1036-1050.

Tiwari, Pragya, Hanhong Bae. "Horizontal gene transfer and Endophytes: An implication for the acquisition of novel traits." *Plants* 9(2020a): 305.

Tiwari, Pragya, Mangalam Bajpai, Lalit Kumar Singh, Shashank Mishra, and Ajar Nath Yadav. "Phytohormones producing fungal communities: Metabolic engineering for abiotic stress tolerance in plants." In *Agriculturally important fungi for sustainable agriculture*. edited by Dr. Vijai Kumar Gupta and Dr. Maria Tuohy, Springer publication, 2020b.

Tiwari, Pragya, Subir Kumar Bose, and Hanhong Bae. "Plant Growth Promoting Soil Microbiomes: beneficial attributes and potential applications." In *Soil Microbiomes for Sustainable Agriculture* - Volume 2: Functional Annotation. edited by Dr. Ajar Nath Yadav, Springer Publication (2021a).

Tiwari, Pragya, Mangalam Bajpai, Lalit Kumar Singh, Ajar Nath Yadav, Hanhong Bae. "Portraying fungal mechanisms in stress tolerance: perspective for sustainable agriculture." In Springer-Nature book on *Recent Trends in Mycological Research*, 2021b.

Tiwari, Pragya, Yashdeep Srivastava, Mangalam Bajpai, Abhishek Sharma. "Bioactive metabolites from natural sources: Prospects and significance in drug discovery and research." *Bioingene PSJ* 1, (2021c):1-14.

Tiwari, Pragya, Yashdeep Srivastava, and Hanhong Bae. "Trends of pharmaceutical design of Endophytes as anti-infective." *Current Topics in Medicinal Chemistry* 21, no. 17 (2021d): 1572-1586.

Tiwari, Pragya, Tushar Khare, Varsha Shriram, Hanhong Bae, Vinay Kumar. "Exploring synthetic biology strategies for producing potent antimicrobial phytochemicals." *Biotechnology Advances* 48, (2021b):107729.

Tiwari, Pragya, Yashdeep Srivastava, Vinay Kumar. "Antimicrobial peptides as effective agents against drug-resistant pathogens." In *Antimicrobial Resistance*; Kumar, V., Shriram, V., Paul, A., Thakur, M., Eds.; Springer: Singapore, 2022a; pp. 289-322.

Tiwari, Pragya, Hanhong Bae. "Endophytic fungi: Key insights, emerging prospects, and challenges in natural product drug discovery." *Microorganisms* 10 (2022b): 360. https://doi.org/10.3390/microorganisms10020360.

Tofern, Britta, Macki Kaloga, Ludger Witte, Thomas Hartmann, and Eckart Eich. "Occurrence of loline alkaloids in *Argyreia mollis* (Convolvulaceae)." *Phytochemistry* 51, no. 8 (1999): 1177-1180.

Tokala, Ranjeet K., Janice L. Strap, Carina M. Jung, Don L. Crawford, Michelle Hamby Salove, Lee A. Deobald, J. Franklin Bailey, and M. J. Morra. "Novel plant-microbe rhizosphere interaction involving *Streptomyces lydicus WYEC108* and the pea plant (*Pisum sativum*)." *Applied and environmental microbiology* 68, no. 5 (2002): 2161-2171.

Turner, J. T., J. S. Lampel, R. S. Stearman, G. W. Sundin, P. Gunyuzlu, and J. J. Anderson. "Stability of the delta-endotoxin gene from *Bacillus thuringiensis subsp. kurstaki* is a recombinant strain of *Clavibacter xyli subsp. cynodontis*." *Applied and Environmental Microbiology* 57, no. 12 (1991): 3522-3528.

van der Lelie, Daniel, Tanja Barac, Safiyh Taghavi, and Jaco Vangronsveld. "Response to Newman: new uses of endophytic bacteria to improve phytoremediation." *Trends in Biotechnology* 1, no. 23 (2005): 8-9.

Verma, Vijay C., Surendra K. Gond, Anuj Kumar, Ashish Mishra, Ravindra N. Kharwar, and Alan C. Gange. "Endophytic actinomycetes from *Azadirachta indica A. Juss.*: isolation, diversity, and anti-microbial activity." *Microbial ecology* 57, no. 4 (2009): 749-756.

Wallwey, Christiane, and Shu-Ming Li. "Ergot alkaloids: structure diversity, biosynthetic gene clusters and functional proof of biosynthetic genes." *Natural product reports* 28, no. 3 (2011): 496-510.

Wang, Qi, Hideyuki Nagao, Yu-Ling Li, Hong-Sheng Wang, and Makoto Kakishima. "*Embellisia oxytropis*, a new species isolated from *Oxytropis kansuensis* in China." *Mycotaxon* (2006).

Werner, Gijsbert DA, Joan E. Strassmann, Aniek BF Ivens, Daniel JP Engelmoer, Erik Verbruggen, David C. Queller, Ronald Noë, Nancy Collins Johnson, Peter Hammerstein, and E. Toby Kiers. "Evolution of microbial markets." *Proceedings of the National Academy of Sciences* 111, no. 4 (2014): 1237-1244.

West, Stuart A., Ashleigh S. Griffin, Andy Gardner, and Stephen P. Diggle. "Social evolution theory for microorganisms." *Nature reviews microbiology* 4, no. 8 (2006): 597-607.

Wilson, Dennis. "Endophyte: the evolution of a term, and clarification of its use and definition." *Oikos* (1995): 274-276.

Yadav, Garima, and Mukesh Meena. "Bioprospecting of endophytes in medicinal plants of Thar Desert: An attractive resource for biopharmaceuticals." *Biotechnology Reports* 30 (2021): e00629.

Yu, Yongtao, Qingmei Zhao, Jianna Wang, Jianhua Wang, Yan Wang, Yumin Song, Guoxia Geng, and Qinfan Li. "Swainsonine-producing fungal endophytes from major locoweed species in China." *Toxicon* 56, no. 3 (2010): 330-338.

Zhang, Peng, Peng-Peng Zhou, and Long-Jiang Yu. "An endophytic taxol-producing fungus from *Taxus media, Cladosporium cladosporioides MD2.*" *Current microbiology* 59, no. 3 (2009): 227-232.

Zhang, Yanyan, Ting Han, Qianliang Ming, Lingshang Wu, Khalid Rahman, and Luping Qin. "Alkaloids produced by endophytic fungi: a review." *Natural Product Communications* 7, no. 7 (2012): 1934578X1200700742.

Chapter 3

Bacterial and Fungal Endophytes from Medicinal Plants and Bio-Prospection of Their Secondary Metabolites

Anupam Maurya[1,*] and V. Gopi[2]
[1]Chemistry Department, Pharmacopoeia Commission for Indian Medicine and Homoeopathy (PCIM&H), Ministry of Ayush, Ghaziabad, India
[2]Microbiology Department, Pharmacopoeia Commission for Indian Medicine and Homoeopathy (PCIM&H), Ministry of Ayush, Ghaziabad, India

Abstract

Endophytes mean "in the plant" (In Greek endon = inside, phyton = plant), which are a diverse group of microorganisms present in plant tissues, including fungi and bacteria. The intracellular spaces within the plant tissues can be colonized by microbes without exerting any adverse or pathological effects. The symptomless endophyte colonization in the plant caught the attention of ecologists, chemists, and researchers for having enormous potential in biotechnological applications. Endophyte microbes are beneficial to plants in many ways, they can carry nutrients from plants, increase the stress tolerance of plants, increase plant disease resistance, modulate plant development, suppress the virulence of pathogens, and suppress the development of competing plant species. Endophytic microbes can help in plant development and yield and can also act as biocontrol agents. Endophytes can also be useful to their host by making a range of natural products that could be harnessed for potential use in medicine, agriculture, or industry. Additionally, they have been shown to have the potential to remove contaminants from soil

*Corresponding Author's Email: manupamk@gmail.com.

In: Endophytes
Editor: Pragya Tiwari
ISBN: 979-8-88697-045-6

by enhancing phytoremediation and play an important role in soil fertility through phosphate solubilization and nitrogen fixation. Currently, the world population is facing disastrous chronic diseases that affect humans. The resistance of pathogens to commercial antibiotics is increasing, thus limiting the therapeutic potential and effectiveness of antibiotics to treat diseases. Inevitably, there is a need to search for novel, affordable and nontoxic natural bioactive compounds from endophytes, that can be used for developing new drugs with multifunctional mechanisms against drug-resistant pathogens. Naturally, plants harbor diverse populations of microbial domains. The independent synthesis of bioactive compounds is very difficult and time-consuming with the emergence of new drug-resistant pathogens. Hence endophytes are being used as good alternative sources of diversified chemical compounds for the treatment of human diseases. Endophytes are well known to produce different primary and secondary metabolites as therapeutic agents. Studies have found that approximately 51% of new drugs developed between 1981 and 2014, synthesized from natural sources. Nowadays, many endophytic natural products are considered new bioactive entities owing a wide range of important biological activities such as antioxidant, antibiotic, antitumor, anti-inflammatory, etc. The endophytes may play an important role in the biotechnological production of bioactive substances as medicinally important agents.

Keywords: bio-prospection, endophytes, secondary metabolites, medicinal values

1. Introduction

From ancient times, people relied on drugs from nature for the treatment of many ailments. Medicinal plants are traditionally used around the world as remedies for the treatment of various diseases, including asthma, gastrointestinal symptoms, skin disorders, respiratory and urinary problems, and liver and cardiovascular diseases (Wyk and Portland 2006; Tian et al. 2014). Traditional systems of medicine continue to be widely practiced due to population rise, inadequate supply of drugs, exorbitant cost of treatments, and development of resistance to currently used drugs. Infectious diseases have led to increased stress on the use of plant materials as a source of medicine and drug leads. Treatment with herbal medicines is considered very safe as there are few or no side effects. India has about 45,000 species of plants and of these, several thousand are said to have medicinal properties. Medicinal plants such as Neem, Turmeric, Tulsi, and Ginger cure several common diseases. The best known and universally used medicine is aspirin (salicylic acid), which possesses the natural origins of the glycoside salicin located in many plant species of the genera *Salix* and *Populus*. Over the last two decades, the use of herbal medicines has considerably increased. However, there is still a significant lack of research data in this area. Recently, the World Health Organization has estimated that 80% of people around the world are dependent on herbal medicines for some aspects of their key health care needs. According to the WHO, around 21,000 plant species have the potential to be used as medicinal plants.

Medicinal plants are used to cure various human diseases as they are sources of various bioactive molecules (Arora et al. 2010). These medicinal properties of plants bioactive molecules include, alkaloids, flavonoids, flavonols, anthraquinones, naphthopyrone glucosides, phenolics, rotenoids, saponins, steroids, terpenes and terpenoids, polyphenols, xanthenes, cytochalasins, benzofurans, polysaccharides, antimicrobials, lignans, alkyl salicylic acids, and carboxylic acids (Bajguz 2007; Cushnie et al. 2014). The bioactive molecules are important for plants to survive and flourish in different environmental conditions (Treutter 2006; Vardhini and Anjhum 2015). Despite the fact that a large number of medicinal plants have been extensively researched in terms of their phytochemical constituents and pharmacological properties (Köberl et al. 2013). The composition of biologically active compounds of medicinal plants varies widely depending on many factors, including their association with microbes (Zhao et al. 2011; Morsy 2014). Normally, plants rely on their microbiome for specific traits and activities, including, induced systemic resistance, nutrient acquisition, growth promotion, and tolerance to abiotic stress factors (Egamberdieva et al. 2010; Malfanova et al. 2011; Sessitsch et al. 2013; Berg et al. 2014). The bioactive secondary metabolites synthesized by medicinal plants can also strongly affect plant-associated microbial communities and their physiological functions (Qi et. al. 2012; Chaparro et al. 2013; Philippot et al. 2013). Novel secondary metabolites from these microbial communities offer opportunities for innovation in leads/drugs development and agrochemical discovery. There is thrilling potential for those wishing to venture into the wild and uncharted terrains of the world to experience the excitement and adventure of engaging in the discovery of endophytes for studying their biology and potential usefulness. In the last two decades, endophytes have reached the top and drawn more attention from researchers because of their recognition and utility in drugs.

The relationship between microbes and plants may have progressed since plants appeared on earth and has allowed these microbes to develop strategic genetic systems and to develop biochemical paths. Nowadays there is a general need for more effective and less toxic compounds, especially considering the multidrug resistance in pathogenic organisms (Magiorakos et al. 2012). Therefore, one way to solve this problem is to explore untapped sources such as the microbes that reside in medicinal plants. Endophytes are potential sources of novel natural products for exploitation in agriculture, medicine, and industry (Kumar and Sagar 2007).

Nowadays, production in the agriculture field is not ecologically balanced and sustainable. The excessive use of synthetic agrochemicals, soil erosion, and irrigation issues, among others in agriculture, are threats to the ecosystem. Considering the use of pesticides, herbicides, and fertilizers, microbial inoculants are a good and reliable option in the agriculture fields, they are capable of acting as biopesticides, bioherbicides, and biofertilizers. Endophytes inhabit the tissues of plants asymptomatically without causing adverse effects. The endophytes benefit host plants, by improving plant nutrient uptake, promoting growth, and stress-related phytohormones. The host plant harbors a unique internal plant environment that is most likely highly bioactive in which the associated microbes, when exposed, produce bioactive secondary metabolites.

In the medicinal field, endophytic actinomycetes are known to produce various chemical entities with unique structures and considerable medicinal importance (Gayathri and Muralikrishnan 2013; Singh and Dubey 2015). Numerous antimicrobial compounds have been reported from various types of endophytic actinomycetes. *Streptomyces* is one of the dominant genera, which is most commonly isolated as endophytic actinomycetes (Zhao et al. 2011; Golinska et al. 2015). Many chemical compounds of biological interest are isolated from *Streptomyces* sp. such as munumbicins (A and B), naphthomycin (A and K), clethramycin, coronamycin, cedarmycin (A and B), saadamycin, and kakadumycin.

In industrial applications, a relatively small number of microbes are used (e.g., cheese, wine, and beer), in environmental sanitation operations, and in the biological control of pests and pathogens. Endophytic microorganisms already established as exceptional sources of bioactive natural products can be explored for the sustainable production of necessary enzymes in different industrial sectors. Antibiotics or hydrolytic enzymes can be released by endophytes to prevent the colonization of pathogens in microbial plants. They are also producing superoxide dismutase, catalase, and hydrogen peroxide reductase to cope with the oxidative stress during colonization (Taghavi et al. 2010). Over the past two decades, endophytes have been studied for their enzymatic potential and many enzymes, including amylase, asparaginase, chitinase, lipases, proteases, xylanase, etc., have been reported so far, from endophytes. Endophytes have recently generated significant interest in the community of microbial chemistry because of their immense potential to contribute to the discovery of new bioactive compounds. It has been suggested that the close biological association between endophytes and their host plant results in the production of a greater number and diversified biological molecules compared to epiphytes or soil-related microbes. The major advantage of using endophytes in drug production is that one can obtain a compound of plant origin without causing much harm to the plant community.

2. Endophytes

The endophytes were first described in Darnel, Germany, in 1904. Endophytes linked with plants can be either eukaryotic or prokaryotic. Prokaryotic endophytes consist of bacteria and archaea, while eukaryotic endophytes consist of fungi, algae, and amoeba. Bacterial endophytes related to plants are often actinomycetes or mycoplasmas. Endophytes are usually present in plant tissues, that are above the surface, but rarely in roots, and are distinguished from mycorrhizae by the absence of external hyphae. There are several epiphytic, rhizospheric and endophytic microorganisms, who live in the different environmental niches of their host plant and form symbiotic associations. Endophytes are the closest and most intimately related to their host plant. They are found in all plant species regardless of their place of origin & ability to enter and prosper in host tissues makes them unique and shows multidimensional interactions in the host plant.

Endophytes (bacteria and fungi) live within healthy plant tissues at a particular time, without causing any immediate overt negative effects. Since, endophyte infections are symptomless, their isolation and identification are done by dissecting the plant tissues through laboratory analysis (Hardoim et al. 2008; Franche et al. 2009). Endophytes are ubiquitous with rich biodiversity, which has been found in all plant species. They form a symbiotic relationship with their plant host. There are approximately three lakh plant species on the planet, and each plant hosts one or more endophytes (Strobel and Daisy 2013). It is estimated that about one million different endophyte species are available. However, only a handful of them have been described so far, which means interesting endophytes are yet to be investigated for their secondary metabolites. It is reported that a considerable proportion of non-plate-culturable endophytes (West et al. 2010; Thomas and Sekhar 2017) recorded a certain degree of difference among the endophytic communities and those obtained from independent cultural approaches, suggesting a non-negative level of non-culture between endophytes. This limitation has been confirmed by other studies, which addressed an important issue of cultivation in grapevine bacterial endophytes.

It is assumed that, in many cases, endophyte microbes act as an organic defense of the plant against foreign plant pathogens. Azevedo et al. (2000) reported that endophytic microorganisms can have the capacity to control plant pathogens. The protection mechanism of the endophytes is exerted directly, by releasing metabolites to attack any antagonists or smooth cells that have affected or indirectly, induced host defense mechanisms or growth. They also promote plant growth by releasing metabolites that activate the host defense response against pathogens, known as the induced systemic resistance process (Azevedo et al. 2000; Kloepper and Ryu 2006). In some cases, the beneficial effect is directly related to microbial metabolisms, such as nitrogen fixation, production of phytohormones, phosphate solubilization, and pathogen suppression or vice-versa. These benefits can be mediated by stimulation of specific activities of the host plant, leading to increased enzymatic catalysis, and enhanced water and nutrient uptake or defense responses (Compant et al. 2013). Many endophytes are able to synthesize bioactive compounds that can be used by plants defense against human pathogens, and some of these compounds have been recognized for the discovery and development of new leads and drugs. Numbers of natural products have been isolated and characterized from endophytes, but the therapeutic properties of endophytes have not been widely explored so far. One endophyte can be capable of producing several bioactive secondary metabolites. The exploration of these bioactive molecules producing endophytes of medicinal systems can provide a basis for lead compounds for human welfare. Endophytes have attracted the attention of the scientists around the world due to their production of novel chemical entities and their exploitation for medicinal purposes (Wang et al. 2000). Both fungi and bacteria are the most common microbes existing as endophytes, but the most frequently isolated endophytes are fungi. It seems that other microbial forms, e.g., mycoplasmas and archaebacteria, most certainly exist in plants as endophytes, but no evidence for them has yet been presented.

2.1. Colonization Mechanism of Endophytes

Endophytes systematically colonize as distinct components of the host. They can input and colonize plants through the vertical seeding technique or through horizontal transmission from the soil to the vegetation. The colonization technique typically begins with a preliminary communication signaling, and the related microorganisms gain initial access through plant root exudates. The rhizospheric zones of the plant and the root exudates produce essential natural compounds that act as chemo-attractants, which assist in facilitating signals or communications between the microbes and plant roots, after which to recruit microbes. This then initiates the technique of endophytic access for the colonization of the tissues of the plant host (López-Ráez et al. 2017) (Table 1).

Table 1. Differentiation of bacterial endophytes from fungal endophytes

Criteria	Bacterial endophyte	Fungal endophyte
Host spectrum	Broad-spectrum host infection	Broad-spectrum host infection usually depends upon the habitat
Mode of infection	Passive colonization through wounded plant cells or tissues and actively via enzymes or using vectors	Mostly through stomata, cell walls, wounded tissue
Root growth	Inter- and/or intra-cellular colonization with slow growth and low densities	Inter- and/or intra-cellular colonization with extensive Growth
Tissue colonized	Primarily intercellular tissues with growth also seen in vascular tissue	Vascular tissues are mostly not colonized
Types of interactions with host	Commensalism, mutualism, pathogenic	Mutualism, commensalism, pathogenic
Specialized structures	Nodule formation	Mycorrhizae, nodules
Benefits to host cell	Growth promotion, N-fixation, secretion of phytohormones metabolites antagonistic to plant pathogens and insects, synthesis of antibiotics and useful compounds	Growth promotion via secretion of phytohormones, improved access to mineral nutrients, combating pathogen stress, synthesis of antibiotics, and useful compounds

2.2. Isolation of Endophytes

1. Plant materials are thoroughly surface sterilized with 70% ethanol and air-dried under a laminar flow hood. This is done to eliminate surface contamination by microbes.
2. Then treat the plant material with 0.5-3.5% sodium hypochlorite for 1-2 minutes, followed by rinsing in sterile double distilled water.
3. Then, with a sterile knife blade, the outer tissues are removed from the specimens and the inner tissues are carefully excised and placed on nutrient agar plates for bacteria and potato dextrose agar for fungi and further incubated at respective temperatures.
4. After the incubation period, growth was observed for both bacteria and fungi.
5. Endophytes are cultured on specific plant materials and are ultimately identified using standard morphological and molecular biological techniques.
6. Eventually, when an endophyte is acquired in pure culture, it is tested for its ability to be cultivated in the form of shock or culture, always with the use of various media and growth conditions.

2.3. Bacterial Endophytes

There are 16 different bacterial phyla that have been reported as endophytes, with most species belonging to the phyla actinobacteria, firmicutes, and proteobacteria (Golinska et al. 2015; Dukare et al. 2022). Bacterial endophytes have been isolated and identified in diverse plant hosts, environments, and different parts of the plant (Dhanya and Padmavathy 2014). However, the occurrence of bacterial endophytes is more prevalent in the root tissues as opposed to the aerial plant tissues (Rosenblueth et al. 2006). Bacterial endophytes confer stress tolerance capabilities on their host plant. They induce allelopathic effects in their host plant and also improve its growth (Table 2) (Mishra et al. 2013). It grants their host plants advantages, which include helping them to mitigate growth-limiting biotic and abiotic influences (Miliute et al. 2015).

Table 2. Overview of medicinal plants and their associated bacterial endophytes (Sharma et al. 2020)

S. No.	Bacterial endophyte	Host plant
	Arthrobacter sp.	*Echinacea purpurea*
	Bacillus amyloliquefaciens	*Scutellaria baicalensis, Oryza sativa*
	Bacillus cereus	*Artemisia annua* L.
	Bacillus subtilis	*Aloe vera, Centella asiatica*
	Bacillus thuringiensis	*Andrographis paniculata, Picrorhiza kurroa*
	Cronobacter sakazakii, Bacillus methylotrophicus	*Tridax procumbens* L.
	Enterobacter sp.	*Sida cordifolia*
	Escherichia coli; Proteus vulgaris	*Curcuma longa*
	Micromonospora	*Maklam phueak*
	Pseudomonas putida	*Berberis lyceum*
	Pseudomonas sp.	*Zingiber officinale*
	Pseudomonas viridiflava	*Lactuca sativa*
	Pseudonocardia sp.	*Artemisia annua*
	Ralstonia sp.	*Aloe vera*
	Rhodococcus bacteria	*Ferula songorica, Arthobacter,*
	Staphylococcus aureus	*Azadirachta indica*
	Streptomyces albidoflavus	*Bruguiera gymnorrhiza*
	Streptomyces aureofaciens CMUAc130	*Zingiber officinale*
	Streptomyces hygroscopicus TP-A0451	*Pteridium aquilinum*
	Streptomyces olivochromogenes	*Tinospora crispa*
	Streptomyces parvulus	*Abutilon indicum*
	Streptomyces sp.	*Allium fistulosum, Maytenus hookeri,*
	Streptomyces sp. *GT 20021503*	*Bruguiera gymnorrhiza*
	Streptomyces sp. *TP-A0556*	*Aucuba japonica, Cryptomeria japonica*
	Streptomyces sp. *YIM 66017*	*Alpiniao xyphylla*
	Streptomyces sp. *Tc 022*	*Alpinia galangal*
	Streptomyces thermoviolaceus NT1	*Catharanthus roseus*

2.4. Fungal Endophytes

Eukaryotic endophytes consist mainly of mushrooms. Mushrooms are heterotrophic organisms, that have symbiotic relationships with a wide variety of body organizations. Based on their phylogeny and taxonomy, the fungal endophytes are either clavicipitaceous or non-clavicipitaceous (Rodriguez et al. 2009). Fungal endophytes are one of the most unexplored and diverse groups of organisms. It creates symbiotic associations with higher life forms and can produce substances beneficial to the host. The diversity of fungal endophytes has been studied based on their numbers, types, colonization rate, and isolation strategy. Fungal diversity varies according to different parts of the plant and the geographical distribution of the host plant in different environments, including temperate (Ganley et al. 2004), tropic (Mohali et al. 2011), aquatic (Debbab et al. 2011), and xerophytic (Suryanarayanan et al.

2005). Fungal endophytes from medicinal plants could be a rich source of functional metabolites (Vaidya and Devasagayam 2007). More than 4,000 bioactive compounds from fungal species have been reported so far.

Commensalism and mutualism denote the balanced phases of plant-microbe interactions. Commensalism delivers an advantage to the fungal endophyte by allowing an uninterrupted nutrient supply and existence without disturbing the host (Table 3). This commensal interrelationship and plant harboring makes the pathogen remain asymptomatic. A wide range of anticancer, antifungal, antimycobacterial, antiviral, antioxidant, insecticidal, antidiabetic, and immunosuppressive bioactive compounds have been isolated from endophytic fungi for their potential ecological role. Hence, it is needed to design a suitable cultivation scheme for their marketable exploitation.

3. Bio-Prospection of Secondary Metabolites Produced by Endophytes

An endophyte is one that exists in the tissues underneath the epidermal cell layers without instigating any adverse effects (Guna tilaka 2006). Both bacteria and fungi are mostly common microbes alive as endophytes, but the most often isolated endophytes are fungi. The term refers to "endophyte," which is generally used for microorganisms that live with plants, but scientists later realized that bacteria could colonize the inner parts of plants as well (Gunatilaka 2006; Ancheeva et al. 2020). The diversity of endophytes empowers them they can naturally produce a number of secondary metabolites. Many secondary metabolites have been isolated from endophytic microbes, such as alkaloids, steroids, terpenes, anthraquinones, cyclic peptides, and flavonoids (Zhao et al. 2018). These secondary metabolites have been used for medicinal purposes for number of years. Natural-based drugs (natural products) have had an enormous impact on modern medicine since about 50% of prescription drugs are based on them. As a matter of fact, the world's first billion-dollar anticancer drug, paclitaxel (Taxol), is a naturally derived medicine from the yew tree. There are a lot of examples demonstrating the value and importance of natural products in modern civilizations (Wani et al. 1971; Tan and Zou 2001; Strobel and Daisy 2003). The fungi are the most common groups of eukaryotic organisms, that produce secondary metabolite which are used as drugs or lead compounds for drug discovery and development. Many significant bioactive compounds have been effectively discovered from fungal endophytes having cytotoxic, antimicrobial, insecticidal, anticancer, and antioxidant activities. The fungal endophytes are used for producing plant-based bioactive compounds such as antimicrobial compounds, pestacin, paclitaxel, isopestacin, podophyllotoxin, camptothecin, hypericin, etc. (Tan and Zou 2001; Ancheeva et al. 2020).

Cryptosporiopsis quercina is a fungal endophyte that is commonly associated with hardwood species in Europe. The fungus was isolated from *Tripterigeum wilfordii*, a medicinal plant native to Eurasia (Wani et al. 1971; Tan and Zou 2001; Strobel and Daisy 2003; Metwalya et al. 2014). The fungus *C. quercina* showed excellent antifungal activity against some important human fungal pathogens such as *Candida albicans* and *Trichophyton spp.* A unique antimycotic lipopeptide, cryptocandin (**1**), was isolated and characterized from *C. quercina* (Bajguz 2007; Malfanova et al. 2011). Cryptocandin (**1**) contains a number of peculiar hydroxylated amino acids and a novel amino acid: 3-hydroxy-4-hydroxy methyl proline. It is related to known bioactive antimycotics, echinocandins, and pneumocandins (Wani et al. 1971; Tan and Zou 2001; Strobel and Daisy 2003). Nevertheless, cryptocandin is being used as a base model to synthesize other antifungal compounds (Strobel and Daisy 2003; Metwalya et al. 2014).

Pestalotiopsis is a genus of ascomycete fungi that is known as the world's most common endophyte of yews. An investigation of the endophytes of *Taxus wallichiana* yielded *P. microspora*, which is a commonly isolated endophyte species, and an initial monoclonal antibody test indicated that it might produce paclitaxel (**2**). Furthermore, several other *P. microspore* isolates were obtained from bald cypress in South Carolina and were also shown to produce paclitaxel. This was the first indication that endophytes residing in plants other than *Taxus* spp. were producing paclitaxel. Therefore, a specific search was conducted for paclitaxel-producing endophytes on continents not known for any indigenous *Taxus spp.* Simply, it appears that the distribution of those fungi making paclitaxel is worldwide and not confined to endophytes of yews (Strobel and Daisy 2003).

Table 3. Overview of medicinal plants and their associated fungal endophytes (Sharma et al. 2020)

S. No.	Fungal endophyte	Host plant
	Alternaria alternate	*Maytenus hookeri, Withania somnifera, Azadirachta indica*
	Alternaria macrospora	*Berberis aristata* DC
	Alternaria sp.	*Ricinus communis, Erythrina variegate*
	Ascomycetes	*Mimosops elengi*
	Aspergillus flaviceps	*Acanthus ilicifolius*
	Aspergillus flavus	*Aegle marmelos*
	Aspergillus flocculus	*Markhamia platycalyx*
	Aspergillus fumigatus	*Justicia beddomei*
	Aspergillus niger	*Azadirachta indica*
	Aspergillus sp.	*Swietenia macrophylla*
	Aspergillus tamarii	*Ficus carica*
	Aspergillus terreus	*Achyranthes aspera*
	Bipolaris sorokiniana	*Costus speciosus*
	Botryosphaeria sp.	*Garcinia atroviridis*
	Chaetomium arcuatum	*Semecarpus anacardium*
	Cochliobolus sp.	*Aerva lanata*
	Colletotrichum gloeosporioides	*Justicia gendarussa, Huperzia serrata, Centella asiatica*
	Colletotrichum sp.	*Tinospora cordifolia*
	Colletotrichum, Fusarium, Phoma and Penicillium	*Pereskia bleo*
	Entrophospora infrequens	*Hypericum japonicum*
	Estalotiopsis microspore	*Artocarpus heterophyllus*
	Eupenicillium sp.	*Erythronium japonicum*
	Fusarium decemcellulare	*Flacourtia inermis*
	Fusarium oxysporum	*Curcuma longa, Juniperus recurva, Dendrobium lindleyi*
	Fusarium solani	*Phaius tankervilleae, Cassia alata*
	Fusarium sp.	*Dysosma versipellis, Maackia chinensis*
	Fusarium sp., *Aspergillus niger*	*Opuntia dilleniid*
	Lasiodiplodia theobromae	*Morinda citrifolia*
	Lecythophora sp.	*Alyxia reinwardtii*
	Neurospora sp.	*Nothapodytes foetida*
	Nigrospora oryzae	*Nyctanthes arbor-tristis*
	Paecilomyces sp.	*Taxus mairei*
	Paraconiothyrium sp.	*Zingiber officinale*
	Penicillium canescens	*Polygonum chinense*
	Penicillium chrysogenum	*Cistanches deserticola*
	Penicillium citrinum	*Tragia involucrate*
	Penicillium fumicalsuri	*Azadirachta indica*
	Penicillium oxalicum	*Gymnema sylvestre*
	Penicillium sp., *Aspergillus* sp.	*Ginkgo biloba*
	Pestalotiopsis sp.	*Azadirachta indica, Terminalia arjuna*
	Phialemonium curvatum	*Passiflora edulis*
	Rosellinia sanctae cruciana	*Albizia lebbeck*
	Spergillus awamori	*Musa acuminate*
	Teratosphaeria sp.	*Pinus clausa*
	Trichoderma koningiopsis, Aspergillus sydowii and *Trichoderma lixii*	*Sonneratia alba*
	Trichoderma lixii	*Bacopa monnieri*
	Xylaria sp.	*Ginkgo biloba* L.

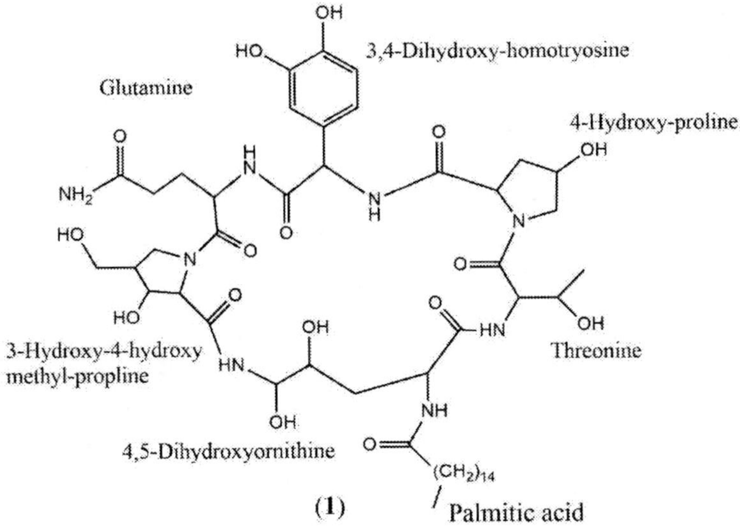

(1)

Figure 1. Cryptocandin.

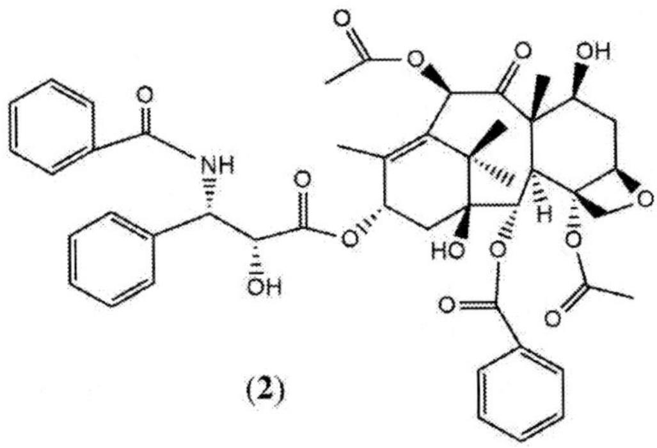

(2)

Figure 2. Paclitaxel.

The filamentous fungus of the phylum Ascomycota genus *Nigrospora* is named *N. sphaerica*, which is an airborne fungus found in soil, air, various plants, and marine organisms. The genus *Nigrospora* has a rich source of bioactive secondary metabolites like nigrosporolides having plant growth-inhibiting activity. The fungus *N. sphaerica* has been reported for the presence of many active secondary metabolites including diterpenes, diketopiperazines, lactones, and nigrosporolides. The endophytic fungus *N. sphaerica* was chemically and biologically investigated by Metwaly et al. (2013), which led to the isolation of nigrosphaerin A (**3**) and 15 new compounds (**4–18**), namely ergosta-6,22-diene-3β,5α,8α-triol (**4**), ergosta-7,22-diene-3β-ol (**5**), ergosta-4,6,8(14),22-tetraene-3β-ol (**6**) ergos-ta-4,6,8(14),22-tetraene-3one (**7**) ergosta-5(6),7,22-triene-3β-ol (**8**), ergosta-7,9(14),22-triene-3β-ol (**9**) ergosta-7,22-epidioxy-3β-ol (**10**), 4-(hydroxymethyl)-3,5-dimethyl dihydro-furan-2(3H)-one (**11**),3-(1-hydro-xyethyl)-4-methyl dihydrofuran-2(3H)-one (**12**), uridine (**13**) adenosine (**14**), 3-benzylhexahydropyrrolo[1,2-a]pyrazine-1,4-dione (**15**) 3-methylhexahy-dropyrrolo[1,2-a]pyrazine-1,4-dione (**16**), methyl 4-hydroxybenzoate (**17**), and 4-(2-hydroxyethyl)phenol (**18**). Apart from the above 15 compounds, 4 fatty acids namely; stearic acid (**19**), oleic acid (**20**), palmitic acid 96 (**21**), and myristic acid (**22**) were also isolated and characterized from *N. sphaerica*. The above-stated compounds, compounds (**3-5, 7,** and **9-18**) were isolated first time from this fungus. All the isolated compounds (**3-22**) from *N. sphaerica* were studied for *in vitro* antileukemic, antileishmanial, antifungal, antibacterial, and antimalarial activities. The studies resulted that compound **7, 9, 11** and **12** exhibited good antileukemic activity against HL60 cells with IC_{50} values of 0.03, 0.39, 0.2, and 0.4 μg/mL, respectively, and against K562 cells with IC_{50} values of 0.35, 0.35, 0.49 and 0.01 μg/mL, respectively. Compounds **5**, **6,** and **8** showed moderate antileishmanial activity with IC_{50} values of **30.2**, **26.4,** and **36.4** μg/mL, respectively. Compound **9** showed moderate antifungal activity against *Cryptococcus neoformans* with an IC_{50} value of 14.8 μg/mL (Metwalya et al. 2014).

Recently, a new endophytic fungus *Curvularia* sp. T12 was isolated from medicinal plant *Rauwolfia macrophylla* which showed the highest homology to *Curvularia* sp. *Curvularia* sp. provides a broad variety of bioactive secondary metabolites including polyketides and steroids (Avinash et al. 2015; Samanthi et al. 2015; Ndjakou et al. 2016; Kaaniche et al. 2019). An investigation of the fungus *Curvularia* sp. T12 led to the isolation of three bioactive compounds; 2'-deoxyribolactone (**23**), hexylitaconic acid (**24**), and ergosterol (**25**) (Kaaniche et al. 2019). All the isolated compounds (**23-25**) were examined for antimicrobial activity against four bacterial strains (*Escherichia coli*, *Micrococcus luteus*, *Pseudomonas agarici,* and *Staphylococcus warneri*), antioxidant and the acetylcholinesterase inhibitory activities. The antibacterial activity test showed that compounds **23** and **24** have antibacterial activities in contrast to all the tried bacterial strains with MIC ranging between 0.17 μg/mL and 0.58 μg/ml. The compounds **23-25** showed good inhibitory potential towards acetylcholinesterase with IC_{50} values of 1.93, 1.54, and 1.52 μM, respectively, while the selected reference galanthamine exhibited an IC_{50} value of 0.5 μM and 0.01 μM for tacrine. Apart from this, compounds **23 & 24** also showed antioxidant activity and exhibited close EC_{50} values of the reference BHT (EC_{50} = 0.46 μM in DPPH assay), reflecting the high scavenging ability of the compounds (Kaaniche et al. 2019).

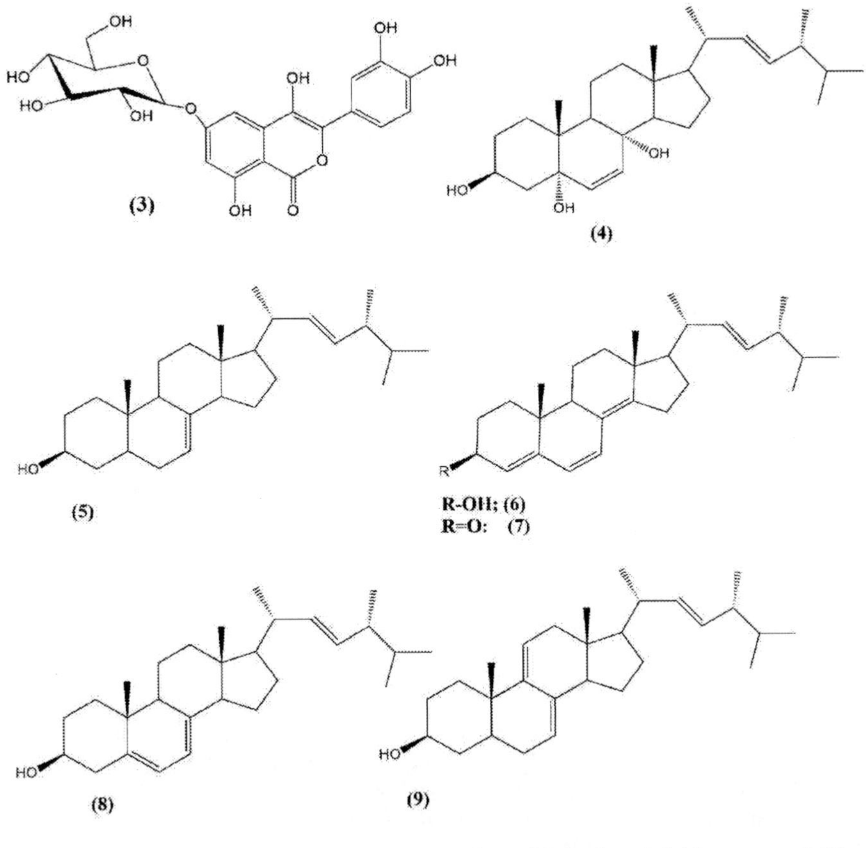

Figure 3. Nigrosphaerin A (**3**), ergosta-6,22-diene-3β,5α,8α-triol (**4**), ergosta-7,22-diene-3β-ol (**5**), ergosta-4,6,8(14),22-tetraene-3β-ol (**6**) ergos-ta-4,6,8(14),22-tetraene-3one (7) ergosta-5(6),7,22-triene-3β-ol (**8**), ergosta-7,9(14),22-triene-3β-ol (**9**).

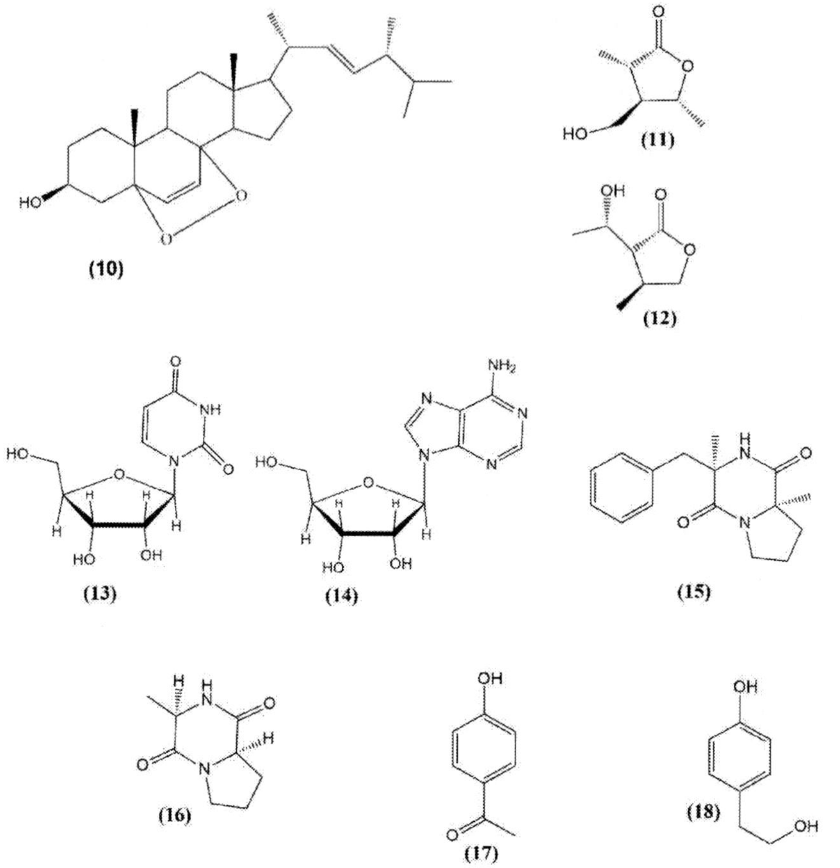

Figure 4. Ergosta-7,22-epidioxy-3β-ol (**10**), 4-(hydroxymethyl)-3,5-dimethyl dihydro-furan-2(3H)-one (**11**),3-(1-hydro-xyethyl)-4-methyl dihydrofuran-2(3H)-one (**12**), uridine (**13**) adenosine (**14**), 3-benzylhexahydropyrrolo[1,2-a]pyrazine-1,4-dione (**15**) 3-methylhexahydropyrrolo[1,2-a]pyrazine-1,4-dione (**16**), methyl 4-hydroxybenzoate (**17**), and 4-(2-hydroxyethyl)phenol (**18**).

Xanthium sibiricum belongs to the family Asteraceae and is an important medicinal plant used in Traditional Chinese Medicine (TCM) for decades. It is commonly distributed in China and its fruit, leaves, and roots have been used for the treatment of several ailments (Zhang et al. 2006; Kan et al. 2011). Li et al. isolated an endophytic fungus, *Eupenicillium* sp. LG41, from the roots of *X. sibiricum* collected from Taiwan, People's Republic of China. The isolated fungal endophyte *Eupenicillium* sp. was chemically investigated and led to the isolation of four new compounds, including two compounds containing the decalin moiety, namely eupenicinicols A and B (**26** and **27**), and two sirenin derivatives, namely eupenicisirenins A and B (**28** and **29**) and four known compounds, namely (2S)-butylitaconic acid (**30**), (2S)-hexylitaconic acid (**31**), xanthomegnin, and viridicatumtoxin (Li et al. 2014). The antibacterial efficacy of the isolated compounds **26–31** was tested against the gram-positive bacterium *Bacillus subtilis* and the gram-negative bacterium *Acinetobacter* sp. BD4. Furthermore, to inspect the possible use of the endophyte as a source of antibacterial drug lead, the clinically important gram-positive bacterium *Staphylococcus aureus subsp. aureus* and the gram-negative bacterium *Escherichia coli* were also included. The compounds **30** and **31** exhibited the most pronounced efficacy against the soil bacteria *Acinetobacter sp.* BD4 with MIC of 1.0 μg/mL for both, and appeared to be more potent than both reference antibiotics (Streptomycin and Gentamicin, MICs of 10 & 5 μg/mL, respectively). Compound 27 was highly active against the clinically relevant *S. aureus*. By comparing, compound **26** (>10.0μg/mL) with 27, it was noticed that changing the substitution at C-11 could radically increase the antibacterial efficacy of compound 26 (1.0μg/mL) (Li et al. 2014).

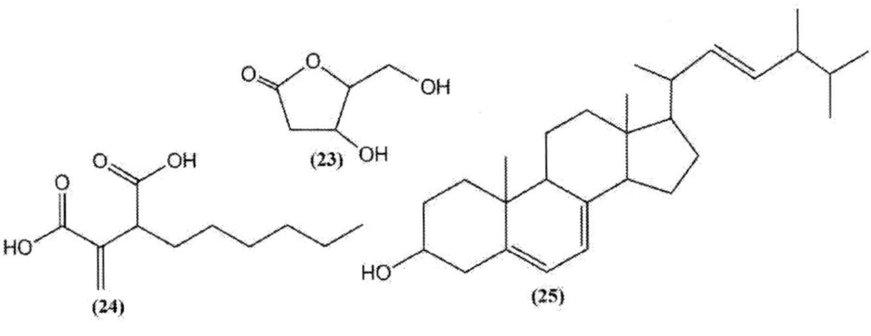

Figure 5. 2'-deoxyribolactone (**23**), hexylitaconic acid (**24**), and ergosterol (**25**).

Artemisia princeps is a perennial herb native to Korea, China, and Japan. Traditionally, it is used for the treatment of inflammation, diarrhoea, bacterial infection, and circular disorders. The fungal endophyte *Phoma* sp. PF2 belongs to the genus of coelomycetous fungi, and was isolated from the leaves of *Artemisia princeps* Pamp, *Phoma* sp., The fungus has been reported to producc various bioactive secondary metabolites, such as cytotoxic α-pyrones (phomones C-F), thiodiketopiperazine (phomazine B), antiangiogenic polyketide (phomaketide A), anti-inflammatory polyketide (phomaketide C), and antibacterial polyketide (barceloneic acid C) (Kong et al. 2014; Sanga et al. 2014; Xia et al. 2015; Lee et al. 2016; Hirano et al. 2017; Kim et al. 2019). The chemical investigation of the cultures of fungal endophytes *Phoma* sp.

PF2 derived from *A. princeps* led to the isolation and identification of eight compounds, including two new isochromanone derivatives viz. (3S,4S)-3,8-dihydroxy-6-methoxy-3,4,5-trimethylisochroman-1-one (**32**) and methyl(S)-8-hydroxy-6-methoxy-5-methyl-4a-(3-oxobutan-2-yl)benzoate (**33**) and six known compounds were allocated as (4S)-clearanol C (**34**), (3R,4S)-3,8-dihydroxy-3-hydroxymethyl-6-methoxy-4,5-dimethylisochroman-1-one (**35**), polygonolide (**36**), (7S,8S)-8-ethyl-7-hydroxy-2,3,8-trimethyl-5,6,7,8-tetrahydro-4H-chromen-4-one (**37**), (7R,8S)-8-ethyl-7-hydroxy-2,3,8-trimethyl-5,6,7,8-tetrahydro-4H-chromen-4-one (**38**), and leptosphaerone (=(7S,8S)-7-ethyl-7, 8-dihydro-8-hydroxy-3,4,7-trimethyl-6H-2-benzopyran-6-one) (**39**). All the isolated molecules were evaluated by measuring the inhibitory activity of nitric oxide (NO) production levels in the lipopolysaccharide (LPS)-induced RAW264.7 macrophage cells. All the isolated compounds showed moderate inhibitory activities on nitric oxide levels in lipopolysaccharide-induced RAW264.7 macrophage cells. Most of the isolates except 6 did not show cell cytotoxicity at 50 μM on RAW264.7 cells in a preliminary MTT assay (Yamamoto et al. 2011; Kim et al. 2019).

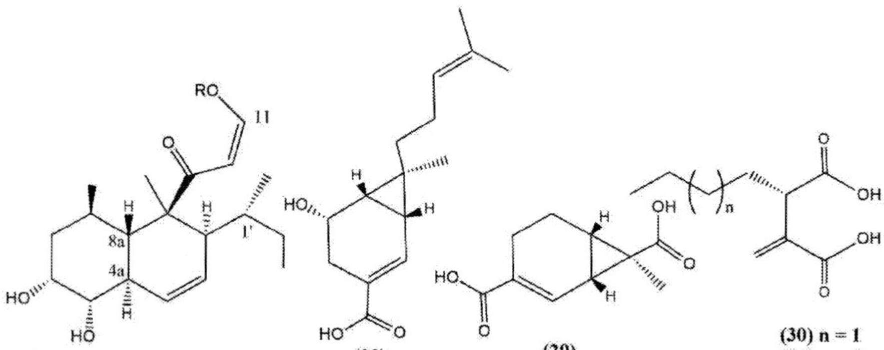

(26) R=H
(27) R=CH$_3$ Δ^{10}=E
(28)
(29)
(30) n = 1
(31) n = 3

Figure 6. Eupenicinicols A and B (**26** and **27**), eupenicisirenins A and B (**28** and **29**), (2S)-butylitaconic acid (**30**) and (2S)-hexylitaconic acid (**31**).

A fungal endophyte, *Aspergillus fumigatus* sp. R7. was isolated from the leaves of sweet potato, i.e., *Ipomoea batatas.* The fungus has been shown to yield shallow, submerged hyphae on potato dextrose agar (PDA) medium. As per pre-screening studies, the endophyte isolates *Aspergillus fumigatus* R7 showed prominent antibacterial activity against gram-positive (*B. subtilis* [16 mm], and *Staphylococcus aureus* [15 mm]) and gram-negative bacteria (*P.*

aeruginosa [19 mm], and *E. coli* [16 mm]). In contrast, the strain extract showed no activity against pathogenic fungi; *A. niger, A. flavus,* and *C. albicans* (Shaaban *et al.* 2011). After harvesting and working up, the afforded crude extract, which was subjected to purification using a series of chromatographic techniques to deliver the ten secondary metabolites namely, linoleic acid (**40**), R(-)-glycerol monolinoleate (**41**) bis-dethio-(bis-methyl-thio)-gliotoxin, (**42**), fumiquinazoline-F (**43**), fumiquinazoline-D (**44**), (Z,Z)-N,N'-[1-[(4-Hydroxy-phenyl)-methylene]-2-[(4-methoxy-phenyl)-methylene]-1,2-ethanediyl]-bis-formamide (**45**), pyrazoline-3-one trimer (**46**), Tricho-9-ene- 2α, 3α, 11α, 16-tetraol (**47**), 2'-deoxy-thymidine (**48**), and cerebroside A (**49**). The *Aspergillus fumigatus* spp. crude extract isolate R7 along with all the secondary metabolites were tested for antimicrobial activity against eleven microbial tests on the basis of the agar diffusion technique (40 μg/disc). The crude extract showed high antibacterial activity, and the isolated molecules from the strain **43** and **44** exhibited antimicrobial activity against the gram-positive *Bacillus subtilis* (12, 15 mm), *S. aureus* (12, 15 mm) and fungi (*C. albicans* [11, 11 mm] and *M. miehi* [12, 13 mm] (Shaaban et al. 2012).

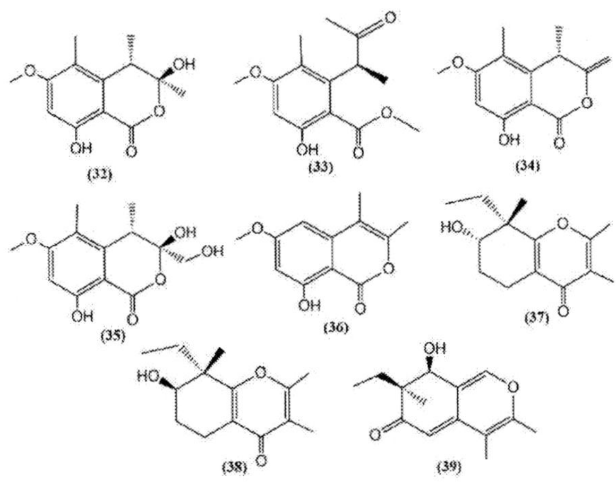

Figure 7. (3S,4S)-3,8-dihydroxy-6-methoxy-3,4,5-trimethylisochroman-1-one (**32**) and methyl(S)-8-hydroxy-6-methoxy-5-methyl-4a-(3-oxobutan-2-yl)benzoate (**33**) and six known compounds were allocated as (4S)-clearanol C (**34**), (3R,4S)-3,8-dihydroxy-3-hydroxymethyl-6-methoxy-4,5-dimethylisochroman-1-one (**35**), polygonolide (**36**), (7S,8S)-8-ethyl-7-hydroxy-2,3,8-trimethyl-5,6,7,8-tetrahydro-4H-chromen-4-one (**37**), (7R,8S)-8-ethyl-7-hydroxy-2,3,8-trimethyl-5,6,7,8-tetrahydro-4H-chromen-4-one (**38**), and leptosphaerone(=(7S,8S)-7-ethyl-7,8-dihydro-8-hydroxy-3,4,7-trimethyl-6H-2-benzopyran-6-one) (**39**).

Mangrove endophytes represent a special group of endophytic microorganisms derived from halophytic plants present in mangrove forests and swamps (Tomlinson 2016; Patil et al. 2018; Ancheeva et al. 2020). Mangrove plants and their endophytes can be viewed as extremophiles due to the fact that the environmental conditions in mangrove ecosystems are associated with many stress factors, such as high salinity (up to 36% salt content), limitations in oxygen and nutrient supply, a high tidal range alternated with periods of drought, and excessively high light exposure (Ancheeva *et al.* 2020). A fungal strain of *Phomopsis longicolla* was derived from the mangrove plant *Sonneratia caseolaris* (Lythraceae), collected in South China (Rönsberg *et al.* 2013; Frank *et al.* 2015). A bioactive compound, phomoxanthone A (**50**), has been reported as one of the main constituents of the fungus *Phomopsis longicolla,* which has also been reported from several mangrove and terrestrial plants. Cytotoxicity screening of phomoxanthone A showed pronounced growth inhibition of cisplatin sensitive and resistant cell lines (IC_{50} values in the range from 0.7 to 5.2 μM), human Burkitt's lymphoma DG75 (IC_{50} = 0.1 μM), and adult lymphoblastic leukemia T Jurkat cells (IC_{50} = 0.5 μM).

The chemical investigation of the endophyte *Embellisia eureka* derived from the plant *Cladanthus arabicus* resulted in the isolation of complex pyrrolidinone alkaloids, embellicines A (**51**) and B (**52**) (Ebrahim et al. 2013; Ancheeva et al. 2020). In a dose- and time-dependent manner, bioactivity screening of these pyrrolidinone alkaloids revealed pronounced cytotoxic activity against human chronic myelogenous leukemia (K562) cells (IC50 less than 10 M), the most potent compound, embellicine B (**52**) exerted IC_{50} values of 0.92, 0.32, 0.25, and 0.21 μM after 8, 24, 48, and 72 h of treatment, respectively. Subsequent evaluation of the activity of 52 toward peripheral blood mononuclear cells (PBMCs) indicated a 40-fold increase in the IC_{50} value, thus confirming cell type selectivity. Further studies specified that the mode of action of both metabolites is most likely via inhibition of TNFα-driven NF-κB activation, which is involved in inflammation and cancer promotion (Ancheeva et al. 2020). Pyrrocidine A (**53**), a close derivative of embellicines A (**51**) and B (**52**), was isolated from the fungus *Neonectria ramulariae* Wollenw KS-246, derived from a dead branch, (plant species was not reported) collected in Japan (Uesugi et al. 2016; Ancheeva et al. 2020). Pyrrocidine A (**53**), originally described as a natural product with antibiotic activity, was shown to exhibit strong cytotoxic properties against human acute promyelocytic leukemia HL60 cells (IC_{50} = 0.12 μM) (Ancheeva et al. 2020).

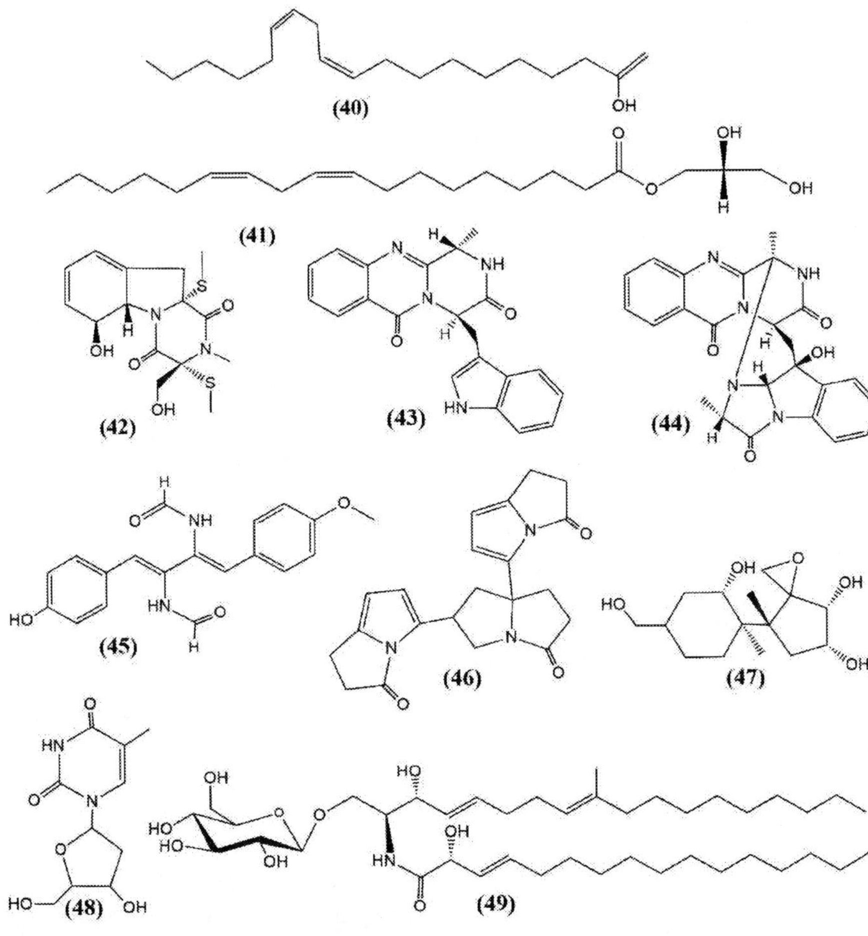

Figure 8. Linoleic acid (**40**), R(-)-glycerol monolinoleate (**41**) bis-dethio-(bis-methyl-thio)-gliotoxin, (**42**), fumiquinazoline-F (**43**), fumiquinazoline-D (**44**), (Z,Z)-N,N'-[1-[(4-Hydroxy-phenyl)-methylene]-2-[(4-methoxy-phenyl)-methylene]-1,2-ethanediyl]-bis-formamide (**45**), pyrazoline-3-one trimer (**46**), Tricho-9-ene- 2α, 3α, 11α, 16-tetraol (**47**), 2'-deoxy-thymidine (**48**), and cerebroside A (**49**).

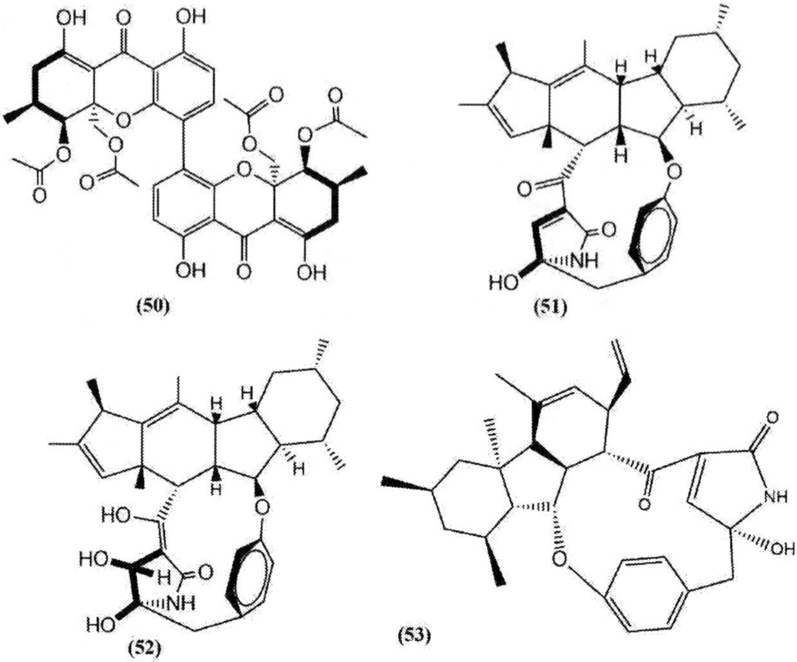

Figure 9. Phomoxanthone A (**50**), embellicines A (**51**) and B (**52**) and Pyrrocidine A (**53**).

Conclusion

The endophytes received considerable attention after being found to protect their guests against insect pests, pathogens, and even domestic herbivores. Endophytes are present in various parts of the plants, from which they are isolated and characterized for the beneficial use of human beings. Entophytes are enormous gems of secondary metabolites with unique structural diversity and biological activities. Endophytes are a feebly investigated cluster of microbes that represent a plentiful and reliable source of bioactive and novel secondary metabolites with potential use in medical, agricultural, and industrial areas. Therefore, researchers all over the world have their focus on endophytes as well as their products. It is also pertinent to mention that a number of endophytic microbes' strains have been much more extensively explored, but a noteworthy number of novel strains are yet to be explored. Both bacterial and fungal endophytes are found in plants, but fungal endophytes are the most common and much explored to date for the bioactive secondary metabolites and biological activities. So far, many bioactive secondary metabolites obtained from the endophytes have been reported for important biological activities like antioxidant, insecticidal, antimicrobial, anticancer, and cytotoxic. These secondary metabolites have become important leads for drug discovery and development, some of them have also been proven as drugs. Many endophytes become vital alternative sources for lots of approved drugs, such as podophyllotoxin obtained from the fungal endophyte *Phialocephala fortinii*, paclitaxel and its analogs obtained from endophytic fungi belonging to the genera *Alternaria, Aspergillus, Botryodiplodia, Botrytis, Cladosporium, Ectostroma, Fusarium, Metarhizium, Monochaetia, Mucor* and *Ozonium,* etc., Camptothecin obtained from endophytic fungal strains of *F. solani*, and vinblastine obtained from fungal endophyte, *Alternaria* sp. isolated from the phloem of *Catharanthus roseus*. Furthermore, bioactivity-guided extraction, fractionation, and isolation (BGEFI) of endophytes secondary metabolites approaches will be very advantageous for unexplored endophytic bacteria and fungi. By using BGEFI, scientist have been given many leads and drugs to the world so far. In recent times, chemo & bioinformatics play a very important role in predicting the compounds drug-likeness and pharmacokinetic features. Computer-aided drug design and discovery may appear to give the most promising leads derived from very common and inexpensive endophyte sources for chemical modifications as well as pre-clinical and clinical trials.

Acknowledgements

The authors are thankful to the Director (I/c) and Section heads, PCIM&H, Ministry of Ayush, for encouragement to write this book chapter.

Conflicts of Interest

The authors declare no conflict of interest.

References

Ancheeva, Elena, Georgios Daletos, and Peter Proksch. "Bioactive Secondary Metabolites from Endophytic Fungi." *Current Medicinal Chemistry* 27 no. 11 (2020): 1836-1854.

Arora, Jaya, Shaily Goyal, and Kishan Gopal Ramawat. "Biodiversity, Biology and Conservation of Medicinal Plants of the Thar Desert." In: Ramawat K. (eds) Desert Plants. *Springer*, Berlin, Heidelberg. 2010.

Avinash K.S., H.S. Ashwini, H.N. Ramesh Babu, and Y.L. Krishnamurthy. Antimicrobial "Potential of Crude Extract of *Curvularia lunata*, an Endophytic Fungi Isolated from *Cymbopogon caesius*." *Journal of Mycology* 4, (2015). https://doi.org/10.1155/2015/185821

Azevedo, João Lúcio, Maccheroni Jr. Walter, Pereira José Odair, and de Araújo, Welington Luiz. "Endophytic microorganisms: a review on insect control and recent advances on tropical plants." *Electronic Journal of Biotechnology* 3, no. 1 (2000): 40–65.

Bajguz, Andrzej. "Metabolism of brassinosteroids in plants." *Plant Physiology and Biochemistry* 45, no. 2 (2007): 95–107.

Berg, Gabriele, Martin Grube, Michael Schloter, and Kornelia Smalla "Unravelling the plant microbiome: looking back and future perspectives." *Frontiers in microbiology* 5, no. 4 (2014): 148.

Chaparro, J.M., Dayakar V. Badri, and Jorge M. Vivanco. "Rhizosphere microbiome assemblage is affected by plant development." *Multidisciplinary Journal of Microbial Ecology* 8, no. 4 (2013): 790–803.

Compant, Stéphane, Günter Brader, Saima Muzammil, Angela Sessitsch, Ahmed Lebrihi, and Florence Mathieu. "Use of beneficial bacteria and their secondary metabolites to control grapevine pathogen diseases." *Biocontrol* 58 no. 4 (2013): 435–455.

Cushnie, T.P. Tim, Benjamart Cushnie and Andrew J. Lamb. "Alkaloids: an overview of their antibacterial, antibiotic-enhancing and Anti virulence activities." *International Journal of Antimicrobial Agents* 44, no. 5 (2014): 377–386.

Debbab, Abdessamad, Amal H. Aly, and Chaidir Chaidir. "Bioactive secondary metabolites from endophytes and associated marine derived fungi." *Fungal Diversity* 49, no. 1 (2011): 1-12.

Dhanya, N. Nair and S. Padmavathy. "Impact of endophytic microorganisms on plants, environment and humans." *Scientific World Journal* 22, no. 2 (2014): 250693.

Dukare, Ajinath, Priyank Mhatre, Hemant S. Maheshwari, Samadhan Bagul, B.S. Manjunatha, Yogesh Khade, and Umesh Kamble. "Delineation of mechanistic approaches of rhizosphere microorganisms facilitated plant health and resilience under challenging conditions." *3 Biotech* 12, no. 3 (2022): 57.

Ebrahim, Weaam, Amal H. Aly, Victor Wray, Attila Mándi, Marie-Hélène Teiten, François Gaascht, Barbora Orlikova, Matthias U. Kassack, WenHan Lin, Marc Diederich, Tibor Kurtán, Abdessamad Debbab, and Peter Proksch. "Embellicines A and B: absolute configuration and NF-κB transcriptional inhibitory activity" *Journal of Medicinal Chemistry* 56, no. 7 (2013): 2991-2999.

Egamberdieva, Dilfuza, Gabriela Berg, Kristina Lindström, and Leena A. Räsänen. "Root colonizing *Pseudomonas* spp. improve growth and symbiosis performance of fodder galega (Galegaorientalis LAM) grown in potting soil." *European Journal of Soil Biology* 46, no. 3–4 (2010): 269-272.

Franche, Claudine, Kristina Lindström, and Claudine Elmerich. "Nitrogen-fixing bacteria associated with leguminous and non-leguminous plants." *Plant and Soil* 321 (2009): 35-59.

Frank, Marian, Hendrik Niemann, Philip Böhler, Björn Stork, Sebastian Wesselborg, Wenhan Lin, and Peter Proksch. "Phomoxanthone A- From Mangrove Forests to Anticancer Therapy." *Current Medicinal Chemistry* 22, no. 30 (2015): 3523-3532.

Ganley Rebecca J., Steven J. Brunsfeld, and George Newcombe. "A community of unknown, endophytic fungi in western white pine." *The Proceedings of the National Academy of Sciences* 101, no. 27 (2004): 10107–10112.

Gayathri, P., and V. Muralikrishnan. "Isolation and characterization of endophytic actinomycetes from mangrove plants for antimicrobial activity." *International Journal of Current Microbiology and Applied Sciences* 2, no. 11 (2013): 78–79.

Golinska, Patrycja, Magdalena Wypij, Gauravi Agarkar, Dnyaneshwar Rathod, Hanna Dahm, and Mahendra Rai. "Endophytic actinobacteria of medicinal plants: diversity and bioactivity." *Antonie Van Leeuwenhoek* 108, no. 2 (2015): 267–289.

Gunatilaka, A.A. Leslie. "Natural products from plant-associated microorganisms: Distribution, structural diversity, bioactivity, and implication of their occurrence." *Journal of Natural Products* 69, no. 3 (2006): 509–526.

Hardoim, Pablo R, Leo S van Overbeek, and Jan Dirk van Elsas. "Properties of bacterial endophytes and their proposed role in plant growth." *Trends in Microbiology* 16 no. 10 (2008): 463-471.

Hirano, Akiko, Masashi Goto, Tsukasa Mitsui, Akiko Hashimoto-Hachiya, Gaku Tsuji, and Masutaka Furue. Antioxidant *Artemisia princeps* extract enhances the expression of filaggrin and loricrin via the AHR/OVOL1 pathway. *International Journal of Molecular Sciences* 18, no. 19 (2017) e1948.

Kaaniche, Fatma, Abdelaaty Hamed, Ahmed S Abdel-Razek, Daniel Wibberg, Negera Abdissa, Imene Zendah El Euch, Noureddine Allouche, Lotfi Mellouli, Mohamed Shaaban, and Nobert Sewald. "Bioactive secondary metabolites from new endophytic fungus Curvularia. sp isolated from *Rauwolfia macrophylla*." *PLoS One* 14 no. 6 (2019): e0217627.

Kan, Suqin, Guangying Chen, Changri Han, Zhong Chen, Xinming Song, Ming Ren, Hong Jiang (2011). "Chemical constituents from the roots of *Xanthium sibiricum*." *Natural Product Research* 25, no. 13 (2011): 1243−1249.

Kim, Jung Wha, Hyun Gyu Choi, Ji Hoon Song, Ki Sung Kang, Sang Hee Shim. "Bioactive secondary metabolites from an endophytic fungus *Phoma* sp. PF2 derived from *Artemisia princeps*. Pamp." *The Journal of Antibiotics* 72 no. 3 (2019): 174-177.

Kloepper, W. Joseph, and Choong-Min Ryu. "Bacterial Endophytes as Elicitors of Induced Systemic Resistance Soil Biology." In: Schulz, C. Boyle, T.N. Sieber (Eds.). *Microbial Root Endophytes*. vol 9. Springer, Berlin, Heidelberg. Pp. 33-52. 2006.

Köberl, Martina, Ruth Schmidt, Elshahat M. Ramadan, Rudolf Bauer, and Gabriele Berg. "The microbiome of medicinal plants: diversity and importance for plant growth, quality and health." *Frontiers in Microbiology* 4, (2013) 400.

Kong, Fandong, Yi Wang, Peipei Liu, Tianhan Dong, and Weiming Zhu. "Thiodiketopiperazines from the marine-derived fungus *Phoma* sp. OUCMDZ-1847." *Journal of Natural Products* 77, no. 1 (2014): 132–137.

Kumar, Shiv, and Anand Sagar. "Microbial Associates of *Hippophae rhamnoides* (Seabuckthorn)." *Plant Pathology Journal* 6, no. 4 (2007): 299-305.

Lee, Ming-Shian, Shih-Wei Wang, Guei-Jane Wang, Ka-Lai Pang, Ching-Kuo Lee, Yueh-Hsiung Kuo, Hyo-Jung Cha, Ruo-Kai Lin, and Tzong-Huei Lee. "Angiogenesis inhibitors and anti-inflammatory agents from *Phoma sp.* NTOU4195." *Journal of Natural Products* 79, no. 12 (2016): 2983–2990.

Li, Gang Li, Souvik Kusari, Marc Lamshöft, Anja Schüffler, Hartmut Laatsch, Michael Spiteller. Antibacterial secondary metabolites from an endophytic fungus, *Eupenicillium* sp. LG41. *Journal of Natural Products* 77 no. 11 (2014): 2335-2341.

López-Ráez, Juan A., Ken Shirasu, and Eloise Foo. "Strigolactones in plant interactions with beneficial and detrimental organisms: The Yin and Yang." *Trends in Plant Science* 22, no. 2 (2017): 527–537.

Magiorakos, P.A., R.B. Carey, Y. Carmeli, M.E. Falagas, C.G. Giske, S. Harbarth, J.F. Hindler, G. Kahlmeter, B. Olsson-Liljequist, D.L. Paterson, L.B. Rice, J. Stelling, M.J. Struelens, A. Vatopoulos, J.T. Weber, and D.L. Monnet. "Multi drug resistant, extensively drug-resistant and pan drug resistant bacteria: an international expert proposal for interim standard definitions for acquired resistance." *Clinical Microbiology and Infection* 18 no. 3 (2012): 268-281.

Malfanova, Natalia, Faina Kamilova, Shamil Validov, Andrey Shcherbakov, Vladimir Chebotar, Igor Tikhonovich, and Ben Lugtenberg "Characterization of Bacillus subtilis HC8, a novel plant-beneficial endophytic strain from giant hogweed." *Microbial Biotechnology* 4, no. 4 (2011): 523–532.

Metwalya, Ahmed M., Hazem A. Kadry, El-HelabAbd-Elsalam, Mohammad Guoyi Maa, Stephen J. Cutler, Samir A. Rossad. "Nigrosphaerin A a new isochromene derivative from the endophytic fungus *Nigrospora sphaerica*." *Phytochemistry Letters* 7, (2014): 1-5.

Miliute, Inga, Odeta Buzaite, Danas Baniulis, and Vidmantas Stanys. "Bacterial endophytes in agricultural crops and their role in stress tolerance: A review." *Zemdirbyste-Agriculture* 102, no. 4 (2015): 465–478.

Mishra, Sandhya, Ram Sanmukh Upadhyay, Chandra Shekhar Nautiyal. "Unravelling the beneficial role of microbial contributors in reducing the allelopathic effect of weeds." *Applied Microbiology and Biotechnology* 97 no. 13 (2013): 5659–5668.

Mohali, S., T.I. Burgess, and M.J. Wingfield. "Diversity and host association of the tropical tree endophyte *Lasiodiplodia theobromae* revealed using simple sequence repeat markers." *Forest Pathology* 35, no. 6 (2011): 385–396.

Morsy, N.M. (2014). "Phytochemical analysis of biologically active constituents of medicinal plants." *Main Group Chemistry* 13, no. 1 (2014): 7-21.

Ndjakou, Lenta Bruno, Jules Rese, Marcel Ladoh-Yemeda, Flora *et. al.* "Purpureone, an antileishmanial ergochrome from the endophytic fungi *Purpureocillium lilacinum.*" *Zeitschrift für Naturforschung B* 71, no. 11 (2016): 1159–1167.

Patil, Ravindra H., Mohini P. Patil, and Vijay L. Maheshwari (2018). "Lead compounds from mangrove-associated microorganisms." *Marine Drugs* 16, no. 9 (2018): e319.

Philippot, Laurent, Jos M. Raaijmakers, Philippe Lemanceau, and Wim H. van der Putten. "Going Back to the Roots: The Microbial Ecology of the Rhizosphere." *Nature Reviews Microbiology* 11, no. 11 (2013): 789–99.

Qi, Xiaojuan, Ensi Wang, Ming Xing, Wei Zhao, and Xia Chen, X. "Rhizosphere and non-rhizosphere bacterial community composition of the wild medicinal plant Rumexpatientia." *World Journal of Microbiology and Biotechnology* 28, no. 5 (2012): 2257–2265.

Rodriguez, R.J., J.F. White, A.E. Arnold, and R.S. Redman. "Fungal endophytes: diversity and functional roles." *New Phytologist* 182, no. 2 (2009): 314–330.

Rönsberg, David, Abdessamad Debbab, Attila Mándi, Vera Vasylyeva, Philip Böhler, Björn Stork, Laura Engelke, Alexandra Hamacher, Richard Sawadogo, Marc Diederich, Victor Wray, Wen Han Lin, Matthias U. Kassack, Christoph Janiak, Stefanie Scheu, Sebastian Wesselborg, Tibor Kurtán, Amal H. Aly, and Peter Proksch. "Pro-apoptotic and immunostimulatory tetrahydroxanthone dimers from the endophytic fungus *Phomopsis longicolla*." *The Journal of Organic Chemistry* 78, no. 24 (2013): 12409- 12425.

Rosenblueth, Mónica, and Esperanza Martínez-Romero. "Bacterial endophytes and their interactions with hosts." *Molecular Plant-Microbe Interactions* 19, no. 8 (2006): 827–837.

Samanthi, K.S., Wickramaarachchi, E.M.K. Wijeratne, and P.A. Paranagama. "Two new bioactive polyketides from *Curvularia trifolii*, an endolichenic fungus isolated from *Usnea* sp." *Journal of the National Science Foundation of Sri Lanka* 43, no. 3 (2015): 217–224.

Sanga, Xia-Nan, Shao-Fei Chena, Ming-XuTanga, Hai-Fen Wanga, Jie Lua, DanZhaoa, Yu-Bo Wanga, Jiao Baia, Hui-Ming Huaa, Gang Chena, Yue-Hu Peia. "α- Pyrone derivatives with cytotoxic activities, from the endophytic fungus *Phoma* sp. YN02-P-3." *Bioorganic & Medicinal Chemistry Letters* 27, no. 11 (2014): 3723–3725.

Sessitsch, Angela, Melanie Kuffner, Petra Kidd, Jaco Vangronsveld, Walter W. Wenzel, Katharina Fallmann, and Markus Puschenreiter. "The role of plant-associated bacteria in the mobilization and phytoextraction of trace elements in contaminated soils." *Soil Biology and Biochemistry* 60, no. 100 (2013): 182–194.

Shaaban, Mohamed, Hamdi Nasr, Amal Z. Hassan, Mohsen S. Asker. "Bioactive secondary metabolites from endophytic *Aspergillus fumigatus*: structural elucidation and bioactivity studies." *Revista latinoamericana de química* 41, no. 1 (2012): 50-60.

Sharma, Ayushi, Bhanu Malhotra, Harsha Kharkwal, Giriraj T. Kulkarni, and Nutan Kaushik. Therapeutic agents from endophytes harbored in Asian medicinal plants. *Phytochemistry* 19, (2020): 691–720.

Singh, Radha, and Ashok K. Dubey. "Endophytic actinomycetes as emerging source for therapeutic compounds." *Indo Global Journal of Pharmaceutical Sciences* 5 no. 2 (2015): 106–116.

Strobel, Gary, and Bryn Daisy. "Bioprospecting for microbial endophytes and their natural products." *Microbiology and Molecular Biology Reviews* 67, no. 4 (2013): 491–502.

Suryanarayanan, Trichur S., Nagamani Thirunavukkarasu, G.N. Hariharan, and P. Balaji. "Occurrence of non-obligate inside lichen thalli." *Sydowia -Horn* 57, no. 1 (2005): 112-130.

Taghavi, Safiyh, Daniel van der Lelie, Adam Hoffman, Yian-Biao Zhang, Michael D. Walla, Jaco Vangronsveld, Lee Newman, and Sébastien Monchy. "Genome sequence of the plant growth promoting endophytic bacterium Enterobacter sp. 638." *PLoS. Genetics* 13, no. 6 (2010): e1000943.

Tan, R.X., and W.X. Zou. "Endophytes: a rich source of functional metabolites." *Natural Product Reports* 18, no. 4 (2001): 448–459.

Thomas, Pious, and Aparna Chandra Sekhar. "Cultivation versus molecular analysis of banana (*Musa* sp.) shoot-tip tissue reveals enormous diversity of normally uncultivable endophytic bacteria." *Microbial Ecology* 73, no. 4 (2017): 885-899.

Tian, Yuan, Séverine Amand, Didier Buisson, Caroline Kunz, François Hachette, Joëlle Dupont, Bastien Nay, and Soizic Prado. "The fungal leaf endophyte *Paraconiothyrium variabile* specifically metabolizes the host-plant metabolome for its own benefit." *Phytochemistry* 108, (2014): 95-101.

Tomlinson, P. Barry. "*Ecology. In: The Botany of Mangroves*," 2nd ed. Cambridge University Press: Cambridge, pp. 11-28. 2016.

Treutter, D. (2006). Significance of flavonoids in plant resistance: a review. *Environ. Chem. Lett.,* 4: 147–157.

Uesugi, Shota, Nozomi Fujisawa, Jun Yoshida, Mitsuru Watanabe, Shingo Dan, Takao Yamori, Yoshihito Shiono, and Ken-ichi Kimura. Pyrrocidine A, a metabolite of endophytic fungi, has a potent apoptosis inducing activity against HL60 cells through caspase activation via the Michael addition. *The Journal of Antibiotics* 69, no. 3 (2016), 133-140.

Vaidya, Ashok D.B., and Thomas P.A. Devasagayam. "Current status of herbal drugs in India: an overview." *Journal of Clinical Biochemistry and Nutrition* 41, no. 1 (2007): 1-11.

Vardhini, Bojjam V., and Naser A. Anjum. "Brassinosteroids make plant life easier under abiotic stresses mainly by modulating major components of antioxidant defense system." *Frontiers in Environmental Science* 2, (2015): 67.

Wang, J., G. Li, H. Lu, Z. Zheng, Y. Huang, and W. Su. "Taxol from *Tubercularia* sp. strain TF5, an endophytic fungus of *Taxus mairei*." *FEMS Microbiology Letters* 193, no. 2 (2000): 249–253.

Wani, Mansukhlal C., Harold Lawrence Taylor, Monroe E. Wall, Philip Coggon, and Andrew T. McPhail. "Plant antitumor agents. VI. The isolation and structure of taxol, a novel antileukemic and antitumor agent from *Taxus brevifolia*." *Journal of the American Chemical Society* 93, no. 9 (1971): 2325-1327.

West, E.R., E.J. Cother, C.C. Steel, and G.J. Ash. "The characterization and diversity of bacterial endophytes of grapevine." *Canadian Journal of Microbiology* 56, no. 3 (2010): 209-2015.

Wyk, Ben-Erik Van, and Michael W. "Portland. Medicinal Plants of the World." *Timber Press,* Pp. 480. 2004.

Xia, Xuekui, Soonok Kim, Sunghee Bang, Hyun-Jung Lee, Changheng Liu, Chan-Il Park and Sang Hee Shim. "Barceloneic acid C, a new polyketide from an endophytic fungus Phoma sp. JS752 and its antibacterial activities." *The Journal of Antibiotics* 68, (2015): 139–141.

Yamamoto, Norio, Yuki Kanemoto, Manabu Ueda, Kengo Kawasaki, Itsuko Fukuda, Hitoshi Ashida. "Anti-obesity and anti-diabetic effects of ethanol extract of *Artemisia princeps* in C57BL/6 mice fed a high-fat diet." *Food & Function* 2, no. 1 (2011): 45–52.

Zhang, Xiao-Qi, Wen-Cai Ye, Ren-Wang Jiang, Zhi-Qi Yin, Shou-Xun Zhao, Thomas C.W. Mak. "Two new eremophilanolides from *Xanthium sibiricum*" *Natural Product Research* 20, no. 13 (2006): 1265−1270.

Zhao, Ke, Petri Penttinen, Tongwei Guan, Jing Xiao, Qiang Chen, Jun Xu, Kristina Lindström, Lili Zhang, Xiaoping Zhang, and Gary A. Strobel. "The diversity and anti-microbial activity of endophytic actinomycetes isolated from medicinal plants in Panxi Plateau China." *Current Microbiology* 62, no. 1 (2011): 182–190.

Zhao, Q., X.M. Zhang, and W.J. Li. "Research progress on bioactive products from endophytes." *Journal of Microbiology* 38, (2018): 103-113.

Chapter 4

Exploring the Multi-Faceted Endophytes to Enhance Plant Growth and Tolerance toward Environmental Stress

Janki K. Patel*

Department of Biological Sciences, P. D. Patel Institute of Applied Sciences, Charotar University of Science and Technology, Changa, Gujarat, India

Abstract

Symbiotic endophytes are known to associate ubiquitously in various plant systems. Among them, some of the endophytes can promote plant growth without causing any negative impact on plant health. Interestingly, at the same time, these microbes are improving the host's stress tolerance against abiotic and biotic factors. The diverse group of bacteria and fungi are colonizing the internal tissues of the different plant parts. The association is expedited upon the exchange of chemical dialogues between the plants and endophytes. Alteration in this association was observed upon the change in the plant growth phase. The core microbiome of the plant is transferred in seed and buffered under different environmental conditions. They promote plant growth using multifarious plant growth-promoting traits similar to rhizobacteria such as nitrogen fixation, siderophore production, photostimulation, P-solubilization, biocontrol activities, etc. Additionally, endophyte colonization plays a significant role in the induction of inducing systemic resistance (ISR) in the plants and can salvage them from future pathogen attacks. The application of endophytes in the field is considered to boost sustainable agricultural production and can reduce the utilization of

*Corresponding Author's Email: jankipatel.bio@charusat.ac.in.

In: Endophytes
Editor: Pragya Tiwari
ISBN: 979-8-88697-045-6

chemical fertilizers. This green strategy can benefit both the environment and human health.

Keywords: antimicrobial compounds, endophytes, Induced Systemic Resistance (ISR), plant growth promotion, root exudates, nutrient availability

1. Introduction

In the modern era, agricultural industries are relying more on chemical fertilizers, pesticides, and growth regulators to fulfill the food demand of the ever-increasing population in the world. The higher usage of these chemicals leads to their entry into the ecosystem and ultimately reaches the higher phylum (Ward 2009). To address this issue, we need to develop the potential biofertilizer and biopesticides while reducing the usage of chemical fertilizers for sustainable agriculture. Endophytes are a suitable alternative to chemical fertilizer as they colonize the internal tissue of the plants and are also known to colonize in the seeds. The application of these microbes can be helpful as they get transferred from one generation to the next without being gets affected by environmental changes (Truyens et al. 2015; Shaik et al. 2019; Tiwari and Bae 2020). The majority of these seed endophytes are known to produce antimicrobial compounds and inhibit phytopathogens (Díaz Herrera et al. 2016; Khalaf et al. 2018; Tiwari et al. 2021; Tiwari et al. 2022). The fungal endophytes are also known to colonize the plants comparably to bacterial endophytes. They promote plant growth similarly to bacterial endophytes (Ripa et al. 2019).

Endophytes can be defined as the group of organisms that colonize the internal tissues of the plant and promote growth without causing any health hazards to the host (Sturz et al. 2000). Importantly, colonization of these endophytes in the internal tissues of the plants was observed comparatively in lower density than the rhizosphere microbial population. Several studies suggest that endophytes are the subset of the rhizosphere microbial population. Both bacteria and fungi are known to colonize the plants and positively promote plant growth via direct or indirect mechanisms. The bacterial endophytes majorly belong to the genera *Arthrobacter, Microbacterium, Streptomyces, Staphylococcus, Pseudomonas, Bacillus, Brevundimonas, Methylobacter, Acinetobacter, Ralstonia, Rhizobium, Achromobacter, Azospirillum, Caulobacter, Chromobacterium, Gluconoacetobacter,*

Azoarcus, Burkholderia, Herbaspirillum, Enterobacter, Klebsiella, Serratia, Paenibacillus, Lysinibacillus, and *Nocardia* sp. (Bhattacharyya et al. 2012; Madhaiyan et al. 2015; Patel et al. 2017; Rosenblueth et al. 2018; Patel et al. 2020).

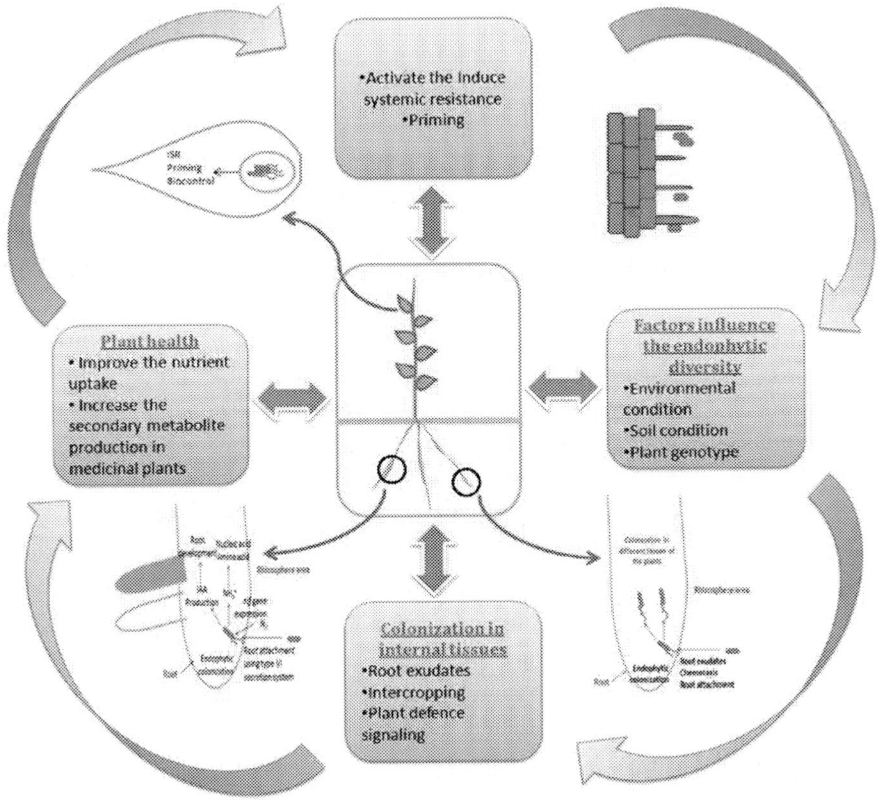

Figure 1. Illustration of influence on endophyte diversity under different conditions along with their ability to promote and protect the plant growth using their multifarious mechanisms.

While the plant growth-promoting fungal endophytes reported till now belongs to the genera *Penicillium, Trichoderma, Hypoxylon, Colletotrichum, Leptosphaerulina, Fusarium, Cladosporium, Cylindrocarpon, Alternaria, Phomopsis, Pleosporales, Trichosphaeriales, Chaetothyriales, Sordariales, Xylariales*, and *Pichia* (Zhou et al. 2014; Khan et al. 2015; Hassan 2017; Ripa et al. 2019; Shah et al. 2019; Xia et al. 2019). This plant growth-promoting endophytes exhibited nitrogen fixation, siderophore production, 1-

aminocyclopropane-1-carboxylic acid deaminase (ACC), indole acetic acid (IAA) production, biocontrol activity, and phosphate solubilization (Patel et al. 2017; Patel et al. 2018; Ripa et al. 2019; Patel et al. 2020; Turbat et al. 2020) (Figure 1).

The genome analysis of the 56 plant-associated bacterial endophytes showed the presence of various genes that contribute to plant growth promotion such as PQS (pyrroloquinoline quinone synthesis) gene cluster, hydrogen cyanide synthesis, nitrogen-fixing genes, auxin synthesis, ACC deaminase production, 2, 3- butanediol synthesis. In comparison to rhizospheric bacteria except 2, 4-diacetylphloroglucinol (DAPG) synthesis gene was not found in any analyzed bacterial endophyte genome (Bruto et al. 2014). Figure 1. Illustration of influence on endophytic diversity under different conditions along with their ability to promote and protect the plant growth using their multifarious mechanisms.

Genomic analysis of endophyte yeast *Rhodotorula mucilaginosa* JGTA-S1 showed the presence of genes related to anti-freezing, fungal hydrolytic enzymes that degrade plant cell wall, genes of the abscisic acid, acetoin, 2, 3-butanediol production and phosphate solubilization. Intriguingly, R. mucilaginosa contains the endosymbiotic diazotrophic bacteria Pseudomonas stutzeri inside their structure. While their genomic comparison with the other fungi reveals their evolutionary relationship with the hemibiotrophic fungi, as it shares the genes related to plant invasion with them (Sen et al. 2019). Comparatively very fewer studies based on molecular approaches are conducted in plant growth-promoting fungal endophytes.

2. Diversity of Endophytes under Various Environmental Conditions

The diversity of fungal endophytes associated with corn, tomato, pepper, and watermelon was studied under different soil conditions such as organic and conventional fields. The high diversity of the culturable fungal endophytes isolated during this study is majorly owned by the organic fields. In all, 550 fungal endophytes were isolated from organic farming while 190 fungal endophytes were isolated from conventional farming (Xia et al. 2019). Earlier the same group of scientists also carried out a similar set of experiments to study the diversity of culturable bacterial endophytes under organic and conventional field conditions. In this study, they obtained similar results as in the case of fungal endophytes. A total of 336 bacterial endophyte isolates were found and in them, 71% of isolates belong to organic farming plants, and only 29% were from conventional farming plants (Xia et al. 2015). These studies indicated that a higher diversity of endophytes was found in organic farming compared to the conventional system.

The effect on the endophytic community upon changes in environmental condition and geographical location was studied by transferring the Mexican breed maize seed to Canadian soil. The young maize plants were harvested and checked whether the original endophyte community was replaced with the native strains due to a change in geographical location and environmental condition. They observed there was partial buffering of endophytes present originally in the maize grown in Mexican soil. The authors attributed it to pre-existing seed endophytes which were transported to the newer location (Johnston-Monje et al. 2014). Intriguingly, another study showed the plant growth stage and tissue type affects the endophyte bacterial communities more compared to the abiotic factors such as elevated CO_2 level and variation in fertilization (Gaidi Ren et al. 2015). Recent studies observed the shift in the endophyte bacterial community depending on the plant growth stage. There were only a few bacterial genera able to colonize throughout the life cycle of the wheat plant (Patel et al. 2021). These studies indicate the plants select their own microbiome (Frank et al. 2017) and govern their successive colonization inside the host (Hassani et al. 2019).

The genotypically different maize and rice plants were analyzed for bacterial endophyte community using the PCR-DGGE method showed the variation in community with the change in host genotype. However, there was also observed the presence of specific bands of DGGE in all the cultivars. This study suggested the presence of specific bacterial endophytes consistently colonizing the plants without being influenced by the host genetic constitution (Hardoim et al. 2011; DAF da Silva et al. 2014). The two different varieties (BAT477 and DOR364) of *Phaseolus vulgaris* were co-inoculated with *Rhizobium-Azospirillum* combination across the different environmental conditions of Cuba. Co-inoculation of *Rhizobium-Azospirillum* on genotype DOR364 resulted in a positive effect by increasing the nitrogen content and promoting the plant growth parameters while genotype BAT477 showed a negative effect on yield and nitrogen fixation compared to singly inoculated *Rhizobium* (Remans et al. 2008).

Bacterial endophytes have been isolated from the plants while they survive under various abiotic stresses. The salt-sensitive and salt-tolerant rice cultivars were analyzed for their endophyte bacterial communities. Strikingly, they observed the presence of common bacterial endophyte *Flavobacterium* sp. in both the cultivars grown under different salt conditions. This endophyte *Flavobacterium* sp. showed the ability to phosphate solubilization, IAA production, and high resistance towards the salt and osmotic stress (Walitang et al. 2017). The diversity of the endophytes was assessed in copper tolerant plants from copper mine Westland and the predominant phyla comprised *Firmicutes*, *Actinobacteria,* and *Proteobacteria.* These bacterial endophytes were found to increase the dry mass of the plant in vermiculate amended with 4 mg Kg^{-1} Cu (Sun et al. 2010).

Comparative analysis of endophyte and rhizospheric diversity was studied by 16S rRNA clonal library in *Stellera chamaejasme* L. showed the presence of identical dominated phylum in decreasing order *Proteobacteria, Firmicutes,* and *Actinobacteria* in both, however, more variable bacteria were present in the rhizosphere (Jin et al. 2014). The plant genotype was not only affecting the bacterial community composition but the age of the plant also plays role in the selective colonization of endophytes inside the plant tissue (Van Overbeek et al. 2008; Gaidi Ren et al. 2015). These reports suggest the bacterial endophyte structure and successful colonization and establishment inside the plant tissue are governed by many varied factors.

3. Endophytic Colonization and Chemical Dialogues between the Plants and Microbes

The endophyte microbial components are helpful for their colonization inside the plant tissue under various environmental conditions such as motility, chemoattraction, adhesion, biofilm formation, cell wall, and outer membrane components as well as enzymes involved in the metabolism of cellulose and trehalose metabolism, secretion systems, and certain others involved in the mitigation of oxidative stress (Pinski et al. 2019; Patel et al. 2021). Bacterial endophyte colonization was studied with *Burkholderia phytofirmans* PsJN (host) and *Pseudomonas syrigae* pv. *pisi* (non-host). Both the strains were inoculated in the grapevine cell suspension and studied for their colonization ability. After the 5 mins of inoculation, their physical interaction with the plant cells was observed and after 24 h, it showed they completely cover the plant cells. In the same experiment, negative control *Escherichia coli* do not show any physical interaction even after the 24 h of inoculation with the plant cell (Bordiec et al. 2011). As plants secrete various root exudates, which play a significant role in attracting endophytes near the roots by chemotaxis mechanism. The rice root exudates effect exerted on endophyte and rhizosphere bacteria was studied in a sterile hydroponic system. Analyses of root exudate showed the presence of major components like carbohydrates and amino acids. The high concentrations of carbohydrates found in this study were glucose, arabinose, mannose, galactose, and glucuronic acid. Along with this, the amino acids found were histidine, proline, valine, alanine, and glycine. The root exudates obtained from the rice plant were tested upon the bacterial endophyte and rhizosphere isolates and showed the highest chemotactic response in the bacterial endophyte compared to rhizosphere isolates (Bacilio-Jiménez et al. 2003). The intercropping of plants affects the composition of the root exudates which ultimately influence the plant-microbe interaction and endophytic colonization. The faba bean and maize intercropping significantly enhance the nodulation. Similarly, the intercropping of faba bean was carried out with wheat and barley does not show a similar effect on nodulation. The nitrogen content was found similar in all the intercropping systems, which indicated the nitrogen concentration does not play a role in the nodulation process in this study. In another set of experiments, maize root exudates inoculated on faba bean root showed enhanced nodulation whereas the wheat and barley root exudates does not show any effect on nodulation. The maize and faba bean intercropping the key signal molecules responsible for the nodulation were screened. The highest production was found in genistein, and it was 73% higher in the intercropping system than in the individual plants' cropping system. The molecular approach was used to find the gene expression in intercropped plants by qRT PCR. The result showed an increase in the expression of the nodulin-like genes and chalcone-flavone isomerase (Li et al. 2016). Thus, the root exudate composition plays a crucial role in defining the endophytic and rhizospheric population in various plant systems.

Apart from the chemical signal through root exudates, studies showed that plant defense signaling pathways control the endophytic colonization inside the plant tissue. The bacterial endophyte *Klebsiella pneumoniae* 342 efficiently colonizes and enhanced the plant growth of *Medicago* spp. under favorable conditions. The introduction of an ethylene precursor molecule in the plant growth medium showed a noticeable reduction in the colonization of *K. pneumoniae* 342 in *Medicago* spp. In reverse, seedlings were grown in the medium containing the ethylene pathway inhibitor (1-methylcyclopropene) again exhibited the efficient endophytic colonization by *K. pneumoniae* 342. In a parallel study, they observed the colonization of human pathogen

Salmonella enterica serovar *typhimurium* in *Medicago* spp. and obtained a similar result to *K. pneumoniae* 342 (Iniguez et al. 2004). The ability of cross-colonization of endophyte isolates from the rice, sorghum, pearl millet, and maize was observed with and without tagging them with fluorescence tags in the wheat plants (Patel et al. 2017; Patel et al. 2020). Microarray analysis of 22,800 genes in *Arabidopsis* was carried out upon colonization of endophyte *Pseudomonas fluorescens* FPT9601-T5. During this study, they note that 95 genes were upregulated, and 105 genes were down-regulated. The gene up-regulated includes the putative auxin-regulated genes and nodulin-like genes downregulation of ethylene-responsive genes was observed (Wang et al. 2005). These studies indicate that plant defense machinery plays a significant role in systematic endophyte colonization in the internal tissues of the plants.

B. phytofirmans PsJN is known for its plant growth promotion ability in *Arabidopsis thaliana.* The mutant strain with impaired quorum sensing (QS) system was used to analyze the endophyte colonization. The strain with defective QS produced a significantly lower level of 3-hydroxy-C8-homoserine lactone compared to the wild-type of strain. The colonization in plants was assessed under gnotobiotic conditions. There was a significantly lower rhizospheric colonization and null colonization of endophytes inside the plant tissue. At the same time, there was a lower ability of root surface adherence by the mutant strain. This was due to the reduction in swimming motility of the mutant strain. The results indicated the importance of the QS system in rhizospheric as well as endophyte colonization in plants (Zúñiga et al. 2013). In the contrary, diazotrophic endophyte *Azoarcus* sp. BH72 does not show the presence of QS genes in the genome, still able to successfully colonize the plant (Krause et al. 2006; Suárez-Moreno et al. 2008).

4. Molecular Signals Involved in Plant Growth Promotion

Plants can absorb nitrogen in the form of nitrate and ammonium ion from the root; however, association with nitrogen-fixing bacteria can make available atmospheric nitrogen for plant metabolism. Nitrogen depletion is the major concern in the case of *Poaceae* plants. Due to this, these plants have developed the mechanism to incur the nitrogen from symbiotic diazotrophic bacteria under the N-limiting soil condition. This mechanism for nitrogen procuring is different compared to other plant-microbe interactions. The plant root tissue secretes the H_2O_2 on diazotrophic bacterial endophytes. This leads to the oxidation of their cell wall, membrane, and cell content. The oxidation of bacterial cells supplies the nitrate for plant growth. The remaining larger molecule was further degraded and absorbed by the plant. Interestingly, diazotrophic endophytes growing in the root tissue itself were used as a source of nitrogen by plants (White et al. 2012).

The microarray analysis of *Arabidopsis* plants after bacterial endophyte colonization showed the up-regulation of certain gene involved in the plant growth namely involved in carbohydrate metabolism like β-galactosidase (Wang et al. 2005). The *Burkholderia kururiensis* M130 strain QS was investigated for role in general metabolic activities by constructing knockout QS gene mutants. The QS mutant strain does not show any negative effect on nitrogen fixation, carbon metabolism, siderophore production, IAA production, lipase activity, and motility (Krause et al. 2006; Suárez-Moreno et al. 2008). On contrary, two QS systems are present in the *Serratia* sp. G3 is known to positively regulate antifungal metabolite production, while, negatively regulating IAA production (B Liu et al. 2009; Xiaoguang Liu et al. 2011). These reports suggest the importance of quorum sensing may differ with the strain and their effect on plant growth promotion.

5. Impact on Secondary Metabolite Production in Plants Upon Endophyte Colonization

Endophytes are known to induce the production of medicinally important metabolites in plants (Tiwari et al. 2022). Fungal endophyte colonization inside *Atractylodes lancea*, a Chinese medicinal plant leads to activation of the jasmonic acid (JA) pathway. This pathway further activates the production of secondary metabolites such as volatile oil. JA pathway and secondary metabolite production was induced by fungal endophyte confirmed upon inoculating the JA inhibitor. Concomitantly, it was observed that the salicylic acid (SA) production was suppressed and due to that higher accumulation of JA occurred (Ren et al. 2012). In the same plant system, colonization of fungal endophyte *Gilmaniella* sp. AL12 induces the biosynthesis of sesquiterpenoids through NO, H_2O_2, ethylene, SA, JA, brassinosteroid, and calcium signaling pathways (Yuan et al. 2016). Bacterial endophyte *E. ludwigii* EnVs6 colonization shifts the metabolite production in the grapevine plants towards cell signaling and plant defense. There was an increased accumulation of hydroxycinnamic acids, and flavonoids and a decrease in the production of phytoalexins was observed (Lòpez-Fernàndez et al. 2016). These studies play a significant role in developing the understanding of secondary metabolite production pathways in plants for the higher production of economically important molecules.

6. Assist in Nutrient Acquisition in Plants

Endophytes assist the plants in acquiring the nutrients from the surrounding environment through nitrogen fixation, siderophore production, phytohormone production, phosphate solubilization, etc. These microbes colonize every tissue of the plant and promote plant growth using multiple PGP traits (Table 1). Therefore, endophyte colonization and manipulation of their diversity play a crucial role in plant growth.

The earlier reports suggested that *in planta* biological nitrogen fixation (BNF) is carried out by *Rhizobium* spp. and it is restricted only to the *Fabaceae* plants. However, recent reports demonstrated the colonization of diverse diazotrophic endophytes in non-legume plants as well (Patel et al. 2017; Ripa et al. 2019; Patel et al. 2020; Turbat et al. 2020). Diazotrophic bacterial endophyte (*Azoarcus* sp. BH72 and *G. diazotrophicus* Pal5) genome sequence showed the presence of similar nitrogen fixing machinery as in other diazotrophs of rhizosphere and legume (Krause et al. 2006; Bertalan et al. 2009). *In vivo* inoculation of 14 diazotrophic endophytes was carried out in pine trees under the N-limiting condition. The total atmospheric nitrogen fixed was up to 23%–53% after one year as estimated through nitrogen-15 (15N) isotope assay as well as it showed the plant growth promotion (Padda et al. 2019). Colonization of foliar fungal endophyte *Colletotrichum tropicale* in leaves of *Theobroma cacao* was observed with and without the presence of the pathogen. Subsequently, in both conditions, the soil was pulsed with the 15N and later traced for the 15N in different plant parts. The colonization of *C. tropicale* significantly improved the N uptake in the plants in both conditions (Christian et al. 2019).

Table 1. A representative list of endophytes carried out the plant growth promotion as well as protection against pathogens attack

Isolate	Traits improve nutrient availability and stress reduction	Biocontrol activity against	Type of biocontrol compound	Source of isolation	Reference
Bacterial endophyte					
Acinetobacter guillouiae	Nitrogen fixation, IAA (Indole Acetic Acid), Phosphate solubilization, Siderophore, Zn-solubilization, Potassium solubilization	-	Hydrogen cyanide producers	Wheat	Rana et al. 2020
Bacillus velezensis LDO2	Phosphate solubilization, Siderophore production	*Aspergillus flavus*	Fengycin, Surfactin, Bacilysin	Peanut root	Deng et al. 2019
Enterobacter roggenkampii ED5	Nitrogen fixation, IAA, Phosphate solubilization, Siderophore production, ACC deaminase, Ammonia Production	*Fusarium moniliforme, Fusarium cubense, Botrytis cinerea, Ceratocystis paradoxa, and Sporisorium scitamineum*	2,3-butanediol, methanethiol, Isoprene, HCN production	Sugarcane	Guo et al. 2020
Lysinibacillus sphaericus	Nitrogen fixation, IAA, ACC deaminase	*Rhizoctonia solani*	-	Rice	Shabanamol et al. 2018
Microbispora sp. GKU 823	ACC deaminase, IAA, nitrogen fixation, phosphate solubilization, siderophore production	-	-	Sugarcane root	Kruasuwan et al. 2016
Paenibacillus spp.	Nitrogen fixation, IAA, Siderophore, Phosphate solubilization, 1-aminocyclopropane-1-decarboxylate (ACC) deaminase	-	-	Cucumber leafstalks	Mahmood et al. 2020
Paraburkholderia sp.	IAA, Siderophore, Phosphate solubilization	*Fusarium* sp. and *Rhizoctonia* sp.	-	*Aloe vera*	Silva et al. 2020

Table 1. (Continued)

Isolate	Traits improve nutrient availability and stress reduction	Biocontrol activity against	Type of biocontrol compound	Source of isolation	Reference
Pseudomonas protegens MP12	IAA, Siderophore, Phosphate solubilization	*Botrytis cinerea, Alternaria alternata, Aspergillus niger, Penicillium expansum and Neofusicoccum parvum*	2,4-diacetylphloroglucinol, Pyoluteorin, and pyrrolnitrin	Grapevine	Andreolli et al. 2019
Pseudomonas stutzeri E25	-	*Botrytis cinerea*	Volatile organic compounds (VOCs)	*Physalis ixocarpa*	Rojas-Solís et al. 2018
Rhizobium radiobacter	Nitrogen fixation, IAA, Siderophore, ACC deaminase	-	-	*Robinia pseudoacacia* in a Pb/Zn mining area	Fan et al. 2018
Streptomyces spp.	Nitrogen fixation, IAA, Siderophore production, Phosphate solubilization	*Magnaporthe oryazae Rhizoctonia Solani*	2-(chloromethyl)-2-cyclopropyloxirane; 2, 4- ditert-butylphenol; 1-ethylthio-3-methyl-1, 3-butadiene	Sorghum	Patel et al. 2017; Patel et al. 2018
Stenotrophomonas maltophilia CR71	-	*Botrytis cinerea*	Volatile organic compounds (VOCs)	*Physalis ixocarpa*	Rojas-Solís et al. 2018
Escherichia coli	-	*Bursaphelenchus xylophilus* (pine wood nematode)	Secondary metabolite extracted from culture supernatant	Pine tree samples	Yunran Liu et al. 2019
Serratia marcescens					
Fungal endophyte					
Trichoderma asperellum	IAA, Phosphate solubilization, Siderophore production	*Fusarium thapsinum, Epicoccum sorghinum, Alternaria alternata,* and *Curvularia lunata*	-	*Sorghum bicolor*	Rajini et al. 2020

Isolate	Traits improve nutrient availability and stress reduction	Biocontrol activity against	Type of biocontrol compound	Source of isolation	Reference
Trichoderma spp.	IAA, Phosphate solubilization, ACC deaminase	-	-	*Triticum aestivum* L.	Ripa et al. 2019
Penicillium spp.	IAA, Phosphate solubilization	*Pseudomonas aeruginosa, Staphylococcus aureus, Bacillus subtilis*	-	*Teucrium polium* L.	Hassan, 2017
Macrophomina phaseolina	IAA, Phosphate solubilization, Siderophore production	*Sclerotium rolfsii, Fusarium oxysporum*	-	Sunchoke and medicinal plants	Suebrasri et al. 2020
Trichoderma spp.					
Fusarium sp.	IAA, Phosphate solubilization, Siderophore production	-	-	*Sophora flavescens*	Turbat et al. 2020
Xylogone sphaerospora					
Alternaria sp.					
idymella sp.					

The accessibility of other nutrients also plays a significant role in plant health such as Zn, Fe, P, and Ca. The fungal endophyte strain *Penicillium roqueforti* was isolated from the root part of the halophytic plant *Solanum surattense*. Inoculation of *P. roqueforti* on the wheat plant in heavy metal contaminated soil and irrigating the soil with wastewater improved the nutrient uptake in wheat. This study showed an increase in the nutrient availability (Mg, K, Na, Ca) in the plant compared to the uninoculated control under the stress conditions (Ikram et al. 2018). The phosphate fertilized and non-fertilized soil was inoculated with the endophyte *Bacillus* spp. showed enhanced uptake of P in both the condition compared to the non-bacterized control pearl millet plants (Ribeiro et al. 2018).

7. Phytohormone Production and Plant Growth Promotion

There are six pathways present in the microbes for the IAA production including the indole-3-acetamide pathway, indole-3-pyruvate pathway, tryptamine pathway, indole-3-acetonitrile pathway, tryptophan side-chain oxidase pathway, and tryptophan-independent pathway (Spaepen et al. 2007). Endophytes are already known to produce the large amount of IAA and influence the plant growth (Ripa et al. 2019; Turbat et al. 2020). The function of the IAA production in the endophytes was evaluated using the *in vitro* plant regeneration techniques. Some of the endophytes isolated from *Withania somnifera* (L.) exhibited the good IAA production. The combination of isolates was inoculated on an endangered plant *Exacum travancoricum* to assess the efficacy of regeneration. The control explants were inoculated with the exogenous IAA and without bacterial inoculation. Co-culturing of *Stenotrophomonas thalpophilum*, *Pseudomonas aeruginosa* and *Enterobacter aerogenes* attributes efficient somatic embryogenesis of *E. travancoricum* in compared to the control plants. Along with that, they noticed that the higher survival rate of plants was due to the inoculation with the IAA producing isolates (Raju et al. 2020). Many fungal endophytes produce the phytohormones. The crude IAA extract of fungal endophytes *Preussia* sp., *Aspergillus japonicus* and *F. oxysporum* promote the root elongation as well as increase the photosynthetic pigments in *Arabidopsis*. In addition, these extracts showed the root growth in rice, and corn plants. Surprisingly, reduction in the biomass of *Arabidopsis* was observed upon the inoculation of the crude IAA extract of the fungal endophytes (Turbat et al. 2020). Application of the *P. roquefortis* on wheat plants that were growing under heavy metal stress improved the stress tolerance. *P. roquefortis* limited the transfer of heavy metal from soil to plants by secreting the IAA (Ikram et al. 2018).

8. Biocontrol Activity against Phytopathogens

Endophytes are known to improve plant fitness by indirect mechanisms such as the secretion of antimicrobial compounds, and by activating the plant defense machinery. Endophytes have been implicated in the suppression of *Fusarium oxysporum* causing wilt disease of banana and pepper (Cao et al. 2005; Yu et al. 2011), *Phytophthora medicaginis* implicated in root rot on chickpea (Misk et al. 2011), *Gaeumannomyces graminis* var. *tritici* (Ggt) responsible for take-all disease in wheat (Coombs et al. 2004), *Rhizoctonia solani* causing damping-off disease (Goudjal et al. 2014), *Pythium aphanidermatum* and *Colletotrichum orbiculare* causing diseases in cucumber (El-Tarabily et al. 2009; Shimizu et al. 2009), and *Alternaria alternata* causing early blight disease in tomato plant (Verma et al. 2011). A total 169 bacterial endophytes were isolated from cucurbits plants. These endophytes were assessed for phytopathogens inhibition against *Rhizoctonia solani*, *Fusarium graminearum*, *Phytophthora capsici*, *Pythium aphanidermatum*, and *Podosphaera fuliginea*. Interestingly, 70% of the endophytes exhibited the antagonistic effect against all the five fungal pathogens. Among them, major genus showed the antagonistic activities belonged to *Bacillus* and *Paenibacillus*. The major metabolite involved in the disease suppression was volatile organic compounds (VOCs) (Khalaf et al. 2018; Tyc et al. 2020). The potential sugarcane endophyte *Bacillus subtilis* showed the *in vitro* inhibitory activity against the *Alternaria, Cochliobolus, Curvularia, Fusarium, Neodeightonia, Phomopsis,* and *Saccharicola* isolated from diseased leaves of sugarcane. Molecular detection of a biosynthetic gene involved in the biocontrol activity showed the presence of *baeR* (bacillaene biosynthesis), *dfnD* (difficidin biosynthesis), *mlnA* (macrolactins biosynthesis), and *fenA* (fengycin biosynthesis) as well as VOCs produced by *B. subtilis* (Hazarika et al. 2019). Diazotrophic endophyte *Streptomyces* spp. has been reported to have an antagonistic effect against the pathogens *Magnaporthe oryzae* and *R. solani* by plate assay as well as by pot experiments in three different cereal plants (Patel et al. 2018a). Endophyte DAPG producing *P. fluorescens* was shown to inhibit the *Ralstonia solanacearum* and reduce the wilt disease in eggplant by more than 70% (Ramesh et al. 2009). Along with this, endophyte has been genetically engineered for better biocontrol activity and these microbes exhibited a similar effect on phytopathogen as the wild-type strains (Patel and Archana 2018).

The fungal endophytes isolated from the various plants of China showed the inhibitory effects on cucumber wilt causing *Fusarium oxysporum*. The highest inhibition of fungal mycelium was perceived using *Penicillium* sp. (66.4%), *Gugnardia mangiferae* (47%), *Hypocrea* sp. (44.2%), *Neurospora* sp. (43.1%), and *Eupenicillium* sp. (43.3%), respectively. Co-inoculation of plant growth-promoting fungus (*Penicillium* sp., *Hypocrea* sp., *Lasiodiplodia theobromae*) and *F. oxysporum* showed the least disease severity in cucumber plants compared to the control (Abro et al. 2019). Biocontrol activity against the cotton leafworm was studied by inoculation of fungal endophyte *Alternaria alternata, Alternaria solani, Curvularia lunata, Nigrospora sphaerica*, and *Stemphylium* sp. Confirmation of pathogenicity of fungal endophytes was observed upon the emergence of fungal mycelium on dead leafworm. The mortality rate of the leafworm inoculated with *C. lunata, A. solani*, and *A. alternata* was 60%, 40%, and 33.3%, respectively (Saad et al. 2019). These studies showed the usefulness of endophytes as biocontrol agents.

9. Bacterial Endophytes Induce Plant Defense against Infection

The induction of induce systemic resistance (ISR) in plants by endophyte is dependent on the efficient colonization (Alquéres et al. 2013). The earlier researcher showed the reduction in plant disease severity in cotton plant upon inoculation of endophyte against *F. oxysporum* f. sp. *vasinfectum* which causes the wilt disease. The plants were injected with the potential endophytes in stem after the 7 d of plantation followed by 10 d of incubation and further infected with fungal pathogen in plants. After 12 d of the fungal treatment, plants were showed to reduction in disease severity compared to the controls. In this study potential bacterial endophyte candidates were *Aureobacterium saperdae*, *Bacillus pumilus*, *phyllobacterium rabiacearum*, *P. putida*, and *Burkholderia solanacearum* (Chen et al. 1995). The Ri-T-DNA transformed pea roots were used as a tool to analyze the visible effect of ISR upon the fungal pathogen attack on endophyte *B. pumilis* inoculated root. The pre-bacterized root challenged with the pathogen strengthens the epidermis and cortical cell wall and restricts the hyphae growth to the outermost root tissue and induces the accumulation of phenolic compounds in intracellular spaces upon pathogen attack (Benhamou et al. 1996). The inoculation of *Streptomyces* sp. EN27 on *Arabidopsis* seed induces the NPR1 dependent pathway gene expression in 7-week-old plants using qPCR (Conn et al. 2008). The volatile compound 2,3-butanediol is produced by bacterial endophyte *Enterobacter aerogenes* and *Bacillus* spp. showed increased disease resistance against the Northern corn leaf blight fungus *Setosphaeria turcica* and bacterial pathogen *Erwinia carotovora* subsp. *carotovora* (Ryu et al. 2004; D'Alessandro et al. 2014). The co-inoculation of endophytic and rhizospheric bacterial formulation in plant reduce the disease infection and induce the high expression of ISR related proteins such as chitinase, β-1,3-glucanase, phenylalanine lyase, peroxidase and polyphenol oxidase in compare to controls (Harish et al. 2008; Sundaramoorthy et al. 2012). The endophyte *B. pumilus* INR7 induced the ISR response against *Xanthomonas axonopodis* pv. *vesicatoria* in pepper plant. This endophyte *B. pumilus* INR7 applied in field along with chemical inducer benzothiadiazole (BTH) showed to higher defense gene expression of CaPR1, CaTin1, and CaPR4 compared to the individual inoculant (Yi et al. 2013). The maize seedlings were treated with the lipopeptide producing endophyte *Bacillus* spp. showed the induction of pathogenesis-related proteins but when the methanolic extract of lipopeptide alone is applied to the plant it does not show any significant effect on ISR gene expression (Gond et al. 2015). The endophyte bacterial colonization in rice and olive plants shows the upregulation of PR10a, NPR1, PAL, and LOX2 which are involved in disease resistance upon pathogen challenge compared to the only fungal challenged plants without endophyte bacterial colonization in the aerial part of the plants (Cabanás et al. 2014; Patel et al. 2018; Patel and Archana 2018). In another study, bacterial endophyte *P. fluorescens* FPT9601-T5 colonize the *Arabidopsis* to induce the up-regulation of PR proteins, WRKY transcription factors, stress response proteins, drought-induced proteins, and peroxidase. Along with this, down-regulation of ethylene response factor, Myb-related transcription factor, and cinnamoyl CoA reductase-like protein (Wang et al. 2005). Comparative analysis of host (*B. phytofirmans* PsJN) and non-host (*P. syringae* pv. *pisi*) plant-microbe was carried out in grapevine plants. In this work, defense-related gene expression was studied after 9 h and 24 h of bacterial endophyte inoculation. The up-regulation of defense-related genes (chit4c, gluc, pin, and lox) was observed in both *B. phytofirmans* PsJN and *P. syringae* pv. *pisi* respectively. Interestingly, *P. syringae* pv. *pisi* (non-host microbe) inoculation causes stronger defense responses and cell death, whereas, *B. phytofirmans* PsJN does not cause cell death in grapevine plants. The mycorrhizal fungus

Piriformospora indica inoculation reduced the *G. orontii* conidia infection in Arwild-type (Col-0), npr1-3 (nonexpressor of PR genes 1-3), and NahG plants. *P. indica* was induced by the JA pathway-dependent defense against the powdery mildew (Stein et al. 2008).

Conclusion

The endophyte population is the subset of the rhizospheric flora. Their diversity is dependent on plant variety as well as prevailing environmental and edaphic conditions. Bacterial endophyte attachment and colonization in the plants are governed by various molecular signals. Moreover, the colonization of many endophytes results in priming of the plants before pathogen attack and prepares them for a more intense and rapid defense upon pathogen attack. Manipulation of endophyte diversity in agricultural land can be a useful strategy for sustainable plant growth promotion.

Acknowledgments

The author acknowledges the research institution for encouragement and support.

Conflict of Interests

No conflict of interest was declared.

References

Abro, Manzoor Ali, Xiang Sun, Xingchun Li, Ghulam Hussain Jatoi, and Liang Dong Guo. "Biocontrol potential of fungal endophytes against *Fusarium oxysporum* f. Sp. *Cucumerinum* causing Wilt in Cucumber." *Plant Pathology Journal* 35, no. 6 (2019): 598–608.

Alquéres, Sylvia, Carlos Meneses, Luc Rouws, Michael Rothballer, Ivo Baldani, Michael Schmid, and Anton Hartmann. "The bacterial superoxide dismutase and glutathione reductase are crucial for endophytic colonization of rice roots by *Gluconacetobacter diazotrophicus* PAL5." *Molecular Plant-Microbe Interactions* 26, no. 8 (2013): 937–945.

Andreolli, Marco, Giacomo Zapparoli, Elisa Angelini, Gianluca Lucchetta, Silvia Lampis, and Giovanni Vallini. "*Pseudomonas protegens* MP12: A plant growth-promoting endophytic bacterium with broad-spectrum antifungal activity against grapevine phytopathogens." *Microbiological Research* 219 (2019): 123–131.

Bacilio-Jiménez, Macario, Sara Aguilar-Flores, Elsa Ventura-Zapata, Eduardo Pérez-Campos, Stephane Bouquelet, and Edgar Zenteno. "Chemical characterization of root exudates from rice (*Oryza sativa*) and their effects on the chemotactic response of endophytic bacteria." *Plant and Soil* 249, no. 2 (2003): 271–277.

Benhamou, Nicole, Joseph W. Kloepper, Andrea Quadt-Hallman, and Sadik Tuzun. "Induction of defense-related ultrastructural modifications in pea root tissues inoculated with endophytic bacteria." *Plant Physiology* 112, no. 3 (1996): 919-929.

Bertalan, Marcelo, Rodolpho Albano, Vânia de Pádua, Luc Rouws, Cristian Rojas, Adriana Hemerly, Kátia Teixeira, et al. "Complete genome sequence of the sugarcane nitrogen-fixing endophyte *Gluconacetobacter diazotrophicus* Pal5." *BMC Genomics* 10, no. 1 (2009): 1-17.

Bhattacharyya, P. N., and D. K. Jha. "Plant growth-promoting rhizobacteria (PGPR): emergence in agriculture." *World Journal of Microbiology and Biotechnology*. 28, no. 4 (2012): 1327-1350.

Bordiec, Sophie, Sandra Paquis, Hélne Lacroix, Sandrine Dhondt, Essaïd Ait Barka, Serge Kauffmann, Philippe Jeandet, et al. "Comparative analysis of defense responses induced by the endophytic plant growth-promoting rhizobacterium B*urkholderia phytofirmans* strain psjn and the non-host bacterium *Pseudomonas syringae* Pv. *pisi* in grapevine cell suspensions." *Journal of Experimental Botany* 62, no. 2 (2011): 595-603.

Bruto, Maxime, Claire Prigent-Combaret, Daniel Muller, and Yvan Moënne-Loccoz. 2014. "Analysis of genes contributing to plant-beneficial functions in plant growth-promoting rhizobacteria and related proteobacteria." *Scientific Reports* 4, no. 1 (2014): 1-10.

Cabanás, Carmen Gómez Lama, Elisabetta Schilirò, Antonio Valverde-Corredor, and Jesús Mercado-Blanco. "The biocontrol endophytic bacterium *Pseudomonas fluorescens* PICF7 induces systemic defense responses in aerial tissues upon colonization of Olive roots." *Frontiers in Microbiology* 5 (2014): 427.

Cao, Lixiang, Zhiqi Qiu, Jianlan You, Hongming Tan, and Shining Zhou. "Isolation and characterization of endophytic *Streptomycete* antagonists of *Fusarium* wilt pathogen from surface-sterilized banana roots." *FEMS Microbiology Letters* 247, no. 2 (2005): 147-152.

Chen, C., E. M. Bauske, G. Musson, R. Rodríguez Kábana, and J. W. Kloepper. "Biological control of *Fusarium* wilt on cotton by use of endophytic bacteria." *Biological Control* 5, no. 1 (1995): 83-91.

Christian, Natalie, Edward Allen Herre, and Keith Clay. "Foliar Endophytic fungi alter patterns of nitrogen uptake and distribution in *Theobroma cacao*." *New Phytologist* 222, no. 3 (2019): 1573-1583.

Conn, V. M., A. R. Walker, and Chris M. M. Franco. "Endophytic actinobacteria induce defense pathways in *Arabidopsis thaliana*." *Molecular Plant-Microbe Interactions* 21, no. 2 (2008): 208-218.

Coombs, Justin T., Philip P. Michelsen, and Christopher M. M. Franco. "Evaluation of endophytic actinobacteria as antagonists of *Gaeumannomyces graminis* Var. *Tritici* in Wheat." *Biological Control* 21, no. 2 (2008): 208-218.

D'Alessandro, Marco, Matthias Erb, Jurriaan Ton, Anna Brandenburg, Danielle Karlen, Jakob Zopfi, and Ted C. J. Turlings "Volatiles produced by soil-borne endophytic bacteria increase plant pathogen resistance and affect tritrophic interactions." *Plant, Cell and Environment* 37, no. 4 (2014): 813-826.

Deng, Zhen-Shan, Bao-Cheng Zhang, Xiang-Ying Qi, Zhi-Hong Sun, Xiao-Long He, Yu-Zhen Liu, Jing Li, Kai-Kai Chen, and Zhan-Xi Lin. "Root-associated endophytic bacterial community composition of *Pennisetum sinese* from four representative provinces in China." *Microorganisms* 7, no. 2 (2019): 47.

Díaz Herrera, Silvana, Cecilia Grossi, Myriam Zawoznik, and María Daniela Groppa. 2016. "Wheat seeds harbor bacterial endophytes with potential as plant growth promoters and biocontrol agents of *Fusarium Graminearum*." *Microbiological Research* 186–187 (May): 37–43. https://doi.org/10.1016/j.micres.2016.03.002.

El-Tarabily, K. A., A. H. Nassar, G. E. St J. Hardy, and K. Sivasithamparam. "Plant growth promotion and biological control of *Pythium aphanidermatum*, a pathogen of cucumber, by endophytic actinomycetes." *Journal of Applied Microbiology* 106, no. 1 (2009): 13-26.

Fan, Miaochun, Zhenshan Liu, Lijun Nan, Entao Wang, Weimin Chen, Yanbing Lin, and Gehong Wei. "Isolation, characterization, and selection of heavy metal-resistant and plant growth-promoting endophytic bacteria from root nodules of *Robinia pseudoacacia* in a Pb/Zn mining area." *Microbiological Research* 217 (2018): 51-59.

Frank, Anna Carolin, Jessica Paola Saldierna Guzmán, and Jackie E. Shay. "Transmission of bacterial endophytes." *Microorganisms* 5, no. 4 (2017): 70.

Gond, Surendra K., Marshall S. Bergen, Mónica S. Torres, and James F. White Jr. "Endophytic *Bacillus* spp. produce antifungal lipopeptides and induce host defence gene expression in maize." *Microbiological Research* 172 (2015): 79-87.

Guo, Dao Jun, Rajesh Kumar Singh, Pratiksha Singh, Dong Ping Li, Anjney Sharma, Yong Xiu Xing, Xiu Peng Song, Li Tao Yang, and Yang Rui Li. 2020. "Complete genome sequence of *Enterobacter roggenkampii* ED5, a nitrogen-fixing plant growth-promoting endophytic bacterium with biocontrol and stress tolerance properties, isolated from sugarcane root." *Frontiers in Microbiology* 11 (2020): 2270.

Hardoim, Pablo Rodrigo, Fernando Dini Andreote, Barbara Reinhold-Hurek, Angela Sessitsch, Leonard Simon van Overbeek, and Jan Dirk van Elsas. "Rice root-associated bacteria: insights into community structures across 10 cultivars." *FEMS Microbiology Ecology* 77, no. 1 (2011): 154-164.

Harish, S., M. Kavino, N. Kumar, D. Saravanakumar, K. Soorianathasundaram, and R. Samiyappan. "Biohardening with plant growth-promoting rhizosphere and endophytic bacteria induce systemic resistance against banana bunchy top virus." *Applied Soil Ecology* 39, no. 2 (2008): 187-200.

Hassan, Saad El Din. "Plant growth-promoting activities for bacterial and fungal endophytes isolated from medicinal plant of *Teucrium polium* L." *Journal of Advanced Research* 8, no. 6 (2017): 687-695.

Hassani, M. Amine, Ezgi Özkurt, Heike Seybold, Tal Dagan, and Eva H. Stukenbrock. "Interactions and co-adaptation in plant metaorganisms." *Annual Review of Phytopathology* 57 (2019): 483-503.

Hazarika, Dibya Jyoti, Gunajit Goswami, Trishnamoni Gautam, Assma Parveen, Pompi Das, Madhumita Barooah, and Robin Chandra Boro. "Lipopeptide mediated biocontrol activity of endophytic *Bacillus subtilis* against fungal phytopathogens." *BMC Microbiology* 1 19, no. 1 (2019): 1-13.

Ikram, Muhammad, Niaz Ali, Gul Jan, Farzana Gul Jan, Inayat Ur Rahman, Amjad Iqbal, and Muhammad Hamayun. "IAA producing fungal endophyte *Penicillium roqueforti* Thom. Enhances stress tolerance and nutrients uptake in wheat plants grown on heavy metal contaminated soils." Edited by Graziella Berta. *PLOS ONE* 13, no. 11 (2018): e0208150.

Iniguez, A. Leonardo, Yuemei Dong, and Eric W. Triplett. "Nitrogen fixation in wheat provided by *Klebsiella pneumoniae* 342." *Molecular Plant-Microbe Interactions* 17, no. 10 (2004): 1078-1085.

Jin, Hui, Xiao Yan Yang, Zhi Qiang Yan, Quan Liu, Xiu Zhuang Li, Ji Xiang Chen, Deng Hong Zhang, Li Ming Zeng, and Bo Qin. "Characterization of rhizosphere and endophytic bacterial communities from leaves stems and roots of medicinal *Stellera chamaejasme* L." *Systematic and Applied Microbiology* 37, no. 5 (2014): 376-385.

Johnston-Monje, David, Walaa K. Mousa, George Lazarovits, and Manish N. Raizada. "Impact of swapping soils on the endophytic bacterial communities of pre-domesticated, ancient and modern maize." *BMC Plant Biology* 14, no. 1 (2014): 1-19.

Khalaf, Eman M., and Manish N. Raizada. "Bacterial seed endophytes of domesticated cucurbits antagonize fungal and oomycete pathogens including powdery mildew." *Frontiers in Microbiology* 9 (2018): 42.

Khan, Abdur Rahim, Ihsan Ullah, Muhammad Waqas, Raheem Shahzad, Sung Jun Hong, Gun Seok Park, Byung Kwon Jung, In Jung Lee, and Jae Ho Shin. "Plant growth-promoting potential of endophytic fungi isolated from *Solanum nigrum* leaves." *World Journal of Microbiology and Biotechnology* 31, no. 9 (2015): 1461-1466.

Krause, Andrea, Adarsh Ramakumar, Daniela Bartels, Federico Battistoni, Thomas Bekel, Jens Boch, Melanie Böhm, et al. "Complete genome of the mutualistic, N2-fixing grass endophyte *Azoarcus* Sp. strain BH72." *Nature Biotechnology* 24, no. 11 (2006): 1384-1390.

Kruasuwan, Worarat, and Arinthip Thamchaipenet. "Diversity of culturable plant growth-promoting bacterial endophytes associated with sugarcane roots and their effect of growth by co-inoculation of diazotrophs and actinomycetes." *Journal of Plant Growth Regulation* 35, no. 4 (2016): 1074-1087.

Li, Bai, Yu Ying Li, Hua Mao Wu, Fang Fang Zhang, Chun Jie Li, Xue Xian Li, Hans Lambers, and Long Li. "Root exudates drive interspecific facilitation by enhancing nodulation and N2 fixation." *Proceedings of the National Academy of Sciences of the United States of America* 113, no. 23 (2016): 6496-6501.

Liu, B, Hongping Qiao, Lili Huang, Heinrich Buchenauer, Qingmei Han, Zhensheng Kang, and Yufei Gong. "Biological control of take-all in wheat by endophytic *Bacillus subtilis* E1R-j and potential mode of action." *Biological Control* 49, no. 3 (2009): 277-285.

Liu, Xiaoguang, Jinli Jia, Roman Popat, Catherine A. Ortori, Jun Li, Stephen P. Diggle, Kexiang Gao, and Miguel Cámara. "Characterisation of two quorum-sensing systems in the endophytic *Serratia plymuthica* strain g3: differential control of motility and biofilm formation according to lifestyle." *BMC Microbiology* 11, no. 1 (2011): 1-12.

Liu, Yunran, Lakshmi Narayanan Ponpandian, Hoki Kim, Junhyun Jeon, Buyng Su Hwang, Sun Keun Lee, Soo Chul Park, and Hanhong Bae. "Distribution and diversity of bacterial endophytes from four *Pinus* species and their efficacy as biocontrol agents for devastating pinewood nematodes." *Scientific Reports* 9, no. 1 (2019): 1-12.

Lòpez-Fernàndez, Sebastiàn, Stéphane Compant, Urska Vrhovsek, Pier Luigi Bianchedi, Angela Sessitsch, Ilaria Pertot, and Andrea Campisano. "Grapevine colonization by endophytic bacteria shifts secondary metabolism and suggests activation of defense pathways." *Plant and Soil* 405, no. 1 (2016): 155-175.

Madhaiyan, Munusamy, Tan Hian Hwee Alex, Si Te Ngoh, Bharath Prithiviraj, and Lianghui Ji. 2015. "Leaf-residing *Methylobacterium* species fix nitrogen and promote biomass and seed production in *Jatropha curcas*." *Biotechnology for Biofuels* 8, no. 1 (2015): 1-14.

Mahmood, Ahmad, and Ryota Kataoka. "Metabolite profiling reveals a complex response of plants to the application of plant growth-promoting endophytic bacteria." *Microbiological Research* 234 (2020): 126421.

Misk, Azza, and Christopher Franco. "Biocontrol of chickpea root rot using endophytic actinobacteria." *BioControl* 56, no. 5 (2011): 811-822.

Overbeek, Leo Van, and Jan Dirk Van Elsas. "Effects of plant genotype and growth stage on the structure of bacterial communities associated with potato (*Solanum tuberosum* L.)." *FEMS Microbiology Ecology*. 64, no. 2 (2008): 283-296.

Padda, Kiran Preet, Akshit Puri, and Chris Chanway. "Endophytic nitrogen fixation - a possible 'hidden' source of nitrogen for lodgepole pine trees growing at unreclaimed gravel mining sites." *FEMS Microbiology Ecology* 95, no. 11 (2019): fiz172.

Patel, Janki K., Rupal Agrawal, and Radhika Sidhdhapara. "Root associated bacterial endophytes from Poaceae plants: identification, characterization and plant growth promotion." *Journal of Microbiology, Biotechnology and Food Sciences* 10, no. 3 (2020): 478-483.

Patel, Janki K., and G. Archana. "Diverse culturable diazotrophic endophytic bacteria from *Poaceae* plants show cross-colonization and plant growth promotion in wheat." *Plant and Soil* 417, no. 1 (2017): 99-116.

Patel, Janki K., and G. Archana. "Engineered production of 2,4-diacetylphloroglucinol in the diazotrophic endophytic bacterium *Pseudomonas* sp. WS5 and its beneficial effect in multiple plant-pathogen systems." *Applied Soil Ecology* 24 (2018): 34-44.

Patel, Janki K., Sheeba Madaan, and G. Archana. "Antibiotic producing endophytic *streptomyces* spp. colonize above-ground plant parts and promote shoot growth in multiple healthy and pathogen-challenged cereal crops." *Microbiological Research* 24 (2018): 34-44.

Patel, Janki K., Krupa Gohel, Hiral Patel, and Tanvi Solanki. "Wheat growth-dependent succession of culturable endophytic bacteria and their plant growth-promoting traits." *Current microbiology* 78, no. 12 (2021): 4103-4114.

Pinski, Artur, Alexander Betekhtin, Katarzyna Hupert-Kocurek, Luis A. J. Mur, and Robert Hasterok. "Defining the genetic basis of plant–endophytic bacteria interactions." *International Journal of Molecular Sciences* 20, no. 8 (2019): 1947.

Rajini, Sollepura B., Murali Nandhini, Arkere C. Udayashankar, Siddapura R. Niranjana, Ole S. Lund, and Harischandra S. Prakash. "Diversity, plant growth-promoting traits, and biocontrol potential of fungal endophytes of *Sorghum bicolor*." *Plant Pathology* 69, no. 4 (2020): 642-654.

Ramesh, R., A. A. Joshi, and M. P. Ghanekar. "Pseudomonads: major antagonistic endophytic bacteria to suppress bacterial wilt pathogen, *Ralstonia solanacearum* in the eggplant (*Solanum melongena* L.)." *World Journal of Microbiology and Biotechnology* 25, no. 1 (2009): 47-55.

Rana, Kusam Lata, Divjot Kour, Tanvir Kaur, Imran Sheikh, Ajar Nath Yadav, Vinod Kumar, Archna Suman, and Harcharan Singh Dhaliwal. "Endophytic microbes from diverse wheat genotypes and their potential biotechnological applications in plant growth promotion and nutrient uptake." *Proceedings of the National Academy of Sciences India Section B - Biological Sciences* 90, no. 5 (2020): 969-979.

Remans, Roseline, Lara Ramaekers, Sarah Schelkens, German Hernandez, Aurelio Garcia, Jorge Luis Reyes, Nancy Mendez, et al. "Effect of *Rhizobium-Azospirillum* co-inoculation on nitrogen fixation and yield of two contrasting *Phaseolus vulgaris* L. genotypes cultivated across different environments in Cuba." In *Plant and Soil* 3 312, no. 1 (2008): 25-37.

Ren, Cheng Gang, and Chuan Chao Dai. "Jasmonic acid is involved in the signaling pathway for fungal endophyte-induced volatile oil accumulation of *Atractylodes lancea* plantlets." *BMC Plant Biology* 12, no. 1 (2012): 1-11.

Ren, Gaidi, Huayong Zhang, Xiangui Lin, Jianguo Zhu, and Zhongjun Jia. "Response of leaf endophytic bacterial community to elevated CO_2 at different growth stages of rice plant." *Frontiers in Microbiology* 6 (2015): 855.

Ribeiro, Vitória Palhares, Ivanildo Evódio Marriel, Sylvia Morais de Sousa, Ubiraci Gomes de Paula Lana, Bianca Braz Mattos, Christiane Abreu de Oliveira, and Eliane Aparecida Gomes. "Endophytic *Bacillus* strains enhance pearl millet growth and nutrient uptake under low-P." *Brazilian Journal of Microbiology* 49 (2018): 40-46.

Ripa, Farhana Alam, Wei Dong Cao, Shuai Tong, and Jian Guang Sun. "Assessment of plant growth promoting and abiotic stress tolerance properties of wheat endophytic fungi." *BioMed Research International* 2019 (2019).

Rojas-Solís, Daniel, Elizabeth Zetter-Salmón, Miguel Contreras-Pérez, Ma del Carmen Rocha-Granados, Lourdes Macías-Rodríguez, and Gustavo Santoyo. "*Pseudomonas stutzeri* E25 and *Stenotrophomonas maltophilia* CR71 endophytes produce antifungal volatile organic compounds and exhibit additive plant growth-promoting effects." *Biocatalysis and Agricultural Biotechnology* 13 (2018): 46-52.

Rosenblueth, Mónica, Ernesto Ormeño-Orrillo, Aline López-López, Marco A. Rogel, Blanca Jazmín Reyes-Hernández, Julio C. Martínez-Romero, Pallavolu M. Reddy, and Esperanza Martínez-Romero. "Nitrogen fixation in cereals." *Frontiers in Microbiology* 9 (2018): 1794.

Ryu, Choong Min, Mohamed A. Farag, Chia Hui Hu, Munagala S. Reddy, Joseph W. Kloepper, and Paul W. Paré. "Bacterial volatiles induce systemic resistance in *Arabidopsis*." *Plant Physiology* 134, no. 3 (2004): 1017-1026.

Saad, Mona M. G., Rehab Y. Ghareeb, and Ali A. Saeed. 2019. "The potential of endophytic fungi as biocontrol agents against the cotton leafworm, *Spodoptera littoralis* (Boisd.) (Lepidoptera: Noctuidae)." *Egyptian Journal of Biological Pest Control* 29, no. 1 (2019): 1-7.

Sen, Tanuka, Colin J. Barrow, and Sunil Kumar Deshmukh. "Microbial pigments in the food industry-challenges and the way forward." *Frontiers in Nutrition.* (2019): 7.

Shabanamol, S., K. Divya, Tijith K. George, K. S. Rishad, T. S. Sreekumar, and M. S. Jisha "Characterization and in planta nitrogen fixation of plant growth-promoting endophytic diazotrophic *Lysinibacillus sphaericus* isolated from rice (*Oryza Sativa*)." *Physiological and Molecular Plant Pathology* 102 (2018): 46-54.

Shah, Sujit, Roshani Shrestha, Sabitri Maharjan, Marc Andre Selosse, and Bijaya Pant. "Isolation and characterization of plant growth-promoting endophytic fungi from the roots of *Dendrobium moniliforme*." *Plants* 8, no. 1 (2018): 5.

Shaik, Sadiq Pasha, and Pious Thomas. "*In vitro* activation of seed-transmitted cultivation-recalcitrant endophytic bacteria in tomato and host–endophyte mutualism." *Microorganisms* 7, no. 5 (2019): 132.

Shimizu, Masafumi, Sachiko Yazawa, and Yusuke Ushijima. "A promising strain of endophytic *Streptomyces* sp. for biological control of cucumber anthracnose." *Journal of General Plant Pathology* 75, no. 1 (2009): 27-36.

Silva, C. F., L. C. Vitorino, M. A. C. Mendonça, W. L. Araújo, M. N. Dourado, L. C. Albuquerque, M. A. Soares, and E. L. Souchie. "Screening of plant growth-promoting endophytic bacteria from the roots of the medicinal plant *Aloe vera*." *South African Journal of Botany* 134 (2020): 3-16.

Silva, DAF da, SR Cotta, RE Vollú, D de Azevedo Jurelevicius, JM Marques, IE Marriel, and L Seldin. "Endophytic microbial community in two transgenic maize genotypes and in their near-isogenic non-transgenic maize genotype." *BMC Microbiology* 14, no. 1 (2014): 1-9.

Soundar Raju, Chellappan, Abubakker Aslam, Devarajan Thangadurai, Jeyabalan Sangeetha, Krishnan Kathiravan, and Appakan Shajahan. "Indole acetic acid (IAA) producing endophytic bacteria on direct somatic embryogenesis and plant regeneration of *Exacum travancoricum Bedd.*" *Vegetos* 33, no. 4 (2020): 690-702.

Spaepen, Stijn, Jos Vanderleyden, and Roseline Remans. 2007. "Indole-3-acetic acid in microbial and microorganism-plant signaling." *FEMS Microbiology Reviews* 31, no. 4 (2007): 425-448.

Stein, Elke, Alexandra Molitor, Karl-Heinz Kogel, and Frank Waller. "Systemic resistance in *Arabidopsis* conferred by the mycorrhizal fungus *Piriformospora indica* requires jasmonic acid signaling and the cytoplasmic function of NPR1." *Plant and Cell Physiology* 4 49, no. 11 (2008): 1747-1751.

Sturz, A. V., B. R. Christie, and J. Nowak. "Bacterial endophytes: potential role in developing sustainable systems of crop production." *Critical Reviews in Plant Sciences* 19, no. 1 (2000): 1-30.

Suárez-Moreno, Zulma Rocí, Jesús Caballero-Mellado, and Vittorio Venturi. "The new group of non-pathogenic plant-associated nitrogen-fixing *Burkholderia* spp. shares a conserved quorum-sensing system, which is tightly regulated by the RsaL repressor." *Microbiology* 154, no. 7 (2008): 2048-2059.

Suebrasri, Thanapat, Hiroyuki Harada, Sanun Jogloy, Jindarat Ekprasert, and Sophon Boonlue. "Auxin-producing fungal endophytes promote the growth of sunchoke." *Rhizosphere* 16 (2020): 100271.

Sun, Le Ni, Yan Feng Zhang, Lin Yan He, Zhao Jin Chen, Qing Ya Wang, Meng Qian, and Xia Fang Sheng. "Genetic diversity and characterization of heavy metal-resistant-endophytic bacteria from two copper-tolerant plant species on copper mine wasteland." *Bioresource Technology* 101, no. 2 (2010): 501-509.

Sundaramoorthy, S., T. Raguchander, N. Ragupathi, and R. Samiyappan. "Combinatorial effect of endophytic and plant growth promoting rhizobacteria against wilt disease of *Capsicum annum* L. caused by *Fusarium solani*." *Biological Control* 60, no. 1 (2012): 59-67.

Tiwari, Pragya, Hanhong Bae. "Horizontal Gene Transfer and Endophytes: An implication for the acquisition of novel traits." *Plants* 2020, 9, 305.

Tiwari, Pragya, Yashdeep Srivastava, and Hanhong Bae. "Trends of pharmaceutical design of Endophytes as anti-infective." *Current Topics in Medicinal Chemistry*, 21, no. 17 (2021): 1572-1586.

Tiwari, Pragya, Hanhong Bae. "Endophytic Fungi: Key insights, emerging prospects, and challenges in natural product drug discovery." *Microorganisms* 10, 360 (2022). https://doi.org/10.3390/microorganisms10020360.

Truyens, Sascha, Nele Weyens, Ann Cuypers, and Jaco Vangronsveld. "Bacterial seed endophytes: genera, vertical transmission and interaction with plants." *Environmental Microbiology Reports* 7, no. 1 (2015): 40-50.

Turbat, Adiyadolgor, Dávid Rakk, Aruna Vigneshwari, Sándor Kocsubé, Huynh Thu, Ágnes Szepesi, László Bakacsy, et al. "Characterization of the plant growth-promoting activities of endophytic fungi isolated from *Sophora flavescens*." *Microorganisms* 8, no. 5 (2020): 683.

Tyc, Olaf, Rocky Putra, Rieta Gols, Jeffrey A. Harvey, and Paolina Garbeva. "The ecological role of bacterial seed endophytes associated with wild cabbage in the United Kingdom." *MicrobiologyOpen* 9, no. 1 (2020): e00954.

Verma, V. C., S. K. Singh, and Satya Prakash. 2011. "Bio-control and plant growth Promotion potential of siderophore producing endophytic *Streptomyces* from *Azadirachta indica* A. Juss." *Journal of Basic Microbiology* 51 (5): 550–56. https://doi.org/10.1002/jobm.201000155.

Walitang, Denver I., Kiyoon Kim, Munusamy Madhaiyan, Young Kee Kim, Yeongyeong Kang, and Tongmin Sa. "Characterizing endophytic competence and plant growth promotion of bacterial endophytes inhabiting the seed endosphere of rice." *BMC Microbiology* 17, no. 1 (2017): 1-13.

Wang, Yanqing, Yumiko Ohara, Hitoshi Nakayashiki, Yukio Tosa, and Shigeyuki Mayama. "Microarray analysis of the gene expression profile induced by the endophytic plant growth-promoting rhizobacteria, *Pseudomonas fluorescens* FPT9601-T5 in *Arabidopsis*." *Molecular Plant-Microbe Interactions* 18, no. 5 (2005): 385-396.

Ward, Mary H. 2009. "Too much of a good thing? Nitrate from nitrogen fertilizers and cancer." *Reviews on Environmental Health* 4, no. 4 (2009): 357-363.

White, James F., Holly Crawford, Mónica S. Torres, Robert Mattera, Ivelisse Irizarry, and Marshall Bergen. 2012. "A proposed mechanism for nitrogen acquisition by grass seedlings through oxidation of symbiotic bacteria." *Symbiosis* 57 (3): 161–71. https://doi.org/10.1007/s13199-012-0189-8.

Xia, Ye, Seth DeBolt, Jamin Dreyer, Delia Scott, and Mark A. Williams. "Characterization of culturable bacterial endophytes and their capacity to promote plant growth from plants grown using organic or conventional practices." *Frontiers in Plant Science* 6 (2015): 490.

Xia, Ye, Mohammad Radhi Sahib, Amna Amna, Stephen Obol Opiyo, Zhenzhen Zhao, and Yu Gary Gao. "Culturable endophytic fungal communities associated with plants in organic and conventional farming systems and their effects on plant growth." *Scientific Reports* 9, no. 1 (2019): 1-10.

Yi, Hwe Su, Jung Wook Yang, and Choong Min Ryu. "ISR meets SAR outside: additive action of the endophyte *Bacillus pumilus* INR7 and the chemical inducer, benzothiadiazole, on induced resistance against a bacterial spot in field-grown pepper." *Frontiers in Plant Science* 4 (2013): 122.

Yu, Xianmei, Chengxiang Ai, Li Xin, and Guangfang Zhou. "The siderophore-producing bacterium, *Bacillus subtilis* CAS15, has a biocontrol effect on fusarium wilt and promotes the growth of pepper." *European Journal of Soil Biology* 47, no. 2 (2011): 138-145.

Yuan, Jie, Kai Sun, Meng Yao Deng-Wang, and Chuan Chao Dai. "The mechanism of ethylene signaling induced by endophytic fungus *Gilmaniella* Sp. AL12 mediating sesquiterpenoids biosynthesis in *Atractylodes Lancea*." *Frontiers in Plant Science* 7 (2016): 361.

Zhou, Zuoping, Changfei Zhang, Wenna Zhou, Wei Li, Long Chu, Jinping Yan, and Haiyan Li. "Diversity and plant growth-promoting ability of endophytic fungi from the five flower plant species collected from Yunnan, Southwest China." *Journal of Plant Interactions* 9, no. 1 (2014): 585-591.

Zúñiga, Ana, María Josefina Poupin, Raúl Donoso, Thomas Ledger, Nicolás Guiliani, Rodrigo A. Gutiérrez, and Bernardo González. "Quorum sensing and indole-3-acetic acid degradation play a role in colonization and plant growth promotion of *Arabidopsis thaliana* by *Burkholderia phytofirmans* PsJN." *Molecular Plant-Microbe Interactions* 26 (5): 546–53. https://doi.org/10.1094/MPMI-10-12-0241-R.

Chapter 5

Some Significant Case Studies on Plant-Associated Endophytes

Avnish Kumar[1,*], Monika Asthana[1] and Pragya Tiwari[2]

[1]Department of Biotechnology, School of Life Sciences,
Dr. Bhimrao Ambedkar University, Agra
[2]Department of Biotechnology, Yeungnam University, Gyeongsan, Gyeongbuk, Republic of Korea

Abstract

Plant-associated endophytes are gaining huge recognition among the scientific community, owing to their emerging prospects in diverse socio-economic applications in the present era. The microorganisms form symbiotic associations with the plants and complete partial or complete life cycles within the host plant. The endophyte population is greatly affected by climatic conditions and the geographical location of the host plant. This symbiotic association has a beneficial impact on both the associated biological organisms. The present chapter discusses an overview of endophytes, their colonization within plants, classification, and the future perspectives in endophyte biology.

Keywords: Classification, Colonization, Endophyte Biology, Plant-Microbe Interactions, Symbiosis

---

*Corresponding Author's Email: akbiotech81@gmail.com.

In: Endophytes
Editor: Pragya Tiwari
ISBN: 979-8-88697-045-6

1. Introduction

Endophytic/endo-symbiotic non-pathogenic microbes (usually fungi or bacteria) are present asymptomatically in the tissues of all plants. Fungal endophytes of grasses perform a defensive mutualism between plants and fungi. Endophytes are microorganisms (bacteria, fungi, or actinomycetes) that have symbiotic inhabitation within healthy plant tissues (Tiwari and Bae 2020; Tiwari et al. 2021). It is considered that endophytes could be isolated from almost all plant species. Mostly, endophytes are from the genera *Enterobacter* sp., *Colletotrichum* sp., *Phomopsis* sp., *Phyllosticta* sp., *Cladosporium* sp., etc. Endophytes when colonizing plant tissues secrete chemicals useful for plant growth, protection, and stability. Some of the secreted compounds are useful to protect plants from herbivorous and as biocontrol agents against plant pathogens. Similarly, many of their products are also of economic importance to humans. These products have importance in the pharmaceutical, food industries, and other industries or research purposes (Tiwari et al. 2021; Tiwari et al. 2022). Endophytic secretions play an important role in nutrient cycling, biodegradation, and bioremediation too. Endophytes are found in the roots, stems, leaves, petioles, inflorescences, fruits, buds, seeds, and other living and dead cells of plants and weeds. The endophyte population is greatly affected by climatic conditions and the location where the host plant grows.

2. Classification of Endophytes

Endophytes are classified based on their cell type as prokaryotic (bacteria and actinomycetes) or eukaryotic (fungi) or based on their source. Fungal endophytes include endophytes of dicots, lichens, bryophytes, ferns as well as fungal endophytes of tree bark, xylem, and roots. According to various studies, fungal endophytes enter the plant through wounds or openings under the interaction of the environment and plants, as well as their hydrolytic enzymes damage the plant cell and make a way inside. In some plant species, bacterial endophytes move from root tissue to reach up to the xylem. During the development of endophytes and host plants, endophyte strains have adapted themselves to extreme environmental conditions and may undergo genetic variations that lead to their ability to synthesize certain phytochemicals. Fungal endophytes are highly diverse in their nature and their presence in plants is dependent on the concentration of plant nutrients, environmental factors, and other microorganisms found in the plant. Some endophytes are found in single histological parts and some endophytes are found in several parts of plants.

3. Endophytes and Their Mechanism of Action

The isolated endophytes from most plants have been employed as producers of natural products in industrial and agriculture setups. In general, they thrive best under a temperature and neutral pH. In the past few decades, researchers have focused on the possibilities and advances in the production of microbial enzymes. For example, toxic chemicals are inactivated by the enzymes of endophytes. Plant microorganisms, whether beneficial or pathogenic, have similar capabilities to survive in the host plant such as their rhizosphere potential, their mobility and penetration into the host plant, mechanisms for entry and propagation inside the plant, and the ability to evade plant immunity. Successful colonization of endophytes in the host plant is influenced by various factors like plant tissue type, plant genotype, microbial taxon and stress conditions, and other biotic and abiotic environmental conditions.

Despite having similar survival steps for endophytes and pathogens, a plant uses a different type of defense mechanism for endophytes than for pathogenic microorganisms. Before colonization, pathogens must survive in the oxidative environment within the host plant. For example, *Enterobacter* sp. produces superoxide dismutase, catalase, and hydroperoxide reductase to counter oxidative stress during the colonization of poplar plants *Populus trichocarpa*. A comparative study of the survival of pathogenic microorganisms of rice plants *Xanthomonas oryzae* pv. pxO99 and the endophyte *Azoarcus olearius* showed that both have differentially expressed genes (DEGs) related to the plant's jasmonate (JA) signaling pathway. It is consistently activated by beneficial endophytes, in contrast to the salicylate (SA) pathway that is only pathogen-infected. Jasmonate is predicted to perform a controlling mechanism for the increase in the number of *Azorcus* endophytes in roots (Mengistu 2020).

Plant pathogens contain enzymes that can degrade the plant cell wall. Similarly, cellulase/hemicellulase and glycoside hydrolase enzymes are found in non-phytopathogens as well. In some endophytes, this enzyme potentially renders the diversity of sugar utilization pathways as one of the factors that enabled the development of the endophyte. The survival of endophytes within the host plant is influenced by various genes with conserved transcriptional regulators. The presence of regulatory proteins is primarily involved in controlling enzyme activity, the most likely function of which occurs when endophytes encounter the host at the time of infection (Ali et al. 2014). The secretion of regulatory proteins defines the plant-microbe interaction. The transport of effector proteins of the bacteria causes the parasite to grow by suppressing the immune system of the host, while the immune system of the plant also recognizes the effector proteins and stimulates the immune response. In contrast, some genes are missing or incomplete in the genome of endophytes. Therefore, it is likely that endophytes adopt symbiotic nature instead of performing their pathogenic action. For example, due to mutation in the T3SS gene *Salmonella enterica* possesses endophytic colonization in *Medicago truncatula.* Jasmonic acid (JA), ethylene (ET), and salicylic acid (SA) signaling are essential for endophyte-mediated induced resistance. In contrast, endophytes can induce disease resistance through the activation of amino acid biosynthesis. Successful endophytes bring many benefits to the plant through their behavior and colonization patterns. Due to these endophytes, plants grow under stress. Endophytes develop a barrier for plant parasites and keep plants healthy. Endophytes help the plants to improve their intake of various minerals, which also kept the plant healthy and increase the crop yield (Mengistu 2020).

As soon as the endophyte meets the plant, communication starts between the endophyte and plants, which manages plant growth under adverse conditions of drought, salinity, temperature, and heavy metal stress through various mechanisms. These mechanisms can be explained by understanding their molecular events. It has been observed that several bio-metabolic pathways or pathogenesis-related gene expression pathways of endophytes and pathogens are executed only when they entered the host. The research on the mechanism behind this is in its initial stages worldwide. The concentration of the plant's free amino acids, homoserine, and asparagine, act as host signals to activate the expression of a lethal gene in virulent strains of *Hematococa* that is expressed only *in planta.* Furthermore, expression of the gene cluster for lolitrem biogenesis in the endophyte is higher in the resident *Neotyphodium lolii* plant in perennial ryegrass, but at least undetectable in fungal cultures grown *in vitro*, lending support to the notion that plant signaling can have been suppressed (Yang et al. 2005). Another example of this symbiotic relationship is the relationship between dicotyledonous plants (Convolvulaceae) and clavipitaceous fungi. Molecular studies based on the 18S rDNA gene and internal transcribed spacer (ITS) have shown that the ergoline alkaloid formation belongs to the epibiotic fungal family

Clavicipitaceae (Ascomycota), whereas no fungal endophyte does so. Fungi associated with this plant and some other unknown fungi are found on various Convolvulaceae plants, which are responsible for the isolated occurrence of ergoline alkaloids in Convolvulaceae (Steiner et al. 2006). Various such citations suggest that there is horizontal gene transfer between the plant and the endophyte, by which the endophytes produce the compounds of the respective plant (Tiwari and Bae 2020).

Endophyte enhances the level of secondary metabolites, which are alkaloids, benzopyrones, chinones, flavonoids, phenolic acids, quinones, steroids, terpenoids, tetralons, xanthones, etc. Additionally, they enhance host growth by promoting nutrient acquisition and protection against abiotic stress. They control pests as well as potentially provide protection to plants from parasitic nematodes and plant disease pathogenesis.

4. Application of Endophytes

The largest group of endophytes, fungal endophytes, produce chemical toxins to protect plants from insects, and nematodes. Other important products are extracellular enzymes like cellulase, proteinase, lipase, esterase, etc. for degradation of dead soil biomass, and solubilizing insoluble phosphates. They are capable to enhance the production of plant growth-promoting hormones including auxins, cytokinins, and gibberellins to promote plant growth. Endophytes spend part of their life span or life cycle within plant tissues. Moreover, the endophytes contribute to the formation of secondary metabolites of the plant.

4.1. Biodegradation of Plastics

Endophytes have been found to be a potent biodegradable agent to degrade plastics. Further, their potential role in plastic biodegradation and management of plant diseases needs to be investigated. For this, it is necessary to identify such species of endophyte first in the ongoing research. To overcome the problem fungal endophytes can be applied in two formats one for the production of biodegradable plastics and another manipulating the fungal genetic system to produce enzymes that can break down plastic material into simpler compounds.

Russell et al. 2011, demonstrated the uses of endophytic fungi for the degradation of synthetic polymer polyester polyurethane (PUR) in both solid and liquid suspensions. The major group of endophytes belongs to the genus Pestalotiopsis, although it was not a universal feature of this genus. Two *Pestalotiopsis microspora* isolates were shown to have the potential to utilize PUR as a carbon source under aerobic as well as anaerobic conditions. The PUR degradation is done by enduophyte with the help of enzyme serine hydrolase.

4.2. Insecticides

Epichloe spp. and *Neotyphodium* spp. may also provide benefits to grasses for insect resistance, increased drought tolerance, and disease suppression. Endophyte contains peramine that appears to prevention to stem weevil (*Listronotus bonariensis*) during the feed. Peramine does not control any pasture insect pests other than Argentine stem weevil (Gaynor and Rowan 1986).

4.3. Feeding Deterrence

Sleepygrass (*Acantherum robustum*) found in the western United States has an endophyte that produces lysergic acid amide. If an animal consumes infected grass, then it keeps them sleeping a little longer. After toxic effects are experienced, animals avoid grazing in areas with sleepy grass including plants in the area. Faeth et al. (2006) reported in a study that the endophyte (*Neotyphodium*) infection frequency of sleepy grass varies from 50 to 100% in various geographic locations and toxic alkaloid production varies from 0 to >150 μg G-1 tissue. Other symbiotics help to deter feeding by grazers via enhancing silicon uptake and increase in foliar levels of phosphorous including induction of secondary metabolites (phenolic compounds, like flavonols isorhamnetin-diglycoside, and rhamnetin derivative).

4.4. Increased Drought Tolerance

Hereme et al. (2020) studied the role of endophytes in plants *Colobanthus quitensis* native to Antarctica. The endophyte reduces oxidative stress and increases the osmoprotective molecules (sugar production, proline) and better relevant gene expression with control over the stomatal opening, photosynthesis, and stoma conduction.

4.5. Increased Heavy Metal Tolerance

Based on the meta-analysis of published studies, Franco-Franklin et al. (2021), has mentioned that bacterial consortia support the presence of bacterial endophyte to reduce heavy metal stress in a variety of hosts. Bacterial endophytes promote the biomass production of host plants under the influence of various heavy metals, thereby playing an important role in effectively reducing the toxicity of various heavy metals. Stress mitigation by bacteria was similar among different plant groups except for non-accumulating plants that benefit most from the symbiotic association.

4.6. Increased Salt Stress Tolerance

Fan et al. (2020) studied ninety-eight endophytes in *Arabidopsis thaliana* under salt stress and without salt stress conditions. *Rhizobacteria* are best among all 98-plant growth-promoting symbionts and their presence is effective in reducing plant salt stress. The action of salt stress derives from regulating osmolytes and antioxidant enzymes. By using biotechnological approaches, one may screen stress tolerance enhancing endophytes may be screened to employed to enhance the capacity of crops to tolerate stresses of marginal lands. Redman et al. (2002) demonstrated endophyte *Curvularia protuberata* generate thermal tolerance in the plants (*Dichanthelium lanuginosum*) to survive in the geothermal soils of Yellowstone.

5. Study of Invasive *Phragmites australis*

Phragmites are wetland plants and play an important role in water purification. These are a natural alternative to wastewater treatment. Despite this work, there is incomplete knowledge of the basic structure of the microbiome at the core of phragmites and the mutual functions of each member of its microbiome. This microbiome could be behind preventing pollution. *P.*

australis contains fungal endosphere communities enriched with a highly diverse range of pathogens. *P. australis* grows plant pathogens on its own tissues, yet there is no evidence of pathogen spread in plant tissues or changes in endosphere structure. An in-depth study of this fact may enrich the understanding of invasion and explain the relevance of widely accepted invasion mechanisms to the invasive ecosystems of *P. australis* (Soares et al. 2016).

Fungal endophytes have been reported for rice plant growth promotion under higher salt regions compared to the low salt region. The plants having fungi showed better growth in NaCl 120 mM concentration. It is supported by the hypothesized mechanisms of increased salt/drought tolerance given by Rodriguez et al. (2008). *Phragmites* genetic makeup and relevant environmental factors like salinity could manipulate the unique community structure and function of the plant. Based on the present knowledge of endophytes, the microbiome in the root of *Phragmites* can be employed with or without using biotechnology to iron-manganese plaques built up in wetlands to help boost their water purification performance (Soares et al. 2016).

5.1. Increase in Oxidative Stress Tolerance

Endophytes are associated with the up-regulation of antioxidants and aquaporin genes in plants and thus can reduce damage in plant tissues due to oxidative stress due to salts including Na^+ ions. The endophyte may enhance NO in plant tissue at the onset of high salt intake. Alnusairi et al. (2021) has shown that exogenous-sourced NO is able to regulate the antioxidant system and thereby limit the NaCl-mediated oxidative damage on membranes. Antioxidant enzymes played an important role in proliferating and increasing photosynthetic efficiency while preserving the structural and functional integrity of the membrane. There are significant effects of NO application on the decrease in harmful Na+ ions and absorption of key mineral elements such as nitrogen (N), potassium (K), and calcium (Ca). Greater K and reduced Na uptake in NO-treated plants lead to a considerable decline in the Na/K ratio. Enhancing of salt tolerance by NO was concomitant with an obvious down-regulation in the relative expression of SOS1, NHX1, AQP, and OSM-34, while D2-protein was up-regulated. Over exposure to salinity may trigger the osmolytes build up in cells for the safety of cellular functioning. Physiologically, Osmo protectants can reduce the cellular osmotic potential, and thereafter a drop in the hydraulic conductivity of the membranes occurs, maybe by decreasing the number of water channels (aquaporins). NO have supposed to be having a regulatory effect on the aquaporin gene and expression of corresponding proteins, mainly proline, and a decrease in its catabolizing enzymes.

6. Microbe-Mediated Nitrogen Scavenging *Phragmites australis* Studies

M. oxydans may cause bacterial infections in human beings. In plants, it can colonize shoots and roots of *Phragmites* as an endophyte. Soares et al. (2016) exposed the plants to ^{15}N-labeled gasses for 10 days. The plants *P. australis* with *M. oxydans* grow significantly even under low soil nitrogen conditions. Soares et al. (2016) further mentioned that bacteria improve their growth by increasing the scavenging of nitrogenous compounds. Further, it was also proposed that the presence of endophyte *M. oxydans* rhizoplane of root hairs enhances microbe-mediated nitrogen scavenging and growth promotion efficiency. Plants exposed to *Microbacterium oxydans* showed double growth than plants not exposed to bacteria. *Microbacterium oxydans* B1 (from *Phragmites*) stimulates root hair development in sterile Bermuda grass seedlings and does not enter the root cells themselves, only colonizing the rhizoplane. The bacteria that colonize in rhizoplanes do not elicit a strong reactive oxygen response in the root hairs of the grass.

7. Bacterial Endophytes Inhibit the Fungal Pathogens

Endophytic microbes are found in a limited area of roots, stems, or leaves. Few endophytes occupy only above-ground plant parts, whereas others are restricted to subterranean organs and tissues. Normally, endophytic microbes are found within the plant in the biotrophic mode and depend on plants only for their nutrition. As a result, many endophytic microbes form obligate associations with plants, most often inside the plant host. Sometimes they establish their relationship with plants even living outside of the host (e.g., arbuscular mycorrhizal fungi or AMF).

Fusarium, *Pythium,* and other water molds cause seed rots during plant germination. *Fusarium* blights are very common and significantly affect seedling development. Verma et al. (2018) demonstrated the protective nature of few bacteria against fungal pathogens using microscopy and fungal inhibition tests. *Pseudomonas* spp. (SLB4- *P. fluorescens*, SLB6-*Pseudomonas* sp., and SY1-*Pseudomonas* sp.) were recognized for promoting seedling development, root and shoot growth, and including stimulation of root hair formation. These bacteria increased phosphate solubilization and production of antifungal volatiles. *Pseudomonas* sp. (SY1) significantly works against *Fusarium, Curvularia* sp. and *Alternaria* sp. in comparing *Pseudomonas* treated plants with non-treated plants. Molecular study for detection of genes associated with pyrrolnitrin synthesis and HCN (hydrogen cyanide) production also recognized corresponding genes in SY1. The dollar spot disease, an economically important disease of *Agrostis stolonifera* (creeping bentgrass relative of *Zea mays*) is caused by the fungus *Sclerotinia homoeocarpa. Sclerotinia homoeocarpa* enters the plant through wounds or other natural openings like stomata. Shehata et al. (2016) experimentally showed that bacterial endophytes *Burkholderia gladioli* endophytes strains could inhibit this fungal pathogen. *B. gladioli* and forty other species of genus *Burkholderia* endophytes could produce antifungal chemicals like cyclic terpene (volatile organic compound), chitinase phenazine, lipopeptides, quinolinones, altericidins, pyrrolnitrin, cepacidines, and siderophores (Vial et al. 2007; Schmidt et al. 2009; Elshafie et al. 2012; Ettinger et al. 2015). Elshafie et al. (2012) had mentioned *B. gladioli* for its antifungal nature against *Botrytis cinerea, Aspergillus flavus, Aspergillus niger, Penicillium digitatum, Penicillium expansum, Sclerotinia sclerotiorum, Rhizoctonia solani, Phytophthora capsici,* and *Phytophthora cactorum.*

The potentially screened endophytes could be used to make effective spray formulations against fungal pathogens. Shehata et al. (2016) suggested an initial formula using qPCR and mentioned that endophyte strains may colonize in shoots or roots to work effectively. Tian et al. (2017), proved the secretion of antifungal protein by *E. festucae* that is highly expressed in the infected plant tissues and may therefore be involved in the disease resistance. *E. festucae* antifungal protein, designated *Efe*-AfpA was not produced by the plant *Epichloë* spp. as it does not contain the gene for corresponding antifungal protein. The antifungal protein was partially purified from the apoplastic proteins of endophyte-infected plant tissue. Further, recombinantly expressed protein expressed in the yeast *Pichia pastoris* had similar antifungal activity against plant pathogen *Sclerotinia homoeocarpa. Efe*-AfpA may therefore be a component of the disease resistance seen in endophyte-infected strong creeping red fescue.

8. Endophytes for Wastewater Treatment

Xenobiotic metabolism and degradation pathways in plants could be altered by using their endophytic community. It is possible to eliminate xenobiotic recalcitrant compounds like carbamazepine (CBZ) from wastewater in CWs. Sauvêtre et al. (2018), studied carbamazepine (CBZ) removal in presence of endophytes in the horseradish plant (*Armoracia rusticana*). Carbamazepine is a recalcitrant pollutant in the aquatic environment. Bacterial endophytes promote the growth of their host and play important role in CBZ metabolism. Sauvêtre et al. (2018), grow the plants with endophytes and without endophytes in spiked Murashige and Skoog (MS) medium. There was up to 5% removal of CBZ by plants without endophytes, but better removal was achieved with bacterial endophyte *Rhizobium radiobacter* (21%) and *Diaphorobacter nitroreductase* (10%). CBZ metabolism is divided into four pathways. Metabolites involving GSH conjugation and 2,3-dihydroxylation, as well as acridine-related compounds are described in plants.

9. Nutrients Move from Microbes to Plants

Macedo-Raygoza et al. (2019) studied Banana (*Musa* spp.)and its endophyte *Enterobacter cloacae* and *Klebsiella pneumoniae,* as antagonists of *Pseudocercospora fijiensis* (causative agent of black Sigatoka disease). These two bacterial endophytes were supporting plant growth in nutrient-limited soils by N-transfer. Both bacteria were also beneficial by producing indole acetic acid, phosphate solubilization, and antifungal to *P. fijiensis.* Maecedo-Raygoza treated the plants with bacterial endophyte and non-endophytic bacteria (*Escherichia coli*) without nutrients and compared them to a water-only treatment, and a mineral nutrients solution treatment. Bacterial endophyte increased the growth and prevented accelerated senescence. *E. coli* treated plants have died when irrigated with only water and weight loss when irrigated only with mineral nutrients solution. Further, it evaluated the transference of organic N from bacteria to plant by using bacteria grown on media containing $^{15}NH_4Cl$ or $Na^{15}NO_3$ as a nitrogen source. Pheophytin isotopomer abundance techniques showed that the ^{15}N was transferred from bacteria to plant tissues. *E. cloacae* could be an alternative biofertilizer to cultivate healthy and better banana crops.

Beltran-Garcia et al. (2014) proposed that the plant gets nitrogen by oxidative degradation of bacterial endophytes. In similar experiments, Beltran-Gracia employed the plant *Agave tequilana* and its endophyte *Bacillus tequilensis* to elucidate organic nitrogen transfer from 15N-labeled bacteria to plants. They observed ^{15}N into tryptophan, deoxynucleosides, and pheophytin derived from chlorophyll a. Plants with colonized endophytes in tissue could get nitrogen by degrading them using reactive oxygen. They showed more nitrogen transfer and greater biomass in comparison to plants grown in soil having heat-killed bacteria. Both the findings presented here support the hypothesis that plants get nitrogen from endophytes under nutrient limitation either by degradation of endophytic microbes or by direct transfer from endophytes.

Rhizophagy is an oxidative process in plants for nutrient extraction from symbiotic microbes. The interaction between plants and the microorganisms found in their soil is called rhizophagy. It is believed that plants cultivate the microorganisms of their interest through this process. The plant cultures the symbiotic microbes in its root's rhizoplane by releasing some exudate from its meristem region. The symbiotic microbes may enter plant tissues through wounds in meristem and other openings like stomata, etc., and grow in periplasmic space between the cell wall and cell membrane. Here they live in wall-less protoplast forms (L form bacteria). During rhizophagy plants either get nutrients by degrading microbes using reactive oxygen or inducing electrolyte leakage from microbes (White et al. 2018).

10. Endophytes are Bodyguards of Plants

English ivy plants harbor a bacterial endophyte (*Bacillus amyloliquefaciens*) in the epidermis and within vascular tissues. Endophytes produce antimicrobial compounds and thus may serve as a 'bodyguard' for plants. The endophyte may reduce the colonization of pathogens in the interior and exterior of the plant (Soares et al. 2015). *Bacillus* endophytes secrete surfactin lipopeptides for direct antibiosis of bacteria and fungi. These lipopeptides may up-regulate the stress resistance genes in plants during the seedlings process (Hazarika et al. 2019). Bacterial endophytes protect the grass *Poaannua* from damping-off disease. The *Pseudomonas* sp. endophyte strains protected the seedlings best by suppressing the hyphae growth (White et al. 2016). The *Pseudomonas* sp. bacteria have inhibitory and altering behavior against pathogenic fungi. The bacterial endophyte can provide protection from some pathogens and may influence soil ecology.

Conclusion

Endophytes (bacteria and fungi) can produce secondary metabolites useful in treating various disorders in humans and useful for plants as supportive molecules like growth regulators and pesticides. The new evolving trends in fungal endophyte-mediated biosynthesis of searching for more bioactive metabolites could be a major demand of the pharmacological industry. It required excellent strategies to overcome the existing challenges in endophyte biology and future direction in endophyte fungi-based drug discovery. Endophytes could be a potential source of novel natural arsenals in the frontiers of the fight against pathogens (Tiwari et al. 2022). The promising ongoing research on economically important endophytes and the metabolites they produce has raised new hopes for finding some solutions of biotechnological and industrial relevance.

Acknowledgments

The authors acknowledge their respective institutions for encouragement and support.

References

Ali, Shimaila, Jin Duan, Trevor C Charles, Bernard R Glick. "A bioinformatics approach to the determination of genes involved in Endophytic behavior in *Burkholderia* spp."*Journal of Theoretical Biology*, 343, (2014), 193–198.

Alnusairi, Ghalia S.H., Yasser S A Mazrou, Sameer H Qari, Amr A Elkelish, Mona H Soliman, Mohamed Eweis, Khaled Abdelaal, GomaaAbd El-Samad, Mohamed F M Ibrahim, Nihal ElNahhas. "Exogenous nitric oxide reinforces photosynthetic efficiency, osmolyte, mineral uptake, antioxidant, expression of stress-responsive genes and ameliorates the effects of salinity stress in Wheat."*Plants*, 10, no. 10 (2021), 1693.

Beltran-Garcia, Miguel J., James F. White, Jr., Fernanda M. Prado, Katia R. Prieto, Lydia F. Yamaguchi, Monica S. Torres, Massuo J. Kato, Marisa H. G. Medeiros & Paolo Di Mascio. "Nitrogen acquisition in *Agave Tequilana* from the degradation of Endophytic bacteria."*Sci Rep*,4,(2014),v6938.

Elshafie, Hazem, S., Ippolito Camele, Rocco Racioppi, Laura Scrano, Nicola S Iacobellis, Sabino A Bufo. "*In vitro* antifungal activity of *Burkholderia gladioli* Pv. *Agaricicola* against some Phytopathogenic fungi."*Int J MolSci*,13,(2012), 16291–16302.

Ettinger, Cassandra, L., Hanan R. Shehata, David Johnston-Monje, Manish N. Raizada, and Jonathan A. Eisen. "Draft genome sequence Of *Burkholderia gladioli* strain UCD-UG_CHAPALOTE (Phylum Proteobacteria)."*Genome Announcement*, 3, (2015), E01462– 14.

Faeth, Stanley H., Dale R Gardner, Cinnamon J Hayes, Andrea Jani, Sally K Wittlinger, Thomas A Jones. "Temporal and spatial variation in alkaloid levels in *Achnatherum robustum*, a native grass infected with the endophyte *Neotyphodium*." *Journal of Chemical Ecology*, 32, (2006), 307–324.

Fan, Di, Sowmyalakshmi Subramanian, and Donald L. Smith. "Plant endophytes promote growth and alleviate salt stress in *Arabidopsis thaliana*."*Sci Rep*, 10, (2020), 12740.

Franco-Franklin, Sandra Moreno-Riascos, and ThauraGhneim-Herrera. "Are endophytic bacteria an option for increasing the heavy metal tolerance of plants? A meta-analysis of the effect size."*Front. Environ. Sci.*, 8, (2021), 603668.

Gaynor, D.L., and D.D. Rowan. "Insect Resistance, Animal Toxicity and Endophyte-Infected Grass."*Proceedings of The New Zealand Grassland Association*, 47, (1986),115-120.

Hazarika, Dibya Jyoti, Gunajit Goswami, Trishnamoni Gautam, Assma Parveen, Pompi Das, Madhumita Barooah & Robin Chandra Boro. "Lipopeptide-mediated biocontrol activity of Endophytic *Bacillus Subtilis* against fungal phytopathogens."*BMC Microbiology*, 19, no. 1 (2019), 71. Https://doi.Org/10.1186/S12866-019-1440-8.

Hereme, Rasme, Samuel Morales-Navarro, Gabriel Ballesteros, Andrea Barrera, Patricio Ramos, Pedro E. Gundel, and Marco A. Molina-Montenegro. "Fungal endophytes exert positive effects on *Colobanthus quitensis* under water stress but neutral under a projected climate change scenario in Antarctica."*Front. Microbiol.*, 11, (2020), 264. Doi: 10.3389/Fmicb.2020.00264.

Macedo-Raygoza, Gloria, M., Benjamín Valdez-Salas, Fernanda M. Prado, Katia R. Prieto, Lydia F. Yamaguchi, Massuo J. Kato, Blondy B. Canto-Canché, Monica Carrillo-Beltrán, Paolo Di Mascio, James F. "White and Miguel J. Beltrán-García. *Enterobacter cloacae*, an endophyte that establishes a nutrient-transfer symbiosis with banana plants and protects against the black Sigatoka pathogen."*Front. Microbiol.*, 10, (2019), 804. Doi: 10.3389/Fmicb.2019.00804.

Mengistu, A.A. Endophytes: Colonization, behavior, and their role in defense mechanism."*International Journal of Microbiology*, Vol. 2020, Article ID 6927219, 8 Pages, 2020. Doi.Org/10.1155/2020/6927219.

Redman, Regina S., Kathy B Sheehan, Richard G Stout, Russell J Rodriguez, Joan M Henson. "Thermotolerance generated by Plant/Fungal symbiosis."*Science*, 298, no. 22 (2002), 1581-1581.

Rodriguez, Rusty J., Joan Henson, Elizabeth Van Volkenburgh, Marshal Hoy, Leesa Wright, Fleur Beckwith, Yong-Ok Kim and Regina S Redman. "Stress tolerance in plants via habitat-adapted symbiosis."*ISME Journal*, 2, (2008), 404–416.

Russell Jonathan R., Huang Jeffrey, Anand Pria, Kucera Kaury, Sandoval Amanda G., Dantzler Kathleen W., Hickman DaShawn, Jee Justin, Kimovec Farrah M., Koppstein David, Marks Daniel H., Mittermiller Paul A., Núñez Salvador Joel, Santiago Marina, Townes Maria A., Vishnevetsky Michael, Williams Neely E., Vargas Mario Percy Núñez, Boulanger Lori-Ann, Bascom-Slack Carol, Strobel Scott A. "Biodegradation of polyester polyurethane by endophytic fungi." Appl Environ Microbiol. 77(17), (2011), 6076-6084.

Sauvêtre, Andres, Robert May, Rudolf Harpaintner, Charlotte Poschenrieder, Peter Schröder. "Metabolism of Carbamazepine in plant roots and endophytic *Rhizobacteria* isolated from *Phragmites Australis*."*J Hazard Mater.*, 342, (2018),85-95.

Schmidt, Silvia, Judith F Blom, JakobPernthaler, Gabriele Berg, Adam Baldwin, Eshwar Mahenthiralingam, Leo Eberl. "Production of the antifungal compound Pyrrolnitrin is quorum sensing-regulated in members of the *Burkholderia cepacia* complex." *Environ Microbiol*, 11, (2009), 1422– 1437.

Shehata, H. R., Lyons, E. M., Jordan, K. S., and Raizada, M. N. "Bacterial endophytes from wild and ancient maize are able to suppress the fungal pathogen *Sclerotinia homoeocarpa*." *J. Appl. Microbiol.*, 120, (2016), 756–769.

Soares MA, Li HY, Kowalski KP, Bergen M, Torres MS, White JF. "Functional role of bacteria from invasive *Phragmites australis* in promotion of host growth."*Microb. Ecol.*, 72, no. 2 (2016), 407-417.

Soares, Marcos Antonio, Hai-Yan Li, Kurt P. Kowalski, Marshall Bergen, Monica S. Torres, and James F. White. "Evaluation of the functional roles of fungal endophytes of *Phragmites australis* from high saline and low saline habitats."*Biol Invasions*, 18, (2016), 2689–2702.

Soares, Marcos Antonio, Jai-Yan Li, Marshall Bergen, Joaquim Manoel da Silva, Kurt P. Kowalski, and James Francis White. "Functional role of an endophytic *Bacillus amyloliquefaciens* in enhancing growth and disease protection pf invasive English Ivy (*Hedera Helix* L.)."*Plant Soil*, 405, (2016), 107–123.

Steiner, Ulrike, Mahalia A. Ahimsa-Müller, Anne Markert, Sabine Kucht, Julia Groß, Nicole Kauf, Monika Kuzma, Monika Zych, Marc Lamshöft, MiroslawaFurmanowa, Volker Knoop, ChristelDrewke and EckhardLeistner. "Molecular characterization of a seed transmitted *Clavicipitaceous* fungus occurring on Dicotyledonous plants (Convolvulaceae)."*Planta*, 224, no. 3 (2006), 533-44.

Tian, Zipeng, Ruying Wang, Karen V. Ambrose, Bruce B. Clarke, and Faith C. Belanger. "The *Epichloëfestucae* antifungal protein has activity against the plant pathogen *Sclerotinia homoeocarpa*, the causal agent of dollar spot disease." *Sci Rep*, 7, (2017), 5643.

Tiwari, Pragya, Hanhong Bae. "Endophytic fungi: Key insights, emerging prospects, and challenges in natural product drug discovery."*Microorganisms*, 10, (2022), 360. https://doi.org/10.3390/microorganisms10020360.

Tiwari, Pragya, Hanhong Bae. "Horizontal gene transfer and Endophytes: An implication for the acquisition of novel traits."*Plants*, 9, (2020), 305.

Tiwari, Pragya, Yashdeep Srivastava, and Hanhong Bae. "Trends of pharmaceutical design of Endophytes as anti-infective."*Current Topics in Medicinal Chemistry*, 21, no. 17 (2021c), 1572-1586.

Verma, Satish K., Kathryn L Kingsley, Marshall S Bergen, Kurt P Kowalski, James F White. "Fungal disease prevention seedlings of Rice (*Oryza sativa*) and other grasses by growth-promoting seed-associated endophytic bacteria from invasive *Phragmites australis*."*Microorganisms*, 6, (2018), 21.

Vial, Ludovic, Marie-Christine Groleau, Valérie Dekimpe, Eric Déziel. "*Burkholderia* diversity and versatility: an inventory of the extracellular products." *J Microbiol Biotechnol*, 17 (2007), 1407–1429.

White, James F., Kathryn L. Kingsley, Satish K. Verma, and Kurt P. Kowalski. "Rhizophagy cycle: an oxidative process in plants for nutrient extraction from symbiotic microbes." *Microorganisms*, 6, no. 3 (2018), 95. Https://Doi.Org/10.3390/ Microorganisms6030095.

White, JF et al. "How do Endophytes function to enhance host plant growth and survival?"*AACC International The Future Of Food sustainability And Safety Symposium*, May 2016 (2016-05-01), pages 1 – 93

Yang, Zhennai, Linda M. Rogers, Yuanda Song, Wenjin Guo, and P. E. Kolattukudy. "Homoserine and asparagine are host signals that trigger in planta expression of a pathogenesis gene in *Nectria haematococca*."*Proceedings of The National Academy Of Sciences Of The United States Of America*, 102, no. 11 (2005), 4197–4202.

Chapter 6

Endophytic Actinomycetes: Overview, Distribution, and Multi-Faceted Applications

Subir Kumar Bose[1], Jyotirmayee Das[2], Mangalam Bajpai[2], Ashutosh Gautam[3] and Pragya Tiwari[4,*]

[1]Department of Biotechnology and Microbiology,
Meerut Institute of Engineering and Technology, Meerut, UP, India
[2]Global Product Compliance (GPC) Regulatory India Private Limited, Nagpur, Maharashtra, India
[3]Spices Board, Regional Office, Srinagar, (J&K UT), India
[4]Department of Biotechnology, Yeungnam University, Gyeongsan, Gyeongbuk, Republic of Korea

Abstract

Endophytic actinomycetes represent a less explored class of microbes with the potential to impact socio-economic areas and considerably impact human lives. This class of microbes is gaining popularity attributed to the biosynthesis of natural, drug-like metabolites, including antibiotics, enzyme inhibitors, and enzymes, among others. The drying pipelines of antibiotics have necessitated the exploration of bioactive metabolites from alternative natural sources, not to forget the importance of plant sources since ancient times. The chapter aims to highlight the forthcoming trends in drug discovery from actinomycetes endophytes with an overview of their classification, potential applications, and future prospects as bio-reservoirs of bioactive metabolites.

*Corresponding Author's Email: pragyatiwari@ynu.ac.kr.

In: Endophytes
Editor: Pragya Tiwari
ISBN: 979-8-88697-045-6

Keywords: bioactive metabolites, bioremediation, endophytic actinomycetes, microbiological methods, phytostimulants, sustainable agriculture

1. Introduction

Natural resources for the production of bioactive metabolites are extensively explored in the present era. The rising incidences of diseases and the discovery of fewer therapeutics necessitate a requirement to explore alternative resources for the discovery of metabolites with pharmacological importance (Tiwari et al. 2021a; Tiwari et al. 2021b; Tiwari et al. 2021c). Actinomycetes (actinobacteria), a category of endophytic microbe, is gaining popularity attributed to the production of natural drug-like metabolites, comprising of enzymes, antibiotics, and enzyme inhibitors (Jiang et al. 2016). The order actinomycetes include filamentous bacteria namely *Pseudonocardia*, *Saccharopolyspora, Amycolatopsis*, *Streptomyces,* and *Micromonospora* sp., and unicellular bacteria. The term 'actinomycete' was derived from Greek ἀκτίς (aktís, 'ray') and μύκης (múkēs, 'mushroom or fungus'), and members highlight considerable diversity in physiology (Ventura et al. 2007; O'Neill and Summers 2015). Moreover, the majority of actinomycetes define an aerobic, saprophytic and complex lifestyle, with few exceptions.

Actinomycetes show the frequency distribution in the environment and may exist as symbiotic partners as well as in solitary habitats. Several studies and patents have documented the identification of more than 22,000 bioactive metabolites from microorganisms, with multi-faceted applications in healthcare, agriculture, and the environment (Tiwari and Bae 2020; Tiwari et al. 2022a; Tiwari et al. 2022b). A few significant antibiotics, produced by actinomycetes and having universal application comprise gentamycin, vancomycin, erythromycin, ivermectin, and vancomycin, among others (Berdy J, 2005; Berdy J, 2012). However, the screening and development of new drugs from microorganisms project challenges, with most of the microbial strains being uncultivable. Several species of bacteria, particularly *Streptomyces*, constitute a key source of bioactive metabolites, showing multi-faceted applications as antibiotics and industrial enzymes (Tiwari et al. 2022b). The complex life cycle of these microbes is closely linked with the biosynthesis of metabolites, progressing from the vegetative stage to hyphal growth, and includes several factors like competition among microbes, abiotic stress, or depletion of nutrients, respectively. While the actinomycetes species are gaining momentum as a biological resource of bioactive metabolites, the research on these microbial species remains challenging (Pachter L, 2007; Abdelnasser SSI, 2012). The prolonged and complex developmental cycles, GC-rich genome hampers molecular characterization and hinders genome sequencing and assembly. In addition, endophytic actinomycetes have gained wide recognition as promoters of plant growth and soil bioremediation and as biocontrol agents to restrict microbial pathogens, finding wider applications in sustainable agriculture (Aamir et al. 2020). Continuous research on endophytes has documented the existence of novel endophytic actinomycetes in bioactive metabolite production.

The endophytic actinomycetes were identified and isolated from a number of plant sources and studied as a promising source of drug discovery (Matsumoto and Takahashi 2017). In *Triticum aestivum*, the technique of restriction fragment length polymorphism (RFLP) was employed for the isolation of 40 genera of endophytic actinobacteria, and validated by scanning electron microscope (SEM) and tissue-specific expression studies (Aamir et al. 2020). Several metabolites of pharmacological significance namely anticancer, antibacterial, and plant growth promotion with multiple applications in agriculture and industries were isolated from endophytic actinomycetes (Passari et al. 2015). Studies have demonstrated the presence of many groups of actinomycetes in different plant tissues having unique structures (Nimnoi et al. 2010). However, limited information is available regarding the association with plants and tissue-specific expression of endophytic actinomycetes. In the last few decades, many species of endophytes were explored and screened for the production of bioactive compounds, while only 20-25% were actinomycetes species, highlighting limited studies (Aamir et al. 2020). At present, this less explored group of endophytic actinomycetes has garnered enormous attention in pharmacological research, as novel candidates for drug discovery.

However, advances in computational biology and genomics continue to address the existing challenges in actinomycetes biology and research despite several limitations. The present chapter focuses on proving key information on the emerging popularity of endophytic actinomycetes as key producers of bioactive metabolites, with multi-faceted applications in environmental bioremediation, healthcare, and industries. Endophytic actinomycetes represent a less explored class of microbes with the potential to impact socio-economic areas and considerably impact human lives.

2. Endophytic Actinomycetes: Overview, Phylogenetic Distribution, and Screening/Isolation Methods

Present in the soil, actinomycetes include free-living microbes, associated with the root of the plants. The plant roots secrete exudates in the rhizosphere, which affects the interaction between a microbe and the plant (Schenk et al. 2012). Endophytic actinomycetes were reported from Chinese cabbage plants and tropical and medicinal plants (Dinesh et al. 2017; Nalini and Prakash 2017). Highlighting ubiquitous occurrence in diverse habitats, including mangroves, halophilic conditions, and arid zones, endophytic actinomycetes are majority present in the roots of plants (Gangwar et al. 2014). Furthermore, the subtropical and temperate regions represent enormous diversity in endophytic actinomycetes species, compared to other habitat zones. Studies have reported that actinomycetes colonization of the plants helps them in adaptation and protection against adverse environmental conditions (Asaf et al. 2017). For example, studies by Mohammadipanah and Wink (2016) showed that the actinobacteria found in desert conditions have genes for biosynthetic pathways for bioactive metabolites and are capable of tolerating extreme conditions. Few studies showed the presence and isolation of actinomycetes from plants in arid zones, suggesting a need for phylogenetic analysis (Thumar et al. 2010). Some key actinomycete species from arid zones comprise *Nocardia*, *Streptomyces*, *Nonomurea*, *Amycolatopsis,* and *Micromonospora* (Huang et al. 2012). Moreover, *Streptomyces* sp. comprises the majority of actinomycetes species found in nutrient-depleted and drought conditions in arid ecosystems. In addition, different habitats have been documented for endophytic actinomycetes. Actinomycetes species like *Isoptericola*, *Nocardiopsis*, *Pseudonocardia*, *Agrococcus*, and *Streptomyces* sp. were reported from mangrove vegetation (Yang et al. 2015a; Yang et al. 2015b), and *Okibacterium endophyticum* from high-salt concentrations (Wang et al. 2015a; Wang et al. 2015b). Moreover, Zhang et al. (2013) discussed the presence and isolation of *Saccharopolyspora dendranthemae* (salt-tolerant actinomycetes) from a plant growing in high salinity conditions. In addition, endophytic actinomycetes species were isolated from the aquatic ecosystems, including actinobacteria from seagrass (Wu et al. 2012) representing diverse habitats, suggesting a key role in the adaptation of marine plants to adverse conditions.

Endophytic actinomycetes colonize different tissues of the plant, namely the stem, bark, leaves, and roots, respectively (Qin et al. 2011). Moreover, the actinomycetes were isolated from diverse kinds of plants, consisting of major crops, medicinal plants, woody species, and halophytes (Qin et al. 2009). The bioactive metabolites from actinomycetes demonstrate multiple pharmacological properties, such as antiviral, anticancer, antimicrobial, and anti-inflammatory, among others (Matsumoto and Takahashi 2017). For example, 46 species of actinomycetes were isolated from different tissues of *Camellia sinensis* (Shan et al. 2018), showing diverse distribution in a wide range of plants.

Table 1. Tabular representation of endophytic actinomycetes, isolated from different parts of the plants

Endophytic actinomycetes (key examples)	Plant species	Source of isolation (plant tissue)	References
Streptomyces sp.	*Centella asiatica*	Stem, Root	Dochhil et al. 2013
Allostreptomyces psammosilenae	*Psammosilene tunicoides*	Roots	Huang et al. 2017
Actinopolyspora sp.	*Aloe vera*	Leaf, stem, root	Gangwar et al. 2014
Streptomyces sp. *Nocardiopsis* sp.	*Dracaena cochinchinensis*	Roots, leaves	Salam et al. 2017
Quadrisphaeraoryzae sp. nov.	*Oryza sativa* L.	Stem, leaves	Muangham et al. 2018 Mingma et al. 2015
Wangellaharbinensis NEAU-J3T *Streptomyces harbinensi NEAU-Da3T*	*Glycine max*	Root	Shen et al. 2013
Marmoricola endophyticus 8BXZ-J1T	*Thespesia populnea*	Plant branches	Jiang et al. 2017
Streptomyces ginkgonis	*Ginkgo biloba*	Seeds	Yan et al. 2018
Saccharomonospora sp.	*Azadirachta indica*	Roots	Verma et al. 2013
Nocardia endophytica KLBMP 1256T	*Jatropha curcas*	Stem	Qin et al. 2011
Plantactinospora solaniradicis	*Solanum lycopersicum*	Root	Li et al. 2018

2.1. Microbiological Methods

Endophytic microbes live in the interior tissues of plants and are symbiotic or mutualistic with them, generating no visible symptoms of illness. Endophytes are found in virtually all plant species on the planet, and it is widely assumed that they contribute to the evolutionary fitness of their hosts by creating a variety of secondary metabolites that aid in disease resistance and survival. Actinomycetes are the most often identified endophytes, followed by gram-positive and gram-negative bacteria, which are being studied as a new source of bioactive chemicals. Actinomycetes are aerobic, gram-positive bacteria that consist of a collection of branching unicellular microorganisms that play an important role in the degradation of organic materials into more easily accessible nutrients. They are also recognized for producing a wide range of secondary metabolites, including antibiotics, anticancer agents, and plant growth hormones, all of which are significant in the pharmaceutical and agricultural sectors.

Endophytic actinomycetes, in particular, those recovered from healthy surface-sterilized tissues, are thought to be a potential source for the synthesis of secondary metabolites, which are natural compounds having antibacterial, antioxidant, and plant growth stimulating properties. There is mounting evidence that new endophytic actinomycetes exist inside diverse tissues of medicinal plants, and some of them create bioactive substances with unique chemical structures. However, data on the tissue distribution and biodiversity of endophytic actinomycetes linked with traditional medicinal plants from diverse habitats is limited.

2.1.1. Isolation of Endophytic Actinomycetes and Their Relative Abundance

By cutting with a scalpel, plant components were split into different organs such as leaves stalks, roots, flowers, and petioles. All samples were thoroughly cleaned in tap water for 3-5 minutes to remove dirt and organic material then cut into pieces and rinsed in Tween 20 for a few seconds before being transferred to clean conical flasks. Surface sterilization was achieved by immersing samples in 70% ethanol for 3 minutes, NaOCl for 1 minute, ethanol for 2 minutes, and three washes with sterile distilled water for 1 minute each. Samples were dried in a laminar airflow chamber before removing the outer tissues using a sterile knife and gently dissecting the interior tissue. To prevent fungal development, kept on Petri plates containing various isolation mediums supplemented with nystatin and cycloheximide (60g/ml). To limit the growth of non-actinomycetes, the media was supplemented with nalidixic acid and $K_2Cr_2O_7$. The plates were incubated for 2–4 weeks in a BOD incubator. Bacterial colonies developing from the plated tissue segments were transferred onto ISP2 medium slants and re-cultured until pure cultures were produced, and the cultures were kept at 4°C.

2.1.2. Isolation of Actinobacteria Strains

Each feces sample was homogenized and swirled for 5 minutes in sterile water. In a Petri plate containing agar CM medium with and without inhibitor, an aliquot of properly diluted supernatant fluid was inserted. KNO_3, K_2HPO_4, $MgSO_4.H_2O$, and glucose were dissolved in deionized water to make the medium, and potassium dichromate was utilized as an inhibitor of other bacteria and fungi (Castillo et al. 2006). Actinomycete growth was examined daily for 20 days after inoculated plates were incubated at 26°C. Plate counts and a stereomicroscope were used to quantify actinomycetes. The appearance of filamentous hyphae and the growth of thick, leathery colonies that clung to the agar surface distinguished actinomycetes. As a result, only mycelium-forming bacteria from the order actinomycetes were included in this study. *Streptomyces* colonies (with aerial mycelium) and other actinomycetes were provisionally ascribed to colonies. Representative actinomycetes colonies were collected and transferred to starch casein agar slants. Glycerol was used to retain pure cultures for identification and characterization (Coombs et al. 2003).

According to prior work, the endophytic actinomycetes were isolated using the dilution technique. 10 gm of sieved soil sample was put in a crucible dish and dried in a 45°C oven for 24 hours. The samples were suspended in 100 mL distilled water in 200 mL triangle bottles, oscillated on the shaking table for 30 minutes, then allowed to stand for 30 minutes before being serially diluted. 0.1 mL of suspension was placed uniformly over the surface of Gauze's medium No.1 agar medium from each dilution, and the plates were incubated at 30°C for 2 weeks. Each serially diluted plate's colonies were purified using potato dextrose agar and cultured at 30°C for 2 weeks (Fiedler et al. 2008).

2.1.3. Preliminary Identification of Actinomycetes

Color of aerial and substrate mycelium, color and features of the colony on the Petri plate, spore mass color, formation of diffusible pigment, consumption of carbon sources, and spore chain morphology were used to identify isolates up to the genus level (Crawford et al. 1993). An oil immersion microscope was used to examine the structure of mycelium. Field Emission Gun-Scanning electron microscopy (FEG-SEM) was used to analyze the spore chain morphology and surface of 10-day-old cultures grown on an ISP1 medium.

2.1.4. Screening of Antagonistic Actinomycetes

According to prior reports, antagonistic actinomycetes were screened utilizing the confronting approach. The second round of screening followed the same method. In brief, one colony plug (6 mm in diameter) of 2 weeks of actinomycetes development was cut in half and inserted equidistantly on both sides of the middle of a PDA-containing plate (Cui et al. 2006). In order to establish a co-culture therapy, two colony plugs (6 mm in diameter) from actinomycetes were symmetrically put 3 cm from one 5-day-old fungal phytopathogen inoculant. Without the actinomycetes, one colony plug of the fungal phytopathogen was deposited on another PDA plate using the same way as the growth control. All treatments and controls were performed in duplicate and incubated at 30°C. When the fungal phytopathogen colony in the growth control had completely reached the center of the Petri dishes, the radius of the relative fungal phytopathogen colony in the treatment dishes was measured. R1 denotes the average radius of each fungal phytopathogen in the treatment, whereas R2 denotes the radius in the growth control. The proportion of fungal phytopathogen growth inhibition in actinomycetes, i.e., Phyto pathogen antagonism, should be determined.

2.1.5. Antibacterial Spectra of Antagonistic Actinomycetes

The antagonistic spectrum of the actinomycetes with strong antagonistic activity against *C. dematium* was examined against the other seven fungal phytopathogens using the co-culture technique.

2.1.6. Identification of Antagonistic Strains

Actinomycetes strains were streaked on agar Gauze's Medium No.1. Plates were then cultured at 30°C for 10 days and then colony morphology was observed under a light microscope (Phenix Optical Technology Co., Ltd, Shangrao, China), to observe their morphological characteristics (Crawford et al. 1993; Cui et al. 2006).

2.2. Molecular and Biochemical Approaches

Traditional Chinese medicinal herbs include physiologically active chemicals and are used as raw materials in the pharmaceutical, cosmetic, and fragrance industries. Medicinal plant endophytes participate in metabolic processes and create comparable or new bioactive chemicals. The Panxi plateau in China's southwest Sichuan province is home to a diverse range of medicinal plants due to its distinct geographical and climatological qualities. Antimicrobial activity analysis, combined with the results of amplifying genes coding for polyketide synthetase (PKS-I, PKS-II), and nonribosomal peptide synthetase (NRPS), revealed that endophytic actinomycetes isolated from medicinal plants in the Panxi plateau possessed broad-spectrum antimicrobial activity and potential natural product diversity, demonstrating that endophytic actinomycetes are valuable reservoirs.

2.2.1. Wild Medicinal Plants

Healthy medicinal plant samples with anticancer, antibacterial, and antiviral activities were obtained in July 2008 from the Panxi plateau in southwest Sichuan, China, according to ethnobotanical qualities. To guarantee that the isolates were endophytes, the cut ends of the root and stem samples were sealed with wax. Within 48 hours, all samples were tested for endophytic actinomycetes.

2.2.2. Surface Sterilization

Surface sterilization was performed with several changes. After drying at room temperature for 48 hours, samples were thoroughly cleaned with running tap water and assessed on size and surface appearance. All content that was clearly damaged was eliminated. To remove dirt and organic debris, samples were washed with sonification for 10 minutes at 150 w. To limit fungal development, tissue pieces were washed in 0.1 percent Tween 20 for 30 seconds before being submerged in 75 percent ethanol for 5 minutes, 2 percent sodium hypochlorite for 5 minutes, and rinsed with 10 percent $NaHCO_3$ for 10 minutes. Samples were cleaned three times in sterile water after each treatment. Finally, in a laminar flow chamber, the samples were fully dried. To ensure that the sterilizing process was successful, aliquots of the sterile distilled water from the final rinse were inoculated on ISP2 medium plates. Surface-sterilized samples were divided into small pieces aseptically, plated onto a selective medium, and cultured at 28°C for a month. The actinomycetes isolates were kept at 4°C on ISP4 slope media.

2.2.3. Identification of Actinomycetes

The isolates were de-replicated by cultural and morphological parameters, including morphology and color of aerial mycelium, characteristics of colonies on the plate, spore mass color, the color of diffusible pigments, and sporophore and spore chain morphology. For further identification, optical microscopy and Gram-straining were used. 16S rRNA Gene Restriction Fragment Length Polymorphism Analysis (ARDRA) (Igarashia et al. 2007).

Isolates' complete genomic DNA was isolated and amplified as previously described (El-Shatoury et al. 2006). The PCR products were digested using the restriction endonucleases Hae-III and Rsa-I, respectively. 19 buffer, 0.1-l gm/1l BSA (Bovine Serum Albumin), 10-U Hae-III/Rsa-I, and 3-ll PCR product were used in the 10-ll experiments. The digestions were carried out at 37°C for 6 hours. Gel electrophoresis was used to separate the digested pieces, which were then visualized with a UV transilluminator. The patterns of combined amplified rDNA restriction analyses were used to classify the isolates.

2.2.4. Evaluation of Antimicrobial Activity

In the antimicrobial activity screening assays, several indicator organisms (*Verticillium dahlia Kleb*., *Fusarium oxysporum f. sp. vasinfectum*, *Aspergillus niger*, *Fusarium oxysporum f. sp. Niveum*, *Colletotrichum orbiculare*, *Fusarium graminearum*, *Exerohilumturcicum*, *Curvularia lunatain* 3 days, an inhibition zone formed, indicating antagonism.

2.3. Computational Methods

Antibacterial antibiotics such as erythromycin, streptomycin, tetracycline, vancomycin, and chemotherapy medications such as doxorubicin are derived from biological sources. Actinobacteria have made a significant contribution to human health (Lee et al. 2012). Due to extensive culture-dependent screening of actinomycete isolates from the terrestrial environment, researchers are currently looking for novel strains from unexplored areas such as the marine ecosystem, which has resulted in the rediscovery of bioactive compounds (Weber et al. 2016).

Chemical screening of effective metabolites with bioactive potential is another technique for drug discovery. It entails sophisticated analytical instrumentation, such as high-performance liquid chromatography, mass spectrometry, or nuclear magnetic resonance, to elucidate the structural features of metabolites (Wahaab et al. 2018). The chemicals employed in this method came from microbiological origins. Various metabolomics tools, such as Cycloquest, GNPS-Genome to Natural Product Platform, NRPquest (Nonribosomal Peptide), and PEP2 path, are now available to characterize BCG in order to improve the metabolic profile of actinomycetes, while compound databases, such as (CHEBI-Chemical Entities of Biological Interest, Chemspider, Novel Antibiotics, PubChem, Antibiticome), are widely used to obtain a complete biochemical profile of metabolite (Galm et al. 2006; Weber et al. 2016).

Bioinformatics tools enable the quick identification of known gene clusters that code for bioactive substances, as well as computer predictions of their chemical structure based on genetic sequence information (Zazopoulos et al. 2003; Farnet et al. 2005; Wahaab et al. 2018). This information can be utilized to either lead to a more targeted drug discovery technique (such as reactivity-guided isolation, peptide, and glycol-genomics techniques, or allow heterologous expression in an optimized expression host) or to allow heterologous expression in an optimal expression host (activation of silent gene cluster) (Castro et al. 2016). There are web-based comprehensive suites like as BAGEL-Bayesian analysis of Gene essentiality that may be used to identify genes encoding precursor peptides, while BLAST and HMMeR provide insight into polyketides formed by type I and type II PKS, both ribosomally and post-ribosomally. Peptides that have been changed during translation (RiPPS). Various bioinformatics approaches for organizing or interpreting massive quantities of MS/MS fragmentation data have recently been developed (Kersten et al. 2011). MAGMa (MS annotation based on *in silico* produced metabolites), for example, allows multistage fragmentation data to be matched against potential molecular substructures and has been effectively used for complicated extracts.

3. Mechanisms of Plant Endophytic Actinomycetes Association and Their Advantages

3.1. Protective and Signaling Mechanisms for Plant Growth Promotion and Soil Health

Endophytic actinomycetes, which are found within the plant tissues, have many beneficial effects on the host plants, like inhibition of pathogens, increase in uptake of nutrients like nitrogen and phosphorus, increasing plant growth, etc. (Hasegawa et al. 2006; El Tarabily et al. 2008). Besides these roles, Actinomycetes are also known to produce secondary metabolites of therapeutic significance (Shukla et al. 2014; Singh and Dubey 2015; Segaran et al. 2017; Fadiji and Babalola 2020). Actinomycetes isolated from various sources like soil, water, etc. are well-known sources of antibiotics (Singh, and Dubey 2015). Here we are discussing endogenous actinomycetes. Endophytes produce several compounds that promote the growth of plants and help them adapt better to the environment (Lee et al. 2004; Das and Varma 2009; Ahmad et al. 2016). Studies have indicated that the organisms and their biotopes that are subject to constant metabolic and environmental interactions should produce even more secondary metabolites. Endophytes are microbes that inhabit such biotopes, namely, higher plants, which is why they are currently considered novel secondary metabolites that have the potential capabilities for applications in medicine, agriculture, and industry (Strobel 2003; Kalidass et al. 2015; Singh and Dubey 2015; Segaran et al. 2017; Pemila 2018). India is a country known for its plant diversity, which could well translate into a unique endophytic microbial diversity as it is considered that nearly every plant is a habitat for one or more endophytes. As evident from the research literature, endophytic actinomycetes are being increasingly explored for new therapeutic molecules, and lots of plants promote activity including biofertilizers (Xin et al. 2009; Kalidass et al. 2015), which may fulfill the current human health needs in the areas of infectious diseases and metabolic disorders as well as plants heaths.

Nowadays, crop production especially organic farming is also facing a lot of challenges including climate change, heavy metals, toxic chemical exposure, etc. As endophytic actinomycetes have the ability to produce nitrogen, phosphate, and other nutrients from the soil, other actinomycetes improve plant health and lead to increase yield and production (Hallmann et al. 1997; Premachandra et al. 2016; Vejan et al. 2016; Benedetto et al. 2017). Amylase, protease, pectinase, lipase, cellulase, and xylanase are industrially significant enzymes that are produced by the actinomycetes. Using terminal restriction fragment length polymorphism (T-RFLP) density of the endophytic actinobacteria in wheat roots can be identified (Shimizu 2011). Both endophytic and rhizospheric are plant-associated bacteria that as valuable effects on the host plants. When compared to rhizospheric bacteria, endophytes are more reliable and specific habitat as it resides within the tissues of the plant (Waheeda and Shyam 2017). Currently, market biofertilizers are available commercially in many brand names among them both bacteria and actinomycetes are available as Gmax PGPR, Galltrol, Diegall, Azo-Green, Nitromax, Custom N_2, BaciGold, Custom GP, Subtilex, Yield Shield, Root Shield Plus and others (Kalidass et al. 2015; Shen et al. 2019). The main factors that may regulate endophyte colonization within a plant or microbial species include the genotype of the plant, the growth stage of the plant, the physiological condition of the plant, the type of plant tissues, the soil health, and the environmental condition in which it is grown, the sampling season as well as different agricultural practices (Gaiero et al. 2013; Golinska et al. 2015).

The most occurring actinomycete found in the soil is *Streptomyces* (Pemila 2018) while *Micromonospora*, *Streptosporangium*, and *Nocardia* (having linear plasmid), are less abundant actinomycetes genera. Actinomycete species grow in close association with plant roots and they are one of the most important root colonizing organisms, frequently performed by *Streptomyces* sp. Their thread-like filamentous colonial morphology helps them to colonize the rhizosphere area efficiently, which in turn enables them to establish host – rhizobacteria symbiosis very effectively, after they establish effective colonization, the actinomycetes can produce many organic compounds and enzymes that are beneficial to plants. The ability to produce these compounds enables the actinomycetes to break down complex organic matter in the soil into simpler forms for the plants to absorb easily (Khamna et al. 2010; Anwar et al. 2016). In addition, they have many pathways to produce plant growth-promoting compounds such as IAA, siderophores, etc. (El Tarabily et al. 2008; Viaene et al. 2016). A great achievement has been found and also reported that around 60% of insecticides and bioactive compounds were discovered in the past 5 years obtained from actinomycetes especially *Streptomyces* sp. (Khamna et al. 2010; Damam et al. 2016). Plant growth-promoting activity of actinomycetes has been reported by many scientists in various plants such as *Pisum sativum*, *Phaseolus vulgaris* L., *Solanum lycopersicum*, *T. aestivum,* and *Oryza sativa* under field conditions (Gopalakrishnan et al. 2013; Chitraselvi and Kalidass 2017).

3.2. Bioactive Metabolite Production and Pharmacological Significance

The secondary metabolites having medicinal properties are produced by actinomycetes. Endophytic actinomycetes are associated with plants in different forms, including bacteria (actinomycetes) or mycoplasma, and also in fungi that have been colonized with plants' different tissues, including root, stem, leaf, and sometimes other parts of seed's actinomycetes, etc. Bacterial endophytes are diverse in nature and are known to produce different bioactive metabolites that act as antimicrobial and anticancer compounds, for example, 76% are reported from the single genus, *Streptomyces* (Berdy 2012; Pemila

2018). More than 200 genera from 16 phyla of bacterial species have been identified to be associated with endophytes and among them, most of the species belong to the phyla *Actinobacteria*, *Proteobacteria*, and *Firmicutes* (Golinska et al. 2015). The diversity of bacterial endophytes broad ranges including gram-positive to gram-negative bacteria, such as *Achromobacter*, *Acinetobacter*, *Agrobacterium*, *Bacillus*, *Brevibacterium*, *Microbacterium*, *Pseudomonas*, *Xanthomonas,* etc. (Sun et al. 2013). Actinomycetes are prokaryotic microorganisms that belong to the phylum *Actinobacteria* and possess mycelium-like fungus and form spores (Chaudhary et al. 2013; Barka et al. 2016).

However, the comparison between actinomycetes and fungi is superficial because most of their properties are like bacteria but unlike bacterial cells, while actinomycetes cells are thin with a chromosome and prokaryotic nucleoid and outer peptidoglycan wall. Endophytic actinomycetes are known to produce various chemical entities with unique structures of considerable medicinal importance (Singh and Dubey 2015; Fadiji and Babalola 2020). Many antimicrobial compounds have been reported from various types of endophytic actinomycetes, *Streptomyces* is one of the dominant genera, which is most isolated as endophytic actinomycetes (Zhao et al. 2011; Golinska et al. 2015). Different compounds isolated from *Streptomyces* sp. include munumbicins A and munumbicins B, naphthomycin (A and K types), cedarmycin, coronamycin, clethramycin, saadamycin, kakadumycins, and saadamycin. Other bioactive compounds isolated from actinomycetes are paclitaxel extracted from *Kitasatospora* sp. associated with *Taxusbaccata*, and tyrosol from *Emblica officinalis*, suggested to inhibit food-borne microbes (Zhao K. 2011; Gangwar et al. 2014; Golinska et al. 2015). Endophytic *mycoplasma* species were found to be in a symbiotic relationship with some red algae, such as *Bryopsis pennata*, *B. hypnoides,* and in *Arcobacter* (Hollants et al. 2011). Fungi are a heterotrophic group of organisms with various life cycles that include symbiotic relationships with a wide variety of autotrophic organisms (Dayleet al. 2001). Fungal endophytes are further classified into two separate groups based on their phylogeny and life-history traits. These include the clavicipitaceous, which infect some grasses confined to cool regions, and non-clavicipitaceous endophytes, which are from asymptomatic tissues of non-vascular plants, ferns and allies, conifers, and angiosperms and are limited to the *Ascomycota* or *Basidiomycota* group (Jalgaonwala et al. 2011; Bhardwaj and Agrawal 2014). Fungal endophytes produce some of the most broadly used antibiotic and anticancer drugs. Penicillenols, extracted from *Penicillium* sp., are cytotoxic to numerous cell lines. Popular anticancer drug Taxol is isolated from fungal endophyte *Taxomyces andreanae.* Clavatol (*Torreya mairei*), sordaricin (*Fusarium* sp.), jesterone (*Pestalotiopsis jesteri*), and javanicin (*Chloridium* sp.) are all known to possess strong antibacterial and antifungal properties against numerous foodborne infectious agents (Jalgaonwala et al. 2011). Pestacin, isolated from *P. microspora*, has excellent antioxidant properties.

3.2.1. Bioactive Compounds from Endophytes

Half of the deaths worldwide are due to infectious and parasitic diseases (Fadiji and Babalola 2020). Scientists are looking to discover novel bioactive compounds that prevent fatal infectious diseases. Endophytic microorganisms are reported to produce a lot of bioactive metabolites in a single plant or microbe which served as a potent source of drugs for the treatment of various diseases and with major applications in agriculture, food, medicine, and cosmetics industries (Strobel and Daisy 2003; Jalgaonwala et al. 2011; Shukla et al. 2014; Fadiji and Babalola 2020). Where in medicine, endophytes are the natural storehouse of novel secondary metabolites that can serve as drugs for anti-arthritic, antimicrobial, anticancer, antidiabetic, anti-insect, and immunosuppressant activities (Jalgaonwala et al. 2011; Pujiyanto et al. 2012; Godstime et al. 2014). secondary metabolites are categorized into several functional groups, alkaloids, terpenoids, flavonoids, quinones, steroids, benzopyranones, chinones, phenolic acids, saponins, tannins, tetralones, and many others (Jalgaonwala et al. 2011; Joseph and Priya 2011; Pimentel et al. 2011; Specian et al. 2012; Godstime et al. 2014). The extraction of secondary metabolites from endophytes is affected by various factors, such as the season of sample collection, climatic conditions, and geographical location (Shukla et al. 2014). However, with a revolutionary synthetic process that has been developed during the past few years, extraction from plants and other natural sources has now become more feasible, efficient, and convenient (Hussain et al. 2012). A bioactive compound produced by endophytes is directly associated with the evolution of the host microorganisms, which may have collected genetic information from higher plants, allowing for better adaptation to the host plant and performing some functions, such as protection from various types of pathogens, insects, and also grazing animals (Strobel 2003; Xin et al. 2009). Some of the commonly found secondary bioactive compounds from endophytes are described. Endophytes are also reported to have the ability to produce novel metabolites, which are used as anticancer agents (Rajamanikyam et al. 2017).

Taxol (paclitaxol), is a diterpene alkaloid produced by the endophyte *Metarhizium anisopliae.* Taxol is naturally found in the bark of *Taxus* tree, and is the most promising anticancer agent developed or synthesized (Liu et al. 2009; Zhang et al. 2009; Jalgaonwala et al. 2011; Rajamanikyam et al. 2017). Camptothecin, from *Nothapodytes foetida*is known to have cytotoxic and antifungal properties (Joseph and Priya, 2011; Han and Rahman, 2012). Huperzine A (HupA), from *Huperzia serrata*, act as a cholinesterase inhibitor (Nair and Padmavathy 2014). Lignans, such as cathartics, emetics, and cholagogue, isolated from endophyte *Podophyllum hexandrum*, are reported to act as anticancer agents (Konuklugil 1995). Resins, such as etoposide and teniposide extracted from *P. emodi*, possess strong anticancer activity (Konuklugil 1995). Compounds such as oxacillin, ampicillin, catechin, gallic acid, and cefalexin are known to possess bactericidal activities (Akiyama et al. 2001). Terpenoids possess antineoplastic, antibacterial, and antiviral effects as well as gastrointestinal stimulation (Jalgaonwala et al. 2011; Godstime et al. 2014). The fungal endophyte, *Cytonaema* sp. produces triterpenoid helvolic acid with strong antibacterial activity (Kumar et al. 2014). The antibiotic production by bacteria is associated with its ability to act as an antagonist to combat plant pathogens (Fadiji and Babalola 2020).

3.3. Application of Endophytic Actinomycetes in Agriculture as Phytostimulants, and Biofertilizers for High Yield and Productivity

A bacterial endophyte is a group of microorganisms also responsible for promoting plant growth, they are phytostimulants, biofertilizers, and biological control and are interrelated in the physiological processes of the plant. Phytostimulation is the production of phytohormones, which regulates plant growth and development. Phytohormones known as auxins, cytokinins, and gibberellins promote morphological changes in roots, influencing the absorption of nutrients and water uptake. Auxins have an enzyme ACC deaminase (1-aminocyclopropane-1-carboxylic acid) that reduces ethylene levels in plants. When the expression of ethylene level is high, it can result in inhibition of cell division, DNA synthesis, growth of roots, and aerial parts of the plant. Endophytes use ACC deaminase enzyme and indoleacetic acid (IAA) to reprogram certain signaling pathways, influencing the endophyte community structure. The modulating effect of ethylene concentration results in root growth and promotes seed germination (Oliveira et al. 2003). Endophytes produce phytohormone (IAA, GA_3, and ethylene) which enhances plant growth promotion and changes the morphology and structure of the plant resulting in increased yield and production of the crops (Gao and Tao 2012). The main function of bio fertilization is the bioavailability of nutrients and results in the best plant growth. The mechanism adopted by endophytes in the production of phytohormones in the host plant is related to the mechanism used by rhizobacteria in plant growth promotion. They help in the growth promotion of non-leguminous plants by the secretion and stimulation of IAA (Patel and Patel 2014), gibberellic acid (Khan et al. 2014), auxins (Dutta et al. 2014), and ethylene (Babalola 2010; Kang et al. 2012).

Endophytes are also alternatives to maximize biological nitrogen fixation by plants because these microorganisms have the ability to colonize structures such as leaves stems and roots, of plants and suffer less competition than the bacteria present in soil (Mendes et al. 2013). In the process of biological nitrogen fixation, the bacterial endophyte assimilates atmospheric nitrogen and converts it into ammonia, transferring this molecule to the plant metabolism (Gaiero et al. 2013). In sugarcane, a 30% increase in the nitrogen content has been reported after inoculation with a bacterial mixture of *Gluconacetobacter diazotrophicus, Herbaspirillum rubrisubalbicans, H. seropedicae, Azospirillum amazonense* and *Burkholderia* sp. (Guimaraes et al. 2013). The phosphorus deficiency is considered a limiting factor to plant growth and development, and this nutritional restriction decreased agricultural income (Xin et al. 2009). For the solubilization of phosphorus, bacterial endophytes synthesize organic acids, increasing the availability of phosphorus to the plant (Gaiero et al. 2013). An increase in the activity of phosphatase and availability of phosphorus content by adopting vermicomposting technique has been analyzed by inoculating *B. silvatlantica, Bacillus* sp., and *H. seropedica.* A study reveals that two organic acids oxalic, citric, and tartaric acids synthesized by *Burkholderia* sp. play the main role involved in the solubilization of phosphorus (Busato et al. 2012). Endophytes also play more resistant to biotic and abiotic stresses that directly regulate plant health and result increase growth and productivity of crops (Gagne-Bourgue et al. 2013).

The cycling of soil nutrients is essential for the sustainability of agricultural systems (Mariano et al. 2004). The benefits of the dynamic of biogeochemical cycles, which result in the solubility of nitrogen, phosphate, and zinc, sulfur oxidation, and nitrate bioavailability, are related to the bacterial community's capacity to provide significant quantities of natural nutrients., Cultivation-based techniques are used for the recovery and testing of isolates, whereas cultivation-independent techniques screen for variations in the total endophytic communities (Menpara and Chanda 2013). Endophytes can be easily isolated from plants or microbial, such as agar, potato dextrose agar, and any nitrogen or carbon-containing media. The method frequently used for the detection of endophytes involves isolation from surface-sterilized host plant tissue.

3.4. Environmental Applications

In nature, endophytes are found to colonize in the different parts of plants' stem, roots, petioles, leaf segments, inflorescences of weeds, fruit, buds, seeds, and dead and hollow hyaline cells of plants (Hata and Sone 2008; Specian et al. 2012; Stepniewska and Kuzniar 2013). If the environmental condition is not good, then endophytes does not properly grow and function. The population of endophytes in a plant species is highly variable and depends on various components, such as host species, host developmental stage, inoculum density, and environmental conditions (Dudeja and Giri 2014). Endophytes also play a significant role in environmental cleaning, including bioremediation, biodegradation agents, biocontrol agents, and nutrient cycling.

3.4.1. Bioremediation/Biodegradation

Bioremediation is a method of which removal of pollutants and wastes from the environment by the application of microorganisms. Endophytes have a peculiar ability to break down complex compounds to their simplest form (Stepniewska, and Kuzniar 2013). It is a biological process in microbes to break down wastes. A group of researchers studied the role of endophytes in bioremediation in *Nicotiana tabacum* plants (Mastretta et al. 2009), Inoculation of *Nicotiana tabacum* with endophytes revealed improved biomass production under Cadmium (Cd) stress, and the total plant Cd concentration was higher compared to non-inoculated plants and the results show that the beneficial effects of seed endophytes on heavy metal toxicity and accumulation. Now a day's manmade synthetic polymer polyester polyurethane (PUR) is a big problem for the environment including water and soil health. To explore the endophytic diversity for the breakdown of plastic, several dozen fungal endophytes were screened for their ability to degrade the synthetic PUR (Russell et al. 2011). Several organisms showed the ability to degrade PUR in both liquid and solid suspensions, robust activity was observed among several isolates in the genus *Pestalotiopsis*. Two *Pestalotiopsis microspore* isolates were able to grow on PUR as the sole carbon source under both aerobic and anaerobic conditions. Molecular characterization of the above activity confirms that an enzyme serine hydrolase is responsible for the degradation of PUR (Russell et al. 2011). An experiment performed by Germain and colleagues explored the degradation of herbicides with bacterial endophytes (*Pseudomonas* sp.), this experiment revealed that there was no sign of accumulation of the herbicide in the plant tissue and no sign of phytotoxicity, unlike the uninoculated plant (Germain et al. 2006).

3.4.2. Biocontrol Agents

Various pathogens that cause infectious diseases are also responsible for the disturbance of the fresh environment. Endophytic microorganisms are considered an effective biocontrol agent that is an alternative to chemical control. Biological control is a mechanism that promotes plant growth by protecting against infectious phytopathogens, the mechanism by the production of siderophores, antibiotics or bacteriocins, etc. (Gaiero et al. 2013). The production of many antibiotics is considered a very efficient mechanism for the control of plant pathogens because it occurs directly in cellular metabolism (Beneduzi et al. 2012).

Fungal endophytes have played an important role in controlling insect herbivory not only in grasses but also in conifers (Posada and Vega 2006). Fungal endophyte has an entomopathogen nature known as *Beauveria bassiana* and was found to control the borer insects in coffee seedlings and sorghum (Posada and Vega 2006; Tefera and Vidal 2009). Another fungal pathogen *Botrytis cinerea* causes the rotting of tomato fruits during storage time. Bacterial endophyte *Bacillus subtilis*, isolated from *Speranskia tuberculate* was found strong antagonistic to the pathogen *B. cinerea* in *vitro* observation (Wang et al. 2009). A new strain of *Burkholderia cepacia,* and *B. pyrrocinia* JK-SH007 were identified as a potential biocontrol agent against canker disease by the research team (Ren et al. 2011). Now a day's genetically engineered endophytes are also used as biocontrol agents and express anti-pest proteins like lectins. In the future, many researchers will be worked on these topics, which has been one of the important studies in endophytes recently. *Chaetomium globosum* YY-11 (a fungal endophyte) having antifungal properties, isolated from rape seedlings, and bacterial endophytes of *Enterobacter* sp. and *Bacillus subtilis* isolated from rice seedlings were used to express *Pinellia ternate* agglutinin (*PtA*) gene (Zhao et al. 2010). These recombinant endophytes expressing *PtA* gene are found to effectively control the population of sap-sucking pests in several crop seedlings. Similarly, in a different study, recombinant bacterial endophyte *Enterobacter cloacae* expressing *PtA* gene proved to be a bioinsecticide against white-backed planthopper, *Sogatella furcifera* (Zhang et al. 2011). The use of recombinant endophytes as biocontrol agents expressing different anti-pest proteins becomes a promising technique for control of plant pests as these endophytes can easily colonize within different crop plants successfully.

3.4.3. Nutrient Cycling

Nutrient cycling is a very important process that happens continuously to balance the existing nutrients existing and make them available for every component of the ecosystem that is directly related to our environments. Biodegradation of the dead biomasses becomes one major step in it to bring the utilized nutrients back to the environment which will be further available to the organisms and this is a cyclic chain process. Many saprophytic organisms (including endophytes) play a major role. Some studies have shown that endophytes have an important role in the biodegradation of the litter of its host plants (Muller et al. 2001; Promputtha et al. 2010). During biodegradation of the litter, the endophytic microbes colonize initially within the plants (Thormann et al. 2003) and facilitate the saprophytic microbes to act on through antagonistic interaction, thus increasing the litter decomposition (Fryar et al. 2001; Terekhova et al. 2005). In another study, it was demonstrated that all endophytes had the ability to decompose organic components, including lignin, cellulose, and hemicellulose however the preferences of various groups of endophytes with respect to organic compounds differed (He et al. 2012).

4. Different "Omics" Approaches in Socio-Economic Applications of Endophytic Actinomycetes

Natural products produced by endophytic actinomycetes are the main source of bioactive molecules (Liu et al. 2009; Genilloud 2017). The development of omics approaches has greatly contributed to the resurgence of antibiotic discovery increasing understanding of complex mechanisms controlling the expression of biosynthetic gene clusters (BGCs) that encode secondary metabolites. This book chapter discussed the application of 'omics' and its integration with genomics, transcriptomics, proteomics, and metabolomics as a powerful tool to discover new antibiotics. Particular actinomycetes possess bioactive natural compounds with a wide range of biological activities. Many well-known bioactive molecules (antibiotics), widely used in medicine for the treatment of infectious diseases, such as tetracyclines, β-lactams antibiotics, aminoglycosides, macrolides, and glycopeptides have been isolated from the actinomycetes genus, *Streptomyces*, as products of the secondary metabolism (Genilloud 2017; Palazzotto and Weber 2018). After the arising of next-generation sequencing (NGS) technologies and data obtained from whole-genome sequences, actinomycetes revealed their potential metabolic landscape and thus the far greater potential to produce specialized metabolites than has been discovered by classic screening-based methods in the previous 50 years (Palazzotto and Weber 2018). Genome sequence data revealed that a typical endophytic actinomycete contains on average 30 secondary metabolite gene clusters and thus theoretically has the genetic potential to produce almost a 10-times higher number of secondary metabolites that have been detected during screening and chemical analytics (Bentley et al. 2002; Challis 2014). In addition, the development of omics techniques, including genomics, transcriptomics, proteomics, metabolomics, and computational resources have increased scientific understanding of the complex cellular mechanism and key pathways affecting secondary metabolites production. In this context, here discussed recent progress and coordination of 'omics' approaches with genomics, transcriptomics, proteomics, metabolomics, and meta-omics (metagenomics, metatranscriptomics, and metaproteomics), using combined above omics approaches as powerful tools to discover new antibiotics from endophytes.

4.1. Genomics and Its Applications of Endophytic Actinomycetes

Now a day's development of sequencing technologies compared with the past decade with potential bioinformatics tools play the metabolic capability of *actinomycetes* triggering a research revolution in the drug discovery domain. Genomic analyses explored that silent gene clusters are present even within the genomes of extensively studied species, for example, *S. avermitilis* or *S. coelicolor* A3 (Bentley et al. 2002; Palazzotto and Weber 2018). Genome mining has been established as a powerful tool to explore the genetic capabilities of a strain by scanning the whole genome of interest and identifying secondary metabolite's biosynthetic gene clusters (BGCs) and Secondary Metabolites Analysis SHell (anti-SMASH) and the connected anti SMASH database, Prediction Informatics for Secondary Metabolomes

(PRISM), Global Alignment for natuRaL-products chemInformatiCs (GARLIC), Generalized Retrobiosynthetic Assembly Prediction Engine (GRAPE) platform and IMG/ABC (Dejong et al. 2016; Hadjithomas et al. 2017) the important tools. Above mentioned bio-informatics tools have been improved with more knowledge to explore the functions of various enzymes that are involved in the biosynthesis of secondary metabolites and make it easy for biosynthetic gene clusters to their possible corresponding products (Kim et al. 2017).

4.2. Transcriptomics Approach for Identification of Secondary Metabolites in Endophytic Actinomycetes

Even though DNA sequencing approaches are also explored and identify the high abundance of BGCs in endophytic *actinomycetes* genomes; a majority of their actual products have not been characterized. In most cases, these biosynthetic gene clusters remain silent under laboratory conditions because of their complex regulation at transcriptional, translational, and post-translational modification levels. Transcriptomics data is the real-time gene expression data that actually express the current need for particular secondary metabolites by actinomycetes. To reveal the mechanisms controlling the metabolic switch, development differentiation, and natural product biosynthesis, transcriptional changes in gene expression levels have been extensively studied (Vockenhuberel et al. 2011; Li et al. 2013). Studies by a scholar have reported the co-cultivation-based elicitation between actinomycetes, with other bacteria or fungi as a powerful method to activate the expression of BGCs for example the co-cultivation of the soil bacterium *S. coelicolor* with the competitor myxobacteria strain *Corallococcus coralloides* B035. Transcriptional analysis revealed that co-cultivation induced an increased expression of red X and red H genes and thus an increment of cylprodigiosin production in *S. coelicolor* (Schaberle et al. 2014; Zhu et al. 2014).

4.3. Impact of Proteomics on Discovery of Antibiotics in Endophytic Actinomycetes

Proteomics represents a complementary approach to transcriptomics and genomics for strain characterization. In addition to genomics and transcriptomics, proteomics has been used as a powerful tool to explore the correlation between metabolic pathways and natural product biosynthesis. Protein expression levels of different samples were compared using a proteomics approach that provides information on differential pathways regulation. An extensive analysis of the proteomic changes associated with antibiotic NAI-107 production was performed on *Microbispora* ATCC- PTA-5024 strain. Adopting differential proteomic analysis it is possible to elucidate many regulatory mechanisms controlling morphological differentiation and antibiotic production during different stages of growth (Gallo et al. 2016). The combined approach of genomic, transcriptomic, and proteomic was carried out on the unusual actinomycete, *Planobispora rosea* the producer of thiopeptide GE2270, and integration of different 'omics' data unveiled the physiology of the *P. rosea* during GE2270 biosynthesis (Tocchetti 2015).

4.4. Metabolomics Tool for the Investigation of Secondary Metabolites in Endophytic Actinomycetes

Genomics-based approaches have revealed that the biosynthetic potential of cultured and uncultured micro-organisms remains inaccessible to date. In fact, a major obstacle is that several BGCs are silent or low expressed under standard laboratory conditions. Different approaches/methodologies have been developed to prompt the expression of such a 'silent' gene cluster, able to explore pathway engineering and heterologous expression. Metabolomics, including the genomic, transcriptomic, and proteomic networks, represents a powerful tool to investigate the response to chemical and biological stimuli of secondary metabolite-producing strains (Covington et al. 2017). Using metabolomics approaches measuring the global level of low molecular weight metabolites, allow metabolic comparison of different biological samples avoiding chemical redundancy.

Lots of unique natural products have been explored by the use of two analytical platforms first one is Nuclear magnetic resonance (NMR) based metabolomics analyses and another one is mass spectrometry (MS).

Multivariate statistical analyses including principal component analysis, partial least-squares regression and the MS/MS database Global Natural Product Social (GNPS) molecular networking platform have greatly improved the analysis of mass spectrometry data allowing a rapid de-replication of known molecules and strain prioritization (Hou et al. 2012; Xie et al. 2014). In addition, VarQuest, a novel algorithm for the identification of peptide natural products variants *via* database search of mass spectra was developed to extend GNPS (Gurevich et al. 2018).

4.5. Meta-Omics Approaches to Explore Peculiar Molecules in Endophytic Actinomycetes

To date, a plethora of natural products remains undiscovered, for the reveal of those peculiar molecules complex omics approaches are used. Metaomics technique (metagenomics, meta-transcriptomics, and metaproteomics) provides culture-independent approaches to explore this hidden potential (Schofield and Sherman 2013). Metagenomics approaches are used for the investigation of the chemistry of uncultivated bacteria. Adopting Single-cell genomics and Metagenome sequencing approaches in the prokaryotic BGC encoding onnamide (a group of polyketides) was identified from a marine sponge *Theonella swinhoei* (source of the onnamides and theopederins). The role of the above symbiotic bacteria as 'artistic' producers of new polyketides and peptides that possess antitumor activity. Another successful example is represented by the culture-independent natural products discovery platform developed by Brady and co-workers (Hover et al. 2018).

Nowadays, different omics tools are available to study biosynthetic gene clusters and pathways among those, Orthogonal Active Site Identification System (OASIS) and Proteomic Interrogation of Secondary Metabolism (PrISM) (Meier et al. 2009; Meier et al. 2011). Using PrISM method from unsequenced environmental *Bacillus* sp. the Cyclic Imine koranimine. With high throughput (meta)-proteome data, it is also possible to investigate the post-translational modifications in *situ* environments and thus shed a light on the mechanisms used by bacteria to regulate physiological mechanisms and natural product production (Li et al. 2014). In 2017, Borriss and his research team demonstrated explored a large-scale lysine malonylation analysis in the model Gram-positive plant growth-promoting rhizobacteria, *Bacillus amyloliquefaciens* FZB42. The bioinformatics analysis revealed that the enzymes involved in rhizobacterium plant interaction and responsible for antibiotic production including polyketide synthases (PKSs) and nonribosomal peptide synthases (NRPSs) were highly malonylated (Fan et al. 2017).

5. Present Scenario and Emerging Importance of Endophytic Actinomycetes

As the pharmacologically important yet less explored class of microbes, endophytic actinomycetes is gaining popularity as prolific producers of diverse bioactive metabolites demonstrating multi-faceted applications in medicine, agriculture, and the environment. With a ubiquitous occurrence, endophytic actinomycetes are identified and isolated from a number of plant sources and studied as a promising source of drug discovery. The different species are found in diverse habitats from aquatic systems to high-saline conditions and marshy areas. The emerging prospects of endophytic actinomycetes have initiated studies on screening and isolation of different species employing microbiological methods, molecular and biochemical approaches, and computational methods. The extensive exploration of endophytic actinomycetes for application in agriculture, environment, and medicine necessitates the analysis and methodical approaches towards screening novel bioactive metabolites and their increased socio-economic applications.

5.1. Endophytic Actinomycetes and Their Role in Plant Defense

Bacterial endophytes are well-known for their active role in the plant defense system and wide range of involvement in the same through different mechanisms. They are the most commonly known source of phenolic compounds which are natural antioxidants (Golinska et al. 2015); such compounds neutralize the free radicals produced during a chain of reactions and help in improving plant health. It has been reported that plants arrest the spread of pathogens through different layers of defense responses and action of bioactive compounds; among them, flavonoids are the major ones (Piasecka et al. 2015; Beck and Stengel 2016). Flavonoids are the largest group of naturally occurring phenolic compounds, which occur in different plant parts in the free-state as well as glycosides.

There are reports suggesting that phenolic compounds and flavonoids in particular are important antimicrobials in plant life. There are many reports on the production of phenolic compounds by actinobacteria. Janardhan et al. (2014) showed total phenolic content of the extract was estimated as μg gallic acid equivalents (GAE)/mg of extract in *Nocardiopsis alba.* Almuhayawi et al. (2021) reported that *Streptomyces* sp. isolated from medicinal plants have antimicrobial activity owing to the presence of high levels of antioxidants like catalase, superoxide dismutase, and glutathione peroxidase which contributes to the development of resistance against different pathogens in such plants. Kemung et al. (2020) studied mangrove-derived *Streptomyces* sp. strain MUSC 14 for its potential in producing antioxidant metabolites and found that strain MUSC 14 displayed strong antioxidant activity against ABTS (2,2'-azino-bis(3-ethylbenzothiazoline-6-sulfonic acid), DPPH (2,2-diphenyl-1-picrylhydrazyl) free radicals and metal chelating activity.

Studies have also demonstrated that antioxidants from plants (Wong et al. 2014; Rolim et al. 2018) and microbes (Tan et al. 2017; Siddharth and Vittal 2018) were effective in chelating with iron and thereby preventing the formation of ROS (reactive oxygen species). Phaniendra et al. (2015) reported that ROS when reacts with electron-rich species damages the biological system at cellular damage through the malfunctioning of structure and function of cell organelles by various means. Since actinobacteria are a rich source of highly active antioxidant compounds; therefore it arrests the activity or neutralizes the effects of ROS and strengthens the plant immune system, which relates to the improved performance of crops in the agricultural system (Tan et al. 2015; Chan et al. 2016; Tang et al. 2016; Tay et al. 2019). So here we see that endophytic actinomycetes play a pivotal and active role in plant defense mechanisms as well as strengthening the immunity against pathogens and diseases.

Conclusion: Endophytic Actinomycetes and Abiotic Stresses Tolerance in Plants

Endophytic actinomycetes contribute not only in the tolerance/resistance against pests and diseases in plants but they are known to play a synergistic role in acclimatization as well as adaptation. The well-known cases come from the mangrove ecosystem, which presents very challenging and harsh conditions to the flora of the niche. Microbial communities of such conditions develop a specific and unique metabolic pathway that allows them to cope with the high fluctuation in salinity, pH, temperature, and tidal gradients (Pramanik et al. 2019). This is possible due to the production of a wide range of biomolecules in microbes. When these microbes have a synergistic relationship with plants, they trigger a similar kind of ability in the host and help them to withstand the harsh conditions (Meena et al. 2017) and it is not very different from endophytic actinomycetes. Ser et al. (2015) demonstrated that *Streptomyces* sp. from the harsh environment has the potential to produce biologically active metabolites that could trigger as well as modulate the metabolic pathways in plants in particularly harsh conditions and helps to withstand such adverse conditions. Altogether, these studies support the notion that *Streptomyces* sp. from the harsh environment has the potential to produce biologically active metabolites that could be of benefit in harsh environments.

A study conducted by Meena et al. (2013) showed that endophytic *Streptomyces* sp. isolated from the marine sources have a considerably high presence of exoenzymes chitinase, xylanase, and amylase, protease, lipase, and cellulose. Studies have shown that there are certain groups of microbe including actinobacteria which modulate the plant response against abiotic stresses (Velasquez et al. 2018) by various means such as turgidity, and viscosity of cytosol, alternating the permeability of cell membrane, and many more. Similarly, Chukwuneme et al. (2019) while studying the effects of drought on the performance of maize crops; screened 7 isolates of *Streptomyces pseudovenezuelae* and investigated their role in the context of drought stress. They found that *S. werraensis* MG547867, *S. luteogriseus* MG669347, *S. indiaensis* MG547868, *S. pseudovenezuelae* MG547870, and *Streptomyces* sp. MG640369 had high tolerance towards the induced drought and isolate *S. pseudovenezuelae* MG547870 could withstand the PEG-8000 concentration as high as 20%. It was found that plants with these isolates were not only protected from the deleterious effects of drought but also showed significant increases in the physiological parameters. The findings of this study suggested that these isolates are important tools capable of being developed into bio-inoculants to effectively improve drought tolerance in plants.

Acknowledgments

The authors acknowledge their respective institutions for encouragement and support.

Conflict of Interests

No conflict of interest was declared.

References

Aamir, Mohd, Krishna Kumar Rai, Andleeb Zehra, Manish Kumar Dubey, Swarnmala Samal, Mukesh Yadav, and Ram Sanmukh Upadhyay. "Endophytic actinomycetes in bioactive compounds production and plant defense system." In *Microbial Endophytes,* pp. 189-229. Woodhead Publishing, 2020.

Abdelnasser S. S. I., Ali A. A. S., Ashraf A. H., Mohammed S. E. S., and Shebl S. S. I: Tapping uncultured microorganisms through metagenomics for drug discovery. *Afr J Biotechnol.* 2012; 11(92):15823–15834. DOI: 10.5897/AJB12.2544.

Ahmad, Sayeed, Syed Salman Abbas, Ravi Prakash, Aftab Alam, and Mohammad Ajmal Husain. "Applications of endophytic Actinomycetes and their role in protection." *Imp. J. Interdiscip. Res* 2 (2016): 854-858.

Anwar, Sumaira, Basharat Ali, and Imran Sajid. "Screening of rhizospheric actinomycetes for various in-vitro and in-vivo plant growth-promoting (PGP) traits and for agroactive compounds." *Frontiers in microbiology* 7 (2016): 1334.

Asaf, Sajjad, Muhammad Aaqil Khan, Abdul Latif Khan, Muhammad Waqas, Raheem Shahzad, Ah-Yeong Kim, Sang-Mo Kang, and In-Jung Lee. "Bacterial endophytes from arid land plants regulate endogenous hormone content and promote growth in crop plants: an example of *Sphingomonas* sp. and *Serratia marcescens*." *Journal of Plant Interactions* 12, no. 1 (2017): 31-38. https://doi.org/10.1080/17429145.2016.1274060.

Babalola, O. O. "Ethylene quantification in three rhizobacterial isolates from *Striga hermonthica*-infested maize and sorghum." *Egyptian Journal of Biology* 12 (2010).

Bascom-Slack, C. A., Ma, C., Moore, E., Babbs, B., Fenn, K., Greene, J. S., Hann, B. D., Keehner, J., Kelley-Swift, E. G., Kembaiyan, V., Lee, S. J., Li, P., Light, D. Y., Lin, E. H., Schorn, M. A., Vekhter, D., Boulanger, L. A., Hess, W. M., Vargas, P. N., Strobel, S. A. "Multiple, novel biologically active endophytic actinomycetes isolated from upper Amazonian rainforests." *Microbial ecology* 58, no. 2 (2009): 374-383.

Beck, Charlotte, Kai Blin, Tetiana Gren, Xinglin Jiang, Omkar Satyavan Mohite, Emilia Palazzotto, Yaojun Tong, Pep Charusanti, and Tilmann Weber. "Metabolic Engineering of Filamentous Actinomycetes." *Metabolic Engineering: Concepts and Applications* 13 (2021): 653-688.

Beneduzi, Anelise, Adriana Ambrosini, and Luciana MP Passaglia. "Plant growth-promoting rhizobacteria (PGPR): their potential as antagonists and biocontrol agents." *Genetics and molecular biology* 35, no. 4 (2012): 1044-1051.

Bentley, S. D., Chater, K. F., Cerdeño-Tárraga, A. M., Challis, G. L., Thomson, N. R., James, K. D., Harris, D. E., Quail, M. A., Kieser, H., Harper, D., Bateman, A., Brown, S., Chandra, G., Chen, C. W., Collins, M., Cronin, A., Fraser, A., Goble, A., Hidalgo, J., Hopwood, D. A. "Complete genome sequence of the model actinomycete *Streptomyces coelicolor A3* (2)." *Nature* 417, no. 6885 (2002): 141-147.

Bérdy J: Thoughts and facts about antibiotics: where we are now and where we are heading. *J Antibiot.* 2012; 65:385–395. DOI: 10.1038/ja.2012.27.

Blin, Kai, Marnix H. Medema, Renzo Kottmann, Sang Yup Lee, and Tilmann Weber. "The antiSMASH database, a comprehensive database of microbial secondary metabolite biosynthetic gene clusters." *Nucleic acids research* (2016): gkw960.

Blin, K., Wolf, T., Chevrette, M. G., Lu, X., Schwalen, C. J., Kautsar, S. A., Suarez Duran, H. G., de los Santos, E. L. C., Kim, H. U., Nave, M., Dickschat, J. S., Mitchell, D. A., Shelest, E., Breitling, R., Takano, E., Lee, S. Y., Weber, T., & Medema, M. H. "antiSMASH 4.0—improvements in chemistry prediction and gene cluster boundary identification." *Nucleic acids research* 45, no. W1 (2017): W36-W41.

Busato, Jader G., Lívia S. Lima, Natália O. Aguiar, Luciano P. Canellas, and Fábio L. Olivares. "Changes in labile phosphorus forms during maturation of vermicompost enriched with phosphorus-solubilizing and diazotrophic bacteria." *Bioresource technology* 110 (2012).

Cao, L., Z. Qiu, J. You, H. Tan, and S. Zhou. "Isolation and characterization of endophytic Streptomyces strains from surface-sterilized tomato (*Lycopersicon esculentum*) roots." *Letters in applied microbiology* 39, no. 5 (2004): 425-430.

Castillo, U. F., Strobel, G. A., Mullenberg, K., Condron, M. M., Teplow, D. B., Folgiano, V., Gallo, M., Ferracane, R., Mannina, L., Viel, S., Codde, M., Robison, R., Porter, H., & Jensen, J. "Munumbicins E-4 and E-5: novel broad-spectrum antibiotics from *Streptomyces NRRL 3052*." *FEMS Microbiology letters* 255, no. 2 (2006): 296-300.

Castro-Falcón, Gabriel, Dongyup Hahn, Daniela Reimer, and Chambers C. Hughes. "Thiol probes to detect electrophilic natural products based on their mechanism of action." *ACS chemical biology* 11, no. 8 (2016): 2328-2336.

Challis, Gregory L. "Exploitation of the Streptomyces coelicolor A3 (2) genome sequence for the discovery of new natural products and biosynthetic pathways." *Journal of Industrial Microbiology and Biotechnology* 41, no. 2 (2014): 219-232.

Chitraselvi, E. R., S. Kalidass, and R. Kant. "Efficiency of rhizosphere bacteria in production of indole acetic acid, siderophore and phosphate solubilization." *Int J Chem Tech Res* 7, no. 6 (2015): 2557-2564.

Chitraselvi, R. P. E., and S. Kalidass. "Biocontrol activity of *Streptomyces violarus* bioformulation against leaf blight disease of groundnut." *Trends in bioscience* 10, no. 39 (2017): 8241-8251.

Coombs, Justin T., and Christopher M. M. Franco. "Isolation and identification of actinobacteria from surface-sterilized wheat roots." *Applied and environmental microbiology* 69, no. 9 (2003): 5603-5608.

Covington, Brett C., John A. McLean, and Brian O. Bachmann. "Comparative mass spectrometry-based metabolomics strategies for the investigation of microbial secondary metabolites." *Natural product reports* 34, no. 1 (2017): 6-24.

Cui, Xiao-Long, Pei-Hong Mao, Min Zeng, Wen-Jun Li, Li-Ping Zhang, Li-Hua Xu, and Cheng-Lin Jiang. "*Streptimonospora salina gen. nov*., sp. nov., a new member of the family Nocardiopsaceae." *International Journal of Systematic and Evolutionary Microbiology* 51, no. 2 (2001): 357-363.

Daguerre, Y., Edel-Hermann, V., & Steinberg, C. (2016). Fungal genes and metabolites associated with the biocontrol of soil-borne plant pathogenic fungi. In J.-M. Mérillon & K. G. Ramawat (Eds.), *Fungal Metabolites* (pp. 1–72). Springer International Publishing. https://doi.org/10.1007/978-3-319-19456-1_27-1.

Damam, Malleswari, Mohd Khaja Moinuddin, and Rana Kausar. "Isolation and screening of plant growth-promoting actinomycetes from rhizosphere of some forest medicinal plants." *International Journal of Chemtech Reesearch* 9, no. 5 (2016): 522-528.

Das, A. *"Symbiosis: the art of living, in symbiotic fungi principles and practice."* Varma A., Kharkwal. eds. 1-28. (2009).

De Oliveira, A. L. M., S. Urquiaga, and J. I. Baldani. "Processos e mecanismos envolvidos na influência de microrganismos sobre o crescimento vegetal." *Embrapa Agrobiologia-Documentos (INFOTECA-E)* (2003).

Dinesh, Raghavan, Veeraraghavan Srinivasan, Sheeja TE, Muthuswamy Anandaraj, and Hamza Srambikkal. "Endophytic actinobacteria: diversity, secondary metabolism and mechanisms to unsilence biosynthetic gene clusters." *Critical reviews in microbiology* 43, no. 5 (2017): 546-566. https://doi.org/10.1080/1040841X.2016.1270895.

Dochhil, Hmingthangziki, Mamtaj S. Dkhar, and D. I. N. A. Barman. "Seed germination enhancing activity of endophytic *Streptomyces* isolated from indigenous ethnomedicinal plant *Centella asiatica*." *Int J Pharm Biol Sci* 4, no. 1 (2013): 256-262.

Dutta, Devanushi, Keshab Chandra Puzari, Robin Gogoi, and Pranab Dutta. "Endophytes: exploitation as a tool in plant protection." *Brazilian archives of Biology and Technology* 57 (2014): 621-629.

El-Shatoury, Sahar, Hesham Abdulla, Omnia El-Karaaly, Waleed El-Kazzaz, and Ahmed Dewedar. "Bioactivities of endophytic actinomycetes from selected medicinal plants in the World Heritage Site of Saint Katherine Egypt." *Int J Bot* 2, no. 3 (2006): 307-312.

El-Tarabily, Khaled A., Amr H. Nassar, and Krishnapillai Sivasithamparam. "Promotion of growth of bean (*Phaseolus vulgaris* L.) in a calcareous soil by a phosphate-solubilizing, rhizosphere-competent isolate of *Micromonospora endolithica*." *Applied soil ecology* 39, no. 2 (2008): 161-171.

Fadiji, Ayomide Emmanuel, and Olubukola Oluranti Babalola. "Elucidating mechanisms of endophytes used in plant protection and other bioactivities with multifunctional prospects." *Frontiers in Bioengineering and Biotechnology* 8 (2020): 467.

Farnet, Chris M., and Emmanuel Zazopoulos. "Improving drug discovery from microorganisms." In *Natural Products*, pp. 95-106. Humana Press, 2005.

Fiedler, H. P., Bruntner, C., Riedlinger, J., Bull, A. T., Knutsen, G., Goodfellow, M., Jones, A., Maldonado, L., Pathom-aree, W., Beil, W., Schneider, K., Keller, S., & Sussmuth, R. D. "Proximicin A, B and C, novel aminofuran antibiotic and anticancer compounds isolated from marine strains of the actinomycete *Verrucosispora*." *The Journal of antibiotics* 61, no. 3 (2008): 158-163.

Fryar, S. C., T. K. Yuen, K. D. Hyde, and I. J. Hodgkiss. "The influence of competition between tropical fungi on wood colonization in streams." *Microbial Ecology* (2001): 245-251.

Gagne-Bourgue, F., K. A. Aliferis, P. Seguin, M. Rani, R. Samson, and S. Jabaji. "Isolation and characterization of indigenous endophytic bacteria associated with leaves of switchgrass (*P anicum virgatum* L.) cultivars." *Journal of applied microbiology* 114, no. 3 (2013): 836-853.

Gaiero, Jonathan R., Crystal A. McCall, Karen A. Thompson, Nicola J. Day, Anna S. Best, and Kari E. Dunfield. "Inside the root microbiome: bacterial root endophytes and plant growth promotion." *American journal of botany* 100, no. 9 (2013): 1738-1750.

Gallo, G., Renzone, G., Palazzotto, E., Monciardini, P., Arena, S., Faddetta, T., Giardina, A., Alduina, R., Weber, T., Sangiorgi, F., Russo, A., Spinelli, G., Sosio, M., Scaloni, A., & Puglia, A. M. "Elucidating the molecular physiology of lantibiotic NAI-107 production in *Microbispora ATCC-PTA-5024*." *BMC genomics* 17, no. 1 (2016): 1-17.

Galm, Ute, and Ben Shen. "Expression of biosynthetic gene clusters in heterologous hosts for natural product production and combinatorial biosynthesis." *Expert opinion on drug discovery* 1, no. 5 (2006): 409-437.

Gangwar, Madhurama, Sonam Dogra, Urmil Phutela Gupta, and Ravindra Nath Kharwar. "Diversity and biopotential of endophytic actinomycetes from three medicinal plants in India." *African Journal of Microbiology Research* 8, no. 2 (2014): 184-191.

Gao, Dawen, and Yu Tao. "Current molecular biologic techniques for characterizing environmental microbial community." *Frontiers of Environmental Science & Engineering* 6, no. 1 (2012): 82-97.

Genilloud, Olga. "Actinomycetes: still a source of novel antibiotics." *Natural product reports* 34, no. 10 (2017): 1203-1232.

Germaine, Kieran J., Xuemei Liu, Guiomar Garcia Cabellos, Jill P. Hogan, David Ryan, and David N. Dowling. "Bacterial endophyte-enhanced phytoremediation of the organochlorine herbicide 2, 4-dichlorophenoxyacetic acid." *FEMS microbiology ecology* 57, no. 2 (2006): 302-310.

Golinska, Patrycja, Magdalena Wypij, Gauravi Agarkar, Dnyaneshwar Rathod, Hanna Dahm, and Mahendra Rai. "Endophytic actinobacteria of medicinal plants: diversity and bioactivity." *Antonie Van Leeuwenhoek* 108, no. 2 (2015): 267-289.

Gopalakrishnan, Subramaniam, Vadlamudi Srinivas, Meesala Sree Vidya, and Abhishek Rathore. "Plant growth-promoting activities of *Streptomyces* spp. in sorghum and rice." *Springer Plus* 2, no. 1 (2013): 1-8.

Guimarães, S. L., and V. L. D. Baldani. "Produção de arroz inoculado com bactérias diazotróficas marcadas com resistência induzida ao antibiótico estreptomicina. *Rev. Ciencias Agrar.* 56, 125–132." (2013).

Gurevich, Alexey, Alla Mikheenko, Alexander Shlemov, Anton Korobeynikov, Hosein Mohimani, and Pavel A. Pevzner. "Increased diversity of peptidic natural products revealed by modification-tolerant database search of mass spectra." *Nature microbiology* 3, no. 3 (2018): 319-327.

Hallmann, J., A. Quadt-Hallmann, W. F. Mahaffee, and J. W. Kloepper. "Bacterial endophytes in agricultural crops." *Canadian journal of microbiology* 43, no. 10 (1997): 895-914.

Hasegawa, Sachiko, Akane Meguro, Masafumi Shimizu, Tomio Nishimura, and Hitoshi Kunoh. "Endophytic actinomycetes and their interactions with host plants." *Actinomycetologica* 20, no. 2 (2006): 72-81.

He, Xingbing, Guomin Han, Yonghui Lin, Xingjun Tian, Changguo Xiang, Qijian Tian, Fangyuan Wang, and Zaihua He. "Diversity and decomposition potential of endophytes in leaves of a *Cinnamomum camphora* plantation in China." *Ecological research* 27, no. 2 (2012): 273-284.

Hou, Bao Chao, En Tao Wang, Ying Li, Rui Zong Jia, Wen Feng Chen, Chao Xin Man, Xin Hua Sui, and Wen Xin Chen. "Rhizobial resource associated with epidemic legumes in Tibet." *Microbial ecology* 57, no. 1 (2009): 69-81.

Hou, Yanpeng, Doug R. Braun, Cole R. Michel, Jonathan L. Klassen, Navid Adnani, Thomas P. Wyche, and Tim S. Bugni. "Microbial strain prioritization using metabolomics tools for the discovery of natural products." *Analytical chemistry* 84, no. 10 (2012): 4277-4283.

Hover, B. M., Kim, S. H., Katz, M., Charlop-Powers, Z., Owen, J. G., Ternei, M. A., Maniko, J., Estrela, A. B., Molina, H., Park, S., Perlin, D. S., & Brady, S. F. . "Culture-independent discovery of the malacidins as calcium-dependent antibiotics with activity against multidrug-resistant Gram-positive pathogens." *Nature microbiology* 3, no. 4 (2018): 415-422.

Huang, Mei-Juan, Manik Prabhu Narsing Rao, Nimaichand Salam, Min Xiao, Hai-Quan Huang, and Wen-Jun Li. "Allostreptomyces psammosilenae gen. nov., sp. nov., an endophytic actinobacterium isolated from the roots of Psammosilene tunicoides and emended description of the family Streptomycetaceae [Waksman and Henrici (1943) AL] emend. Rainey et al. 1997, emend. Kim et al. 2003, emend. Zhi et al. 2009." *International Journal of Systematic and Evolutionary Microbiology* 67, no. 2 (2017): 288-293.

Huang, X. L., L. Zhuang, H. P. Lin, J. Li, M. Goodfellow, and K. Hong. "Isolation and bioactivity of endophytic filamentous actinobacteria from tropical medicinal plants." *African Journal of Biotechnology* 11, no. 41 (2012): 9855-9864. https://doi.org/10.5897/AJB11.3839.

Igarashi, Yasuhiro, Martha E. Trujillo, Eustoquio Martínez-Molina, Saeko Yanase, Satoshi Miyanaga, Takamasa Obata, Hiroaki Sakurai, Ikuo Saiki, Tsuyoshi Fujita, and Tamotsu Furumai. "Antitumor anthraquinones from an endophytic actinomycete *Micromonospora lupini sp. nov*." *Bioorganic & medicinal chemistry letters* 17, no. 13 (2007): 3702-3705.

Jalgaonwala, Ruby Erach, Bhavna Vishwas Mohite, and Raghunath Totaram Mahajan. "A review: natural products from plant-associated endophytic fungi." *J Microbiol Biotechnol Res* 1, no. 2 (2011): 21-32.

Jiang, Yi, Qinyuan Li, Xiu Chen, and Chenglin Jiang. "Isolation and cultivation methods of Actinobacteria." *Actinobacteria-basics and biotechnological applications* (2016): 39-57. http://dx.doi.org/10.5772/61457.

Jiang, Zhong-Ke, Zhen Pan, Fei-Na Li, Xiao-Jun Li, Shao-Wei Liu, Li Tuo, Ming-Guo Jiang, and Cheng-Hang Sun. "Marmoricola endophyticus sp. nov., an endophytic actinobacterium isolated from *Thespesia populnea*." *International Journal of Systematic and Evolutionary Microbiology* 67, no. 11 (2017): 4379-4384.

Kang, Jun Won, Zareen Khan, and Sharon L. Doty. "Biodegradation of trichloroethylene by an endophyte of hybrid poplar." *Applied and environmental microbiology* 78, no. 9 (2012): 3504-3507.

Kersten, Roland D., Nadine Ziemert, David J. Gonzalez, Brendan M. Duggan, Victor Nizet, Pieter C. Dorrestein, and Bradley S. Moore. "Glycogenomics as a mass spectrometry-guided genome-mining method for microbial glycosylated molecules." *Proceedings of the National Academy of Sciences* 110, no. 47 (2013): E4407-E4416.

Kersten, Roland D., Yu-Liang Yang, Yuquan Xu, Peter Cimermancic, Sang-Jip Nam, William Fenical, Michael A. Fischbach, Bradley S. Moore, and Pieter C. Dorrestein. "A mass spectrometry–guided genome mining approach for natural product peptidogenomics." *Nature chemical biology* 7, no. 11 (2011): 794-802.

Khamna, Sutthinan, Akira Yokota, John F. Peberdy, and Saisamorn Lumyong. "Indole-3-acetic acid production by *Streptomyces* sp. isolated from some Thai medicinal plant rhizosphere soils." *EurAsian Journal of BioSciences* 4 (2010).

Khan, A. L., Waqas, M., Kang, S. M., Al-Harrasi, A., Hussain, J., Al-Rawahi, A., Al-Khiziri, S., Ullah, I., Ali, L., Jung, H. Y., & Lee, I. J. "Bacterial endophyte *Sphingomonas sp. LK11* produces gibberellins and IAA and promotes tomato plant growth." *Journal of Microbiology* 52, no. 8 (2014): 689-695.

Lee, Jonathan A., Mark T. Uhlik, Christopher M. Moxham, Dirk Tomandl, and Daniel J. Sall. "Modern phenotypic drug discovery is a viable, neoclassic pharma strategy." *Journal of medicinal chemistry* 55, no. 10 (2012): 4527-4538.

Lee, Sunhee, M. Flores-Encarnacion, M. Contreras-Zentella, L. Garcia-Flores, J. E. Escamilla, and Christina Kennedy. "Indole-3-acetic acid biosynthesis is deficient in *Gluconacetobacter diazotrophicus* strains with mutations in cytochrome c biogenesis genes." *Journal of Bacteriology* 186, no. 16 (2004): 5384-5391.

Li, Wenchao, Xiaowei Guo, Linlin Shi, Junwei Zhao, Liangliang Yan, Xiaotong Zhong, Chen Zhang, Yufei Chen, Xiangjing Wang, and Wensheng Xiang. "*Plantactinospora solaniradicis sp. nov*., a novel actinomycete isolated from the root of a tomato plant (*Solanum lycopersicum* L.)." *Antonie van Leeuwenhoek* 111, no. 2 (2018): 227-235.

Li, Zhou, Yingfeng Wang, Qiuming Yao, Nicholas B. Justice, Tae-Hyuk Ahn, Dong Xu, Robert L. Hettich, Jillian F. Banfield, and Chongle Pan. "Diverse and divergent protein post-translational modifications in two growth stages of a natural microbial community." *Nature communications* 5, no. 1 (2014): 1-11.

Liu, Gang, Keith F. Chater, Govind Chandra, Guoqing Niu, and Huarong Tan. "Molecular regulation of antibiotic biosynthesis in *Streptomyces*." *Microbiology and molecular biology reviews* 77, no. 1 (2013): 112-143.

Liu, Kaihui, Xiaowei Ding, Baiwan Deng, and Wenqiang Chen. "Isolation and characterization of endophytic taxol-producing fungi from *Taxus chinensis*." *Journal of Industrial Microbiology and Biotechnology* 36, no. 9 (2009): 1171.

Mariano, Rosa de Lima Ramos, Elineide Barbosa da Silveira, Sayonara Maria Paulino de Assis, Andréa Maria André Gomes, Ana Rosa Peixoto Nascimento, and Virginia Maria Tenório Sabino Donato. "Importância de bactérias promotoras de crescimento e de biocontrole de doenças de plantas para uma agricultura sustentável." *Anais da Academia Pernambucana de Ciência Agronômica* 1 (2004): 89-111.

Mastretta, Chiara, Safiyh Taghavi, Daniel Van Der Lelie, Alessio Mengoni, Francesca Galardi, Christina Gonnelli, Tanja Barac, Jana Boulet, Nele Weyens, and Jaco Vangronsveld. "Endophytic bacteria from seeds of *Nicotiana tabacum* can reduce cadmium phytotoxicity." *International Journal of Phytoremediation* 11, no. 3 (2009): 251-267.

Matsumoto, A., Takahashi, Y., 2017. Endophytic actinomycetes: a promising source of novel bioactive compounds. *J. Antibiot.* 70 (5), 514.

McCarthy, Alan J., and Stanley T. Williams. "Actinomycetes as agents of biodegradation in the environment—a review." *Gene* 115, no. 1-2 (1992): 189-192.

Medema, Marnix H., Rainer Breitling, Roel Bovenberg, and Eriko Takano. "Exploiting plug-and-play synthetic biology for drug discovery and production in microorganisms." *Nature Reviews Microbiology* 9, no. 2 (2011): 131-137.

Meier, Jordan L., and Michael D. Burkart. "Proteomic analysis of polyketide and nonribosomal peptide biosynthesis." *Current opinion in chemical biology* 15, no. 1 (2011): 48-56.

Meier, Jordan L., Sherry Niessen, Heather S. Hoover, Timothy L. Foley, Benjamin F. Cravatt, and Michael D. Burkart. "An orthogonal active site identification system (OASIS) for proteomic profiling of natural product biosynthesis." *ACS chemical biology* 4, no. 11 (2009): 948-957.

Mendes, Rodrigo, Paolina Garbeva, and Jos M. Raaijmakers. "The rhizosphere microbiome: significance of plant beneficial, plant pathogenic, and human pathogenic microorganisms." *FEMS microbiology reviews* 37, no. 5 (2013): 634-663.

Menpara, Disha, and Sumitra Chanda. "Endophytic bacteria-unexplored reservoir of antimicrobials for combating microbial pathogens." *Microbial pathogens and strategies for combating them: Science, technology and education* (2013): 1095-1103.

Mingma, Ratchanee, Kannika Duangmal, Arinthip Thamchaipenet, Savitr Trakulnaleamsai, Atsuko Matsumoto, and Yoko Takahashi. "*Streptomyces oryzae sp. nov.*, an endophytic actinomycete isolated from stems of rice plant." *The Journal of Antibiotics* 68, no. 6 (2015): 368-372.

Mohammadipanah, Fatemeh, and Joachim Wink. "Actinobacteria from arid and desert habitats: diversity and biological activity." *Frontiers in microbiology* 6 (2016): 1541. https://doi.org/10.3389/fmicb. 2015.01541.

Muangham, Supattra, Kenika Lipun, Atsuko Matsumoto, Yuki Inahashi, and Kannika Duangmal. "*Quadrisphaera oryzae sp. nov.*, an endophytic actinomycete isolated from leaves of rice plant (Oryza sativa L.)." *The Journal of Antibiotics* 72, no. 2 (2019): 93-98.

Nair, D. N., and S. Padmavathy. "Impact of endophytic microorganisms on plants, environment and humans. *Sci World J* 2014: 250693." (2014).

Nalini, M. S., and H. S. Prakash. "Diversity and bioprospecting of actinomycete endophytes from the medicinal plants." *Letters in applied microbiology* 64, no. 4 (2017): 261-270.

Nimnoi, Pongrawee, Neelawan Pongsilp, and Saisamorn Lumyong. "Endophytic actinomycetes isolated from *Aquilaria crassna Pierre ex Lec* and screening of plant growth promoters production." *World Journal of Microbiology and Biotechnology* 26, no. 2 (2010): 193-203. https://doi.org/10.1007/s11274-009-0159-3.

O'Neill M, Summers E. Collins (Firm: Bishopbriggs Scotland) *Collins, English dictionary*; (2015).

Omojate Godstime, C., O. Enwa Felix, O. Jewo Augustina, and O. Eze Christopher. "Mechanisms of antimicrobial actions of phytochemicals against enteric pathogens–a review." *J Pharm Chem Biol Sci* 2, no. 2 (2014): 77-85.

Pachter L. "Interpreting the unculturable majority." *Nat Methods*. 2007; 4:479–480. DOI: 10.1038/nmeth0607-479.

Palazzotto, Emilia, and Tilmann Weber. "Omics and multi-omics approaches to study the biosynthesis of secondary metabolites in microorganisms." *Current opinion in microbiology* 45 (2018): 109-116.

Passari, Ajit K., Vineet K. Mishra, Ratul Saikia, Vijai K. Gupta, and Bhim P. Singh. "Isolation, abundance and phylogenetic affiliation of endophytic actinomycetes associated with medicinal plants and screening for their in vitro antimicrobial biosynthetic potential." *Frontiers in microbiology* 6 (2015): 273. https://doi.org/10.3389/fmicb.2015.00273.

Patel, Monita V., and Rajesh K. Patel. "Indole-3-acetic acid (IAA) production by endophytic bacteria isolated from saline desert, the little Runn of Kutch." *CIBTech J. Microbiol* 3 (2014): 17-28.

Posada, Francisco, and Fernando E. Vega. "Inoculation and colonization of coffee seedlings (Coffea arabica L.) with the fungal entomopathogen *Beauveria bassiana* (Ascomycota: Hypocreales)." *Mycoscience* 47, no. 5 (2006): 284-289.

Promputtha, Itthayakorn, Kevin D. Hyde, Eric H. C. McKenzie, John F. Peberdy, and Saisamorn Lumyong. "Can leaf degrading enzymes provide evidence that endophytic fungi becoming saprobes?" *Fungal Diversity* 41, no. 1 (2010): 89-99.

Pujiyanto, Sri, Y. U. L. I. N. Lestari, Antonius Suwanto, S. Budiarti, and Latifah K. Darusman. "Alpha-glucosidase inhibitor activity and characterization of endophytic actinomycetes isolated from some Indonesian diabetic medicinal plants." *Int J Pharm Pharm Sci* 4, no. 1 (2012): 327-333.

Qin, Sheng, Jie Li, Hua-Hong Chen, Guo-Zhen Zhao, Wen-Yong Zhu, Cheng-Lin Jiang, Li-Hua Xu, and Wen-Jun Li. "Isolation, diversity, and antimicrobial activity of rare actinobacteria from medicinal plants of tropical rain forests in Xishuangbanna, China." *Applied and environmental microbiology* 75, no. 19 (2009): 6176-6186.

Qin, Sheng, Ke Xing, Ji-Hong Jiang, Li-Hua Xu, and Wen-Jun Li. "Biodiversity, bioactive natural products, and biotechnological potential of plant-associated endophytic actinobacteria." *Applied Microbiology and Biotechnology* 89, no. 3 (2011): 457-473.

Rajamanikyam, Maheshwari, Varahalarao Vadlapudi, and Suryanarayana Murty Upadhyayula. "Endophytic fungi as novel resources of natural therapeutics." *Brazilian Archives of Biology and Technology* 60 (2017).

Ren, Jia Hong, Jian Ren Ye, Hui Liu, Xu Ling Xu, and Xiao Qin Wu. "Isolation and characterization of a new *Burkholderia pyrrocinia strain JK-SH007* as a potential biocontrol agent." *World Journal of Microbiology and Biotechnology* 27, no. 9 (2011): 2203-2215.

Russell, J. R., Huang, J., Anand, P., Kucera, K., Sandoval, A. G., Dantzler, K. W., Hickman, D., Jee, J., Kimovec, F. M., Koppstein, D., Marks, D. H., Mittermiller, P. A., Núñez, S. J., Santiago, M., Townes, M. A., Vishnevetsky, M., Williams, N. E., Vargas, M. P. N., Boulanger, L. A., Strobel, S. A.. "Biodegradation of polyester polyurethane by endophytic fungi." *Applied and environmental microbiology* 77, no. 17 (2011): 6076-6084.

Salam, N., Khieu, T. N., Liu, M. J., Vu, T. T., Chu-Ky, S., Quach, N. T., Phi, Q. T., Narsing Rao, M. P., Fontana, A., Sarter, S., & Li, W. J. "Endophytic actinobacteria associated with *Dracaena cochinchinensis Lour.*: isolation, diversity, and their cytotoxic activities." *BioMed Research International* 2017 (2017).

Schäberle, Till F., Annika Orland, and Gabriele M. König. "Enhanced production of undecylprodigiosin in Streptomyces coelicolor by co-cultivation with the corallopyronin A-producing myxobacterium, *Corallococcus coralloides*." *Biotechnology letters* 36, no. 3 (2014): 641-648.

Schenk, Peer M., Lilia C. Carvalhais, and Kemal Kazan. "Unraveling plant–microbe interactions: Can multi-species transcriptomics help?" *Trends in biotechnology* 30, no. 3 (2012): 177-184.

Schulz, Barbara, Christine Boyle, Siegfried Draeger, Anne-Katrin Römmert, and Karsten Krohn. "Endophytic fungi: a source of novel biologically active secondary metabolites." *Mycological research* 106, no. 9 (2002): 996-1004.

Segaran, Gayathri, Ranjitha Dhevi V. Sundar, Sugashini Settu, Saranya Shankar, and Mythili Sathiavelu. "A review on endophytic actinomycetes and their applications." *J Chem Pharm Res* 9, no. 10 (2017): 152-158.

Segaran, Gayathri, Ranjitha Dhevi V. Sundar, Sugashini Settu, Saranya Shankar, and Mythili Sathiavelu. "A review on endophytic actinomycetes and their applications." *J Chem Pharm Res* 9, no. 10 (2017): 152-158.

Shan, Wenna, Ying Zhou, Huihui Liu, and Xiaomin Yu. "Endophytic actinomycetes from tea plants (*Camellia sinensis*): isolation, abundance, antimicrobial, and plant-growth-promoting activities." *BioMed research international* 2018 (2018).

Shen, Fo-Ting, Jui-Hung Yen, Chien-Sen Liao, Wen-Ching Chen, and Yi-Ting Chao. "Screening of rice endophytic biofertilizers with fungicide tolerance and plant growth-promoting characteristics." *Sustainability* 11, no. 4 (2019): 1133.

Shen, Yue, Chongxi Liu, Xiangjing Wang, Junwei Zhao, Feiyu Jia, Yuqin Zhang, Liang Wang, Deguang Yang, and Wensheng Xiang. "*Actinoplanes hulinensis sp. nov*., a novel actinomycete isolated from soybean root (*Glycine max* (L.) Merr)." *Antonie van Leeuwenhoek* 103, no. 2 (2013): 293-298.

Shimizu, Masafumi. "Endophytic actinomycetes: biocontrol agents and growth promoters." In *Bacteria in agrobiology: Plant growth responses*, pp. 201-220. Springer, Berlin, Heidelberg, 2011.

Shukla, S. T., P. V. Habbu, V. H. Kulkarni, K. S. Jagadish, A. R. Pandey, and V. N. Sutariya. "Endophytic microbes: a novel source for biologically/pharmacologically active secondary metabolites." *Asian J Pharmacol Toxicol* 2, no. 3 (2014): 1-6.

Singh, Radha, and A. K. Dubey. "Endophytic actinomycetes as emerging source for therapeutic compounds." *Indo Global J Pharm Sci* 5, no. 2 (2015): 106-116.

Skinnider, Michael A., Chris A. Dejong, Philip N. Rees, Chad W. Johnston, Haoxin Li, Andrew L. H. Webster, Morgan A. Wyatt, and Nathan A. Magarvey. "Genomes to natural products prediction informatics for secondary metabolomes (PRISM)." *Nucleic acids research* 43, no. 20 (2015): 9645-9662.

Skinnider, Michael A., Nishanth J. Merwin, Chad W. Johnston, and Nathan A. Magarvey. "PRISM 3: expanded prediction of natural product chemical structures from microbial genomes." *Nucleic acids research* 45, no. W1 (2017): W49-W54.

Specian, Vânia, Maria Helena Sarragiotto, João Alencar Pamphile, and Edmar Clemente. "Chemical characterization of bioactive compounds from the endophytic fungus *Diaporthe helianthi* isolated from *Luehea divaricata*." *Brazilian Journal of Microbiology* 43, no. 3 (2012): 1174-1182.

Stępniewska, Z., and A. Kuźniar. "Endophytic microorganisms-promising applications in bioremediation of greenhouse gases." *Applied microbiology and biotechnology* 97, no. 22 (2013): 9589-9596.

Strobel, Gary A. "Endophytes as sources of bioactive products." *Microbes and infection* 5, no. 6 (2003): 535-544.

Strobel, Gary, and Bryn Daisy. "Bioprospecting for microbial endophytes and their natural products." *Microbiology and molecular biology reviews* 67, no. 4 (2003): 491-502.

Sun, Xinhua, Xianglan Kong, Huquan Gao, Tianjiao Zhu, Guangwei Wu, Qianqun Gu, and Dehai Li. "Two new meroterpenoids produced by the endophytic fungus *Penicillium sp. SXH-65*." *Archives of pharmacal research* 37, no. 8 (2014): 978-982.

Tefera, Tadele, and Stefan Vidal. "Effect of inoculation method and plant growth medium on endophytic colonization of sorghum by the entomopathogenic fungus *Beauveria bassiana*." *BioControl* 54, no. 5 (2009): 663-669.

Terekhova, V. A., and T. A. Semenova. "The structure of micromycete communities and their synecologic interactions with basidiomycetes during plant debris decomposition." *Microbiology* 74, no. 1 (2005): 91-96.

Thormann, Markus N., Randolph S. Currah, and Suzanne E. Bayley. "Succession of microfungal assemblages in decomposing peatland plants." *Plant and Soil* 250, no. 2 (2003): 323-333.

Thumar, J. T., Dhulia, K., Singh, S. P., 2010. Isolation and partial purification of an antimicrobial agent from halotolerant alkaliphilic *Streptomyces aburaviensis strain Kut-8. World J. Microbiol. Biotechnol.* 26, 2081–2087. https://doi.org/10.1007/s11274-010-0394-7.

Tiwari, Pragya, Hanhong Bae. "Endophytic fungi: Key insights, emerging prospects, and challenges in natural product drug discovery." *Microorganisms* 10 (2022b): 360. https://doi.org/10.3390/microorganisms10020360.

Tiwari, Pragya, Hanhong Bae. "Horizontal gene transfer and Endophytes: An implication for the acquisition of novel traits." *Plants* 9(2020): 305.

Tiwari, Pragya, Tushar Khare, Varsha Shriram, Hanhong Bae, Vinay Kumar. "Exploring synthetic biology strategies for producing potent antimicrobial phytochemicals." *Biotechnology Advances* 48, (2021b):107729.

Tiwari, Pragya, Yashdeep Srivastava, and Hanhong Bae. "Trends of pharmaceutical design of Endophytes as anti-infective." *Current Topics in Medicinal Chemistry* 21, no. 17 (2021c): 1572-1586.

Tiwari, Pragya, Yashdeep Srivastava, Mangalam Bajpai, Abhishek Sharma. "Bioactive metabolites from natural sources: Prospects and significance in drug discovery and research." *Bioingene PSJ* 1, (2021a):1-14.

Tiwari, Pragya, Yashdeep Srivastava, Vinay Kumar. "Antimicrobial peptides as effective agents against drug-resistant pathogens." In *Antimicrobial Resistance*; Kumar, V., Shriram, V., Paul, A., Thakur, M., Eds.; Springer: Singapore, 2022a; pp. 289-322.

Tocchetti, A., Bordoni, R., Gallo, G., Petiti, L., Corti, G., Alt, S., Cruz, J. C. S., Salzano, A. M., Scaloni, A., Puglia, A. M., De Bellis, G., Peano, C., Donadio, S., & Sosio, M. A genomic, transcriptomic and proteomic look at the GE2270 producer *Planobispora rosea*, an uncommon actinomycete. *PLoS One* 2015, 10:e0133705.

Ventura, Marco, Carlos Canchaya, Andreas Tauch, Govind Chandra, Gerald F. Fitzgerald, Keith F. Chater, and Douwe van Sinderen. "Genomics of Actinobacteria: tracing the evolutionary history of an ancient phylum." *Microbiology and molecular biology reviews* 71, no. 3 (2007): 495-548. DOI.1038/ja.2005.1

Verma, Vijay C., Swechha Anand, Christian Ulrichs, and Santosh K. Singh. "Biogenic gold nanotriangles from *Saccharomonospora* sp., an endophytic actinomycetes of *Azadirachta indica A. Juss*." *International Nano Letters* 3, no. 1 (2013): 1-7.

Viaene, Tom, Sarah Langendries, Stien Beirinckx, Martine Maes, and Sofie Goormachtig. "*Streptomyces* as a plant's best friend?" *FEMS microbiology ecology* 92, no. 8 (2016).

Vockenhuber, Michael-Paul, Cynthia M. Sharma, Michaela G. Statt, Denis Schmidt, Zhenjiang Xu, Sascha Dietrich, Heiko Liesegang, David H. Mathews, and Beatrix Suess. "Deep sequencing-based identification of small non-coding RNAs in *Streptomyces coelicolor*." *RNA biology* 8, no. 3 (2011): 468-477.

Wahaab, Femina, and Kalidass Subramaniam. "Bioprospecting marine actinomycetes for multidrug-resistant pathogen control from Rameswaram coastal area, Tamil Nadu, India." *Archives of microbiology* 200, no. 1 (2018): 57-71.

Waithaka, Paul Njenga, Eliud Mugu Gathuru, Benson Muriuki Githaiga, and Calvince Omoro Ouma. "Microbial degradation of maize waste materials using actinomycetes isolated from Egerton University Soils, Njoro in Kenya." *Intl Res J Biol Sci* 1, no. 1 (2019): 31-36.

Wang, Hong-Fei, Yong-Guang Zhang, Ji-Yue Chen, Jian-Wei Guo, Li Li, Wael N. Hozzein, Yuan-Ming Zhang, Mohammed AM Wadaan, and Wen-Jun Li. "*Frigoribacterium endophyticum sp. nov*., an endophytic actinobacterium isolated from the root of *Anabasis elatior* (CA Mey.) Schischk." *International journal of systematic and evolutionary microbiology* 65, no. Pt_4 (2015): 1207-1212.

Wang, Hong-Fei, Yong-Guang Zhang, Juan Cheng, Wael N. Hozzein, Wei-Hong Liu, Li Li, Ji-Yue Chen, Jian-Wei Guo, Yuan-Ming Zhang, and Wen-Jun Li. "*Labedella endophytica sp. nov*., a novel endophytic actinobacterium isolated from stem of *Anabasis elatior* (CA Mey.) Schischk." *Antonie van Leeuwenhoek* 107, no. 1 (2015): 95-102.

Wang, Shutong, Tongle Hu, Yanling Jiao, Jianjian Wei, and Keqiang Cao. "Isolation and characterization of *Bacillus subtilis EB-28*, an endophytic bacterium strain displaying biocontrol activity against *Botrytis cinerea Pers*." *Frontiers of Agriculture in China* 3, no. 3 (2009): 247-252.

Weber, Tilmann, and Hyun Uk Kim. "The secondary metabolite bioinformatics portal: computational tools to facilitate synthetic biology of secondary metabolite production." *Synthetic and Systems Biotechnology* 1, no. 2 (2016): 69-79. Wohlleben, Wolfgang, Yvonne Mast, Evi Stegmann, and Nadine Ziemert. "Antibiotic drug discovery." *Microbial biotechnology* 9, no. 5 (2016): 541-548.

Wu, Houbo, Wen Chen, Guanghua Wang, Shikun Dai, Danyan Zhou, Hengzhi Zhao, Yatao Guo, Yongchang Ouyang, and Xiang Li. "Culture-dependent diversity of Actinobacteria associated with seagrass (*Thalassia hemprichii*)." *African Journal of Microbiology Research* 6, no. 1 (2012): 87-94.

Xie, P., Ma, M., Rateb, M. E., Shaaban, K. A., Yu, Z., Huang, S. X., Zhao, L. X., Zhu, X., Yan, Y., Peterson, R. M., Lohman, J. R., Yang, D., Yin, M., Rudolf, J. D., Jiang, Y., Duan, Y., & Shen, B. "Biosynthetic potential-based strain prioritization for natural product discovery: a showcase for diterpenoid-producing actinomycetes." *Journal of natural products* 77, no. 2 (2014): 377-387.

Xin, Gang, Gengyun Zhang, Jun Won Kang, James T. Staley, and Sharon L. Doty. "A diazotrophic, indole-3-acetic acid-producing endophyte from wild cottonwood." *Biology and Fertility of Soils* 45, no. 6 (2009): 669-674.

Yan, Xia, Yanfang Li, Nana Wang, Yue Chen, and Li-li Huang. "*Streptomyces ginkgonis sp. nov*., an endophyte from *Ginkgo biloba*." *Antonie Van Leeuwenhoek* 111, no. 6 (2018): 891-896.

Yang, GuiLiu, LiFang Yang, MingGuo Jiang, JiaFa Wu, GuangHua Gan, Li Tuo, and ChengHang Sun. "Isolation, identification, and bioactivity of endophytic Actinomycetes from mangrove plants in Beilun River." *Journal of Agricultural Biotechnology* 23, no. 7 (2015): 894-904.

Yang, Xueqiong, Tianfeng Peng, Yabin Yang, Wei Li, Jie Xiong, Lixing Zhao, and Zhongtao Ding. "Antimicrobial and antioxidant activities of a new benzamide from endophytic *Streptomyces sp. YIM 67086*." *Natural product research* 29, no. 4 (2015): 331-335.

Zazopoulos, Emmanuel, Kexue Huang, Alfredo Staffa, Wen Liu, Brian O. Bachmann, Koichi Nonaka, Joachim Ahlert, Jon S. Thorson, Ben Shen, and Chris M. Farnet. "A genomics-guided approach for discovering and expressing cryptic metabolic pathways." *Nature biotechnology* 21, no. 2 (2003): 187-190.

Zhang, Peng, Peng-Peng Zhou, and Long-Jiang Yu. "An endophytic taxol-producing fungus from Taxus media, *Cladosporium cladosporioides MD2*." *Current microbiology* 59, no. 3 (2009): 227-232.

Zhang, XianFang, Jingjing Li, Gaofu Qi, Kai Wen, Jinzhong Lu, and Xiuyun Zhao. "Insecticidal effect of recombinant endophytic bacterium containing *Pinellia ternata* agglutinin against white backed planthopper, *Sogatella furcifera*." *Crop Protection* 30, no. 11 (2011): 1478-1484.

Zhang, Yue-Ji, Wen-Di Zhang, Sheng Qin, Guang-Kai Bian, Ke Xing, Yun-Feng Li, Cheng-Liang Cao, and Ji-Hong Jiang. "*Saccharopolyspora dendranthemae sp. nov*. a halotolerant endophytic actinomycete isolated from a coastal salt marsh plant in Jiangsu, China." *Antonie van Leeuwenhoek* 103, no. 6 (2013): 1369-1376.

Zhao, Ke, Petri Penttinen, Tongwei Guan, Jing Xiao, Qiang Chen, Jun Xu, Kristina Lindström, Lili Zhang, Xiaoping Zhang, and Gary A. Strobel. "The diversity and antimicrobial activity of endophytic actinomycetes isolated from medicinal plants in Panxi plateau, China." *Current microbiology* 62, no. 1 (2011): 182-190.

Zhao, Xiuyun, Gaofu Qi, Xianfang Zhang, Nan Lan, and Xinrong Ma. "Controlling sap-sucking insect pests with recombinant endophytes expressing plant lectin." *Nature precedings* (2010): 1-1.

Zhu, Hua, Jasper Swierstra, Changsheng Wu, Genevieve Girard, Young Hae Choi, Willem Van Wamel, Stephanie K. Sandiford, and Gilles P. van Wezel. "Eliciting antibiotics active against the ESKAPE pathogens in a collection of actinomycetes isolated from mountain soils." *Microbiology* 160, no. 8 (2014): 1714-1725.

Ziemert, Nadine, Mohammad Alanjary, and Tilmann Weber. "The evolution of genome mining in microbes–a review." *Natural product reports* 33, no. 8 (2016): 988-1005.

Chapter 7

Bio-Prospection of Endophytes and Yield Improvement Strategies for the Production of High-Value Metabolites

Vyoma Mistry[1], Siddhi Darji[1], A. Sankarnarayanan[2] and Abhishek Sharma[1,3,*]

[1]C. G. Bhakta Institute of Biotechnology, Uka Tarsadia University, Gopal-Vidyanagar, Maliba Campus, Surat, India

[2]Sri Sathya Sai University for Human Excellence, Navanihal, Okali Post, Kamalapur, Kalaburagi, Karnataka, India

[3]Department of Biotechnology and Bioengineering, Institute of Advanced Research, Koba Institutional Area, Gandhinagar, India

Abstract

Endophytes are endosymbiont bacterium and fungus species that spent at least a part of their life cycle residing and growing asymptomatically intra- and intercellularly in the plant tissues. Being in close association with plants, endophytes gain the metabolic advantages of plants and becomes the potential source of novel natural products of medicinal and pharmaceutical importance. The giant biological diversity coupled with the ability to synthesize pharmaceutically active secondary metabolites has put endophytes in the center stage of microbial cell factory research. The present review highlights the strategies of endophytes used to biosynthesize the novel biologically active natural products and bio-prospection of endophytes and yield of metabolites. Endophytes help plants to sustain biotic and abiotic stress, and biotransformation and

*Corresponding Author's Email: abhi19ind@gmail.com.

In: Endophytes
Editor: Pragya Tiwari
ISBN: 979-8-88697-045-6

contribute to plant growth, carbon sequestration, phytoremediation, and protection from heavy metal toxicity. Moreover, yield-improvement strategies for the production of high-value metabolites in endophytes have been extensively discussed.

Keywords: bio-products, endophytes, plant-endophytes association, secondary metabolites, yield improvement strategies

1. Introduction

Since ancient times, plants appeared as the most promising source of pharmaceutically important biomolecules. Paradoxically in the present scenario, the microbes associated with plants are having a high impact on the production of vital biomolecules (Subbulakshmi et al. 2012; Tiwari et al. 2021a). These microbes called endophytes - are a group of organisms that subsists inter/intracellularly in plants with no symptoms of disease to the host. De Bary, 1866 coined the term Endophytes which discriminate the epiphytic organisms located on the plant surface. Endophytes are pervasive and found almost in all species of plants (Nair and Padmavathy 2014; Tiwari and Bae 2020). The endophytes can be described taxonomically as they are of different classes like bacteria, viruses, fungi, archaea, and protozoans. They show complex interaction with their host mostly mutualism and rarely parasitism. Numerous studies show that the growth of endophytes is strictly constrained by plants but, several mechanisms of endophytes maintain their symbiosis. Endophytes benefit plants to acclimatize to the environment in several ways: 1) increase stress tolerance for plants, 2) produce plant growth-promoting molecules, 3) guard host plant from pathogens, 4) control plant growth by increasing nutrient uptake, 4) suppress weed growth (White et al. 2019).

According to fossil records, endophytes were originally associated with land plants for > 400 million years (Krings et al. 2007; Tiwari and Bae 2020). In a single host plant, its endophytes can be of one or more types (Strobel and Daisy 2003). Colonization of endophytes can be observed in various parts of the host plant like- leaves, stems, roots, fruit, buds, seeds, etc. (Hata and Sone 2008; Specian et al. 2012; Stępniewska and Kuzniar 2013). The variable population of endophytes in plants depends mostly on host type and environmental conditions (Dudeja and Giri 2014). The symbiotic interactions between organisms and their bioactive metabolites are less explored by researchers. In the last few years, studies showed that novel chemicals are been less discovered from natural sources. And on the other hand, plant sources are explored vastly to discover novel chemical entities with a curative purpose (Gauda et al. 2016). Endophyte microorganisms perform a vital role in this pursuit for bioactive curative compounds, with potential use in drug discovery (Lam 2007; Tiwari et al. 2021b; Tiwari et al. 2021c). The present chapter highlights the diversity of endophytes, their association with host plants, and yield improvement approaches for the production of pharmaceutically important biomolecules. Such studies may improve the understanding of endophytes and address the need for new and useful compounds necessary to combat various pathogens associated with human health and other possible medicinal uses.

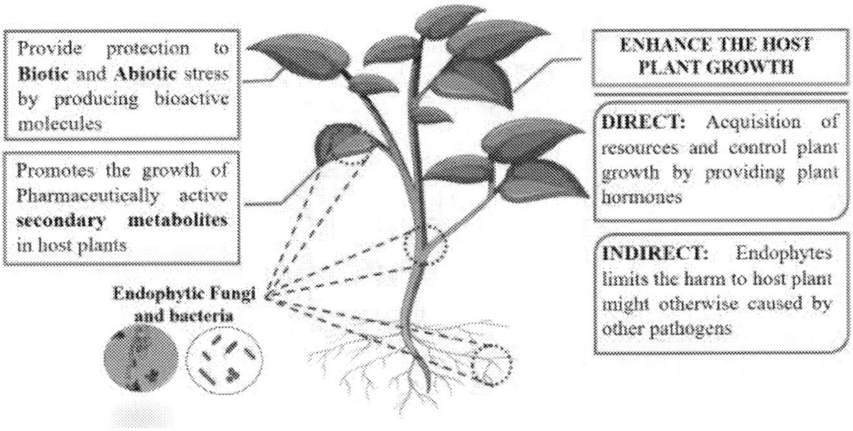

Figure 1. Plant endophytes association and their applications.

2. Endophytic Diversity

Endophytes represent huge diversity both biologically and ecologically. They can be classified as bacteria, fungi, protozoans, algae, archaea, etc. Depending on their function with the host plant they are categorized into two types- obligate and facultative. Obligate endophytes are those that depend on plant metabolism for existence, plants spread these obligate endophytes via vectors or by vertical transmission (Hardoim et al. 2008). Whereas, the facultative endophytes are those that live exterior to the host plant during a certain stage of their life cycle mostly linked with plants through the soil environment (Abreu-Tarazi et al. 2010).

Table 1. List of major bacterial and fungal endophytes benefiting their host plant in various biotic and abiotic stress conditions

Host plant	Endophytic species	Function	Reference
Bacterial endophytes			
Oryza sativa Rice	*Azoarcus* sp. *Pseudomonas stuzeri* *Azospirillium* sp. *Burkholderia* sp.	Nitrogen fixation Phytohormone secretion nif gene cluster, antifungal action, and indirect Plant growth promotion activity	Krause et al. 2006 Kaneko et al. 2011 Kwak et al. 2012
Wheat	*Klebsiella pneumoniae* *Azospirillium lipoferum*	Plant growth promotion by Nitrogen fixation	Fotus et al. 2008 Wisniewsky-Dye et al. 2011
Maize	*Klebsiella pneumoniae* *Burkholderia phytofirmans*	Nitrogen fixation IAA synthesis	Fotus et al. 2008 Weilharter et al. 2011
Pepper	*Pseudomonas* sp. *Bacillus* sp. *Enterobacter clocae*	Unknown role in Plant growth promotion	Liu et al. 2012 Arvind et al. 2009
Poplar	*Enterobacter* sp. *Stenotrophomonas maltophilia* *Serratia proteamaculans*	Sederophore, IAA, acetoin and butanediol synthesis ACC deaminase	Taghavi et al. 2009
Sugarcane	*Gluconacetobacter diazotrophicus*	Nitrogen fixation, auxin synthesis	Bertalan et al. 2009
Potato	*Variovorax paradoxus* *Burkholderia phytofirmans*	IAA synthesis ACC deaminase	Han et al. 2011 Weilharter et al. 2011
Tomato	*Brebacillus brevis* *Pseudomonas* sp.	Plant growth-promoting activity	Patel et al. 2012 Yang et al. 2011
Fungal endophytes			
Chrysanthemum morifolium	*Chaetomium globosum*, *Botrytis* sp.	Tolerance against salt stress by increased POD activity	Liu et al. 2011
Salvia miltiorrhiza	*Arbuseular mycorrhiza*	Resistance to drought increases nutrient absorption	Wang et al. 2009
Cordia alliodora	*Leucocoprinus gongylophorus*	Protection against insects by producing antagonistic chemicals	Bittleston et al. 2011
Salvia miltiorrhiza	*Arbuseular mycorrhiza*	Drought resistant	Meng and He 2011
Phragmites australis	*Choiromyces aboriginum*, *Stachybotrys elegans*, *Cylindrocarpon* sp.	Resistance against biotic stress by producing cell wall degrading enzymes	Cao et al. 2009
Oryza sativa	*Sordariomycetes* sp.	Resistance against heavy metal like Pb^{2+}	Li and Zhang 2015
Capsicum annuum	*Penicillium resedanum*	Protection against heat stress	Jia et al. 2016

In this chapter, the efforts are made to highlight major bacterial and fungal endophytes that are of advantage to their host plant and guard them against various abiotic and biotic stress conditions (Table 1). Bacterial endophytes are the most prominent and contain a diverse group of organisms from more than 200 genera. Amongst them, the dominant genera include *Probacteria*, *Actinobacteria*, and *Firmicutes* (Golinska et al. 2015). The biological and ecological diversity of bacterial endophytes ranges from gram-negative to gram-positive organisms. Active biomolecules produced by bacterial endophytes include mostly antimicrobial and anticancer and more than 75% are from single genera of *Streptomyces* sp. (Gauda et al. 2015; Tiwari et al. 2022).

The fungal endophytes are classified into four groups based on taxonomy and six other criteria: host range, host tissues colonized, *in planta* colonization, mode of transmission, and *in planta* biodiversity. These four groups are divided into clavicipitaceous endophytes (Class 1) and non-clavicipitaceous endophytes (Class 2, 3, and 4). The clavicipitaceous, which infects some grasses confined to cool regions, and the non-clavicipitaceous endophytes, which are from asymptomatic tissues of non-vascular plants, ferns and allies, conifers, and angiosperms and are limited to the *Ascomycota* or *Basidiomycota* group (Jalgaonwala et al. 2011; Bhardwaj and Agrawal 2014).

3. Endophyte – Plant Associations and Applications

Endophytic organisms can invade the host plant tissue at any stage of their life cycle and helps them with biotic and abiotic stress tolerance. Rhizospheric bacteria that live in close contact with plants have the potential to invade the plant tissue and it is the primary source of colonizing endophytes. Fungal endophytes have several mechanisms for plant growth development. Initially, fungi produce hormones that enhance the host plant's growth. On the other hand, some fungal endophytes produce bioactive molecules like flavonoids, alkaloids, and terpenoids. These bioactive compounds increase resistance to biotic and abiotic stress and some would enhance the production of secondary metabolites, produced by both endophytes and host plant species. By all these mechanisms endophyte are associated with plants and help plants to sustain biotic and abiotic stress, biotransformation, carbon sequestration, phytoremediation, and protection from heavy metal toxicity.

Endophytic organisms promote plant growth directly or indirectly by various mechanisms. Direct promotion of plant growth happens when either

(i) the plant growth-promoting endophytes facilitates the acquisition of resources from the environment including phosphorous, nitrogen and iron; or (ii) control plant growth by providing various plant hormones like auxin, cytokinin, ethylene. Indirect promotion of plant growth by PGPB occurs when an organism limits the harm to host plants caused by various pathogenic agents including nematodes, fungi, and bacteria (Santoyo et al. 2016). Some endophytes help to enhance the resistance of host plants against biotic and abiotic stresses through the production of bioactive molecules (Nejad and Johnson 2000). In symbiotic stress tolerance, the fungal endophyte performs a biological trigger that stimulates the defense systems of the host plant. As a result, the activated defense mechanism in plants can sustain against stress conditions and in the case of pathogenic organisms direct competition occurs between endophytes and pathogens. These bioactive molecules can be used as therapeutic compounds against particular pathogens. The chapter focuses on the bio-prospection of endophytes to improve the yield of pharmaceutically important metabolites.

Still, plants are a major source of medicine, but the over-harvesting, endangered scenario, and extinction of medicinal plants define some major challenges (Brower 2008; Urban 2015). Hence, it is necessary to find out alternative resources that can save by increasing desirable benefits. Being in close association with the plants, endophytes gain the metabolic advantage of plants. Henceforth, endophytes have attracted the researcher for their potential and prospects as a valuable source of secondary metabolites, plant hormones (phytohormones), confers biotic-abiotic stress, metabolic advantages, and plant growth promotion, among others.

3.1. Secondary Metabolites

Plants produce bioactive compounds as a defense response to environmental stress known as "secondary metabolites." They are pharmaceutical active secondary metabolites that have the highest role in various medicinal applications. Flavonoids, steroids, peptides, quinols, alkaloids, terpenoids, phenols, polyketones, podophyllotoxin, camptothecin, vincristine, and vinblastine are the example of the plant secondary metabolites which have been identified as anticancer, antimicrobial, antioxidant, immunosuppressive, anti-inflammatory (Korkina 2007). Stierle et al. (1993) have mentioned the first report on the fungal endophyte *Taxomyces andreanae* which produces taxol. This fungus was isolated from *Taxus brevifolia* in the early 1990s. Culture supernatant of endophytes increased the production of secondary metabolites in comparison to control (Li and Tao 2009) evidenced by increased secondary metabolites production and becoming plant resistant when co-cultured with endophytes. Endophytes reduced the growth of fungi by triggering the hydrolase (hydroxylation) for the plant cell through endophyte colonization. Endophytes effectively reduced the attack of the pathogens by secreting the elicitors like glycoprotein, polysaccharides, and lipopolysaccharides that prompt plant defense mechanisms and increase secretion of plant secondary metabolites. However, there were tapered data on which endophytes can survive in the host plant when an enormous range of secondary metabolites is produced (Gao et al. 2010; Fadiji and Babalola 2020).

3.2. Plant Growth and Its Mechanisms

For plant growth, endophytes colonize the internal tissue of every part of the plant (Santoyo et al. 2016). Endophytes can enter through various ways in various parts. For example, if it has to enter the leaves or young steam it goes from the stomata (Roos and Hattingh 1983; Santoyo et al. 2016). These endophytes help in the growth of the plants through various mechanisms like phytostimulation, phytoimmobilization, phytostabilization, phytotransformation, phytovolatilization, phytofilterastion, biofertilization and biocontrol (Conesa et al. 2012; Sahoo et al. 2017). Each of the mechanism terms is described in the below (Table 2).

Table 2. Various mechanism and description of the endophytes for the promotion of plant growth

Mechanism	Description	References
Phytostimulation	Upregulation of phytohormnes (Plant hormones or bioactive compounds)	Bloemberg and Lugtenberg 2001; Sahoo et al. 2017
Phytoimmobilization	Removing of the contamination and further accelerated with the symbiotic endophytes	Arthur et al. 2005; Ma et al. 2011; Perez-de-Mora et al. 2011
Phytostabilization	Control of the contaminant in the soil through the accumultation/adsorption onto roots	Defined by USEPA

Table 2. (Continued)

Mechanism	Description	References
Phytotransformation/ Phytodegradation	Degradation of the organic compounds either in plants or through the compounds like enzymes in simple molecular form that could be used by plant tissue further if it's necessary	Newman and Reynolds et al. 2004; Etim 2012; defined by USEPA
Phytovolatilization	Releasing of the contaminants in an atmosphere (Volatile form)	Schiavon and Pilon-Smits 2017
Phtofilterasation	It anchorage phytocompounds, contaminants or nutrients to absorbed or bound to non-living surface (like roots, seeds) through the help of certain surfaces	Conesa et al. 2012
Biofertilization	By establish the symbiotic relationship its co-ordinate with the nutrients and minerals and enhance the absorption without altering the nature of soil.	Roychowdhury et al. 2017
Bio control	Protection from the phytopathogens	Eljounaidi et al. 2016

Since the past years, researchers are working on isolation of the bioactive compounds from the medicinal plant with the aim of using drugs or templates which can be further modified (Balunas and Kinghorn 2005; Atanasov et al. 2015; Chen et al. 2015). 15,000 out of 50,000-80,000 plant species are documented for medicinal use (Cordell and Colvard 2005). However, from several years, plant-based bioactive medicinal compounds brings numerous disadvantages that is extremely low yields which also vary with environmental conditions, over-harvesting leads to the limited supply and extremely complex structure with unknown mechanisms give the challenges to synthesis or semi-synthesis of bioactive molecules (Venugopalan and Srivastava 2015; Tiwari et al. 2021b). Hence, there is vital need to discover alternative routes that can protect by increasing in desirable benefit. There are numeral approaches to increase the yield of bioactive secondary metabolites like, genetically modification of the plants, transgenic cell lines, pathways engineering, cell suspension culture, hairy root cultures, callus culture, plant metabolic engineering by overexpression of the enzymes, increasing the activity of rate-limiting enzymes. While the genome instability, optimization of suitable cell lines, low yield, and scale-up are some difficulties; except for some cases like taxol production by phyton biotech, shikonin, ginseng, and berberine production by Mitsui Chemicals are the limitations with the plant cell, tissue culture, cell lines, hairy root culture, suspension cultures and callus culture (Linden 2006). Jalgaonwala et al. (2011) and Omojate godstime et al. (2014) have reported that endophytic microorganisms are the storehouse of bioactive secondary metabolites that have anti-cancer, anti-diabetic, anti-microbial, anti-arthritic, anti-insect, and immunosuppressant activities. These studies defined new possibilities and fulfilled the demands of life-saving drugs (Tiwari et al. 2022). Moreover, endophytes simplify the bioprocessing and evolved novel resistance mechanisms to toxic production (Mu et al. 1999; Kusari et al. 2011a; Gouda et al. 2016). Hence, the endophytic microorganisms are the pre-eminent alternative to overcome drug resistance in various pathogenic microorganisms (Godstime et al. 2014). Bioactive compounds such as vinblastine, paclitaxel, podophyllotoxin, hypericin, camptothecin have been commercially produced by various fungal endophytes in the respective plants (Joseph and Priya 2011; Zhao et al. 2011a; Tiwari et al. 2022).

4. Yield Improvement Strategies for Pharmaceutical Metabolites

Genetic and metabolic engineering and/or bioprocessing strategies may be used separately or in combination for enhancing the production of high-value metabolites. There are various strategies used for yield improvement and discussed as follows.

4.1. Strain Improvement

Strain improvement provides ultimate opportunities in cost reduction without affecting the significant capital outlay. Microbial production strains are the heart of the fermentation industry (Parekh et al. 2000). Strain improvement is not only for the yield enhancement, but it also improves other characteristics like removal of unwanted metabolites, improved utilization of carbon-nitrogen sources, improved oxygen transfer in the fermenter (Barrios-Gonzalez et al. 2003; Aarthi and Smita 2015). Strain improvement can be performed by the classical genetic method or the molecular genetic method.

4.1.1. Classical Genetic Method

In this method, the strain development process relies on mutation and is followed by random/rational screening, which can be further tested and performed using new improved mutants. The mutation can be employed physically (UV, X-ray, γ-rays, laser, microwaves, ion beams, etc.) or chemically (ethyl methyl sulfonate (EMS), Sodium Nitrite ($NaNO_2$), Diethyl sulphate (DES), Nitrosoguanidine (NTG)). This experimental approach has a long history of achievement such as Taxol production by Hygromycin resistant mutants (Kai et al. 2009). Mutagenesis in fungal endophytes is performed by protoplast mutagenesis or fusion, mycelium, or spores. Mycelium mutagenesis has the disadvantage of genetic separation in offspring and is difficult to control, whereas protoplast mutation does not have this limitation. Hence, protoplast mutagenesis is the foremost for yield improvement in fungal endophyte (Kai et al. 2009). Zhao et al. (2011) obtained a high level of taxol using HDF-68 strain. This HDF-68 strains were obtained by protoplast fusion of $UV_{40\text{-}19}$ and $UV_{50\text{-}6}$ *Nodulisporium sylviforme* strains.

4.1.2. Molecular Genetic Method

Biochemical and molecular genetic tools (for identify biosynthetic pathways, adequate vectors, and effective transformation) are used followed by cloning and transformation. Efficiently fungal DNA transformation is not yet known. However, the research on efficient fungal DNA transformation will give a strong platform on the regulation of genes and insertion of key biosynthetic genes/pathways. Recently published reports of genetic manipulation are mostly on the taxol producing endophytes. Restriction enzyme-mediated integration increases the transformation rates comparable to conventional transformation (Bölker et al. 1995; Manivasakam et al. 2001). Spiering et al. (2005) performed a gene-silencing technique, RNAi in fungal endophytes. Liu et al. (2013) reported *Agrobacterium* tumefaciens-mediated genetic transformation (ATMT) for taxol production. Taxol production was not only reported in fungal endophytes but also in *E. coli* (Ajikumar et al. 2010) and yeast episomal vectors (Dejong et al. 2006).

4.2. Bioprocess Optimization

Bioprocess optimization is well-developed in plant cell or tissue culture cultivation whereas researchers are still working on bioprocess optimization through various strategies in endophytes (Aarthi and Smita 2015).

4.2.1. Optimization of Fermentation Parameters

Every microorganism has its own parameters for cultivation. Therefore, it is very important to provide or optimize the correct parameters like media composition, pH, temperature, photoperiod, etc. Bode et al. (2002) invented OSMAC (One strain many compounds) where they describe the single strain can also produce various compounds by changing parameters. Henceforth, statistic and single factor optimization using variable parameters can provide the paramount output. Grond et al. (2002) and Wang et al. (2014a) stated that the OSMAC approach has been directed to the discovery of novel metabolites.

4.2.2. Elicitor/Inhibition Addition

Elicitor is the signaling molecule that triggers the formation of bio-compound/secondary metabolite. Badi et al. (2009) reported that in cell culture elicitor triggers the secondary metabolites in plant defense, hypersensitive response, and/or pathogen-related proteins. The mechanism and effective strategies using elicitor/inhibition are well established in plant cell culture (Radman et al. 2003; Zhao et al. 2005a; Namdeo 2007; Vasconsuelo and Boland 2007). In the case of endophytes, fungal endophytes are also widespread. Interestingly fungal endophytes have seen significant enhancement in the product yield by ethanol-like small molecules which can support the idea that how elicitation can help in cost-effective, high product yield in a small and large scale of fungal fermentation (Zhao et al. 2013c; Venugopalan and Srivastava 2015). In addition to the elicitors, inhibitors can be used. This inhibition will help in diverting the efflux towards the product of interest. In studies on taxol production by endophytes in 1985, large amounts of ergosterol were also produced. Li et al. (1998b) used sterol inhibitors which diverted the pool towards the taxol production and reported 50 fold product enhancement.

4.2.3. Precursor Feeding

To increase the desirable products in plant cell culture/tissue culture increasing biosynthetic precursor supply in the initial on intermediate steps will lead to increase in desirable product yield in the culture media. With increasing the precursor feeding reaction rate, yield production, and toxicity in cells are increased (Mulabagal and Tsay 2004). Hence, the concentration of precursor feeding needs to be optimized. These days precursor feeding is explored in fungal endophytes (Gaosheng and Jingming 2012). Taxol and camptothecin biosynthesis is well studied in fungal endophytes. Feeding of sodium benzonate has a positive effect in *Pestalotiopsis microspore Ne32* taxol production (Li et al. 1998b) and feeding of tryptamine and tryptophan has a positive effect in *Trichoderma atroviride* LY357 and *Entrophosppora infrequens* RJMEF001 respectively (Amna et al. 2012; Pu et al. 2013).

4.2.4. Use of Adsorbent Resin/Solid Supports

Toxicity of the product affects the growth of an organism. To decrease the toxicity and increase the yield *in situ* product removal (ISPR) is used. It's an integrated bioprocess of product recovery, elimination of feedback inhibition, avoiding product auto-toxicity, reducing cost, yield improvement of product, and overcoming product degradation (Schügerl and Hubbuch 2005; Luo et al. 2014). ISPR can be employed in two ways either with liquid solvent or adsorption with a solid sorbent (Xu et al. 2009). Recently, resin has been used extensively in fungal endophyte fermentation as it has enhanced novel bioactive molecules (González-Menéndez et al. 2014). Polystyrene resin DM-301 have used in *Hyalodendriella* sp. *ponipodef 12* which enhanced the yield of botrallin and TMC-264, 2.29 and 11.76 folds respectively (Luo et al. 2014). The presence of solid support enhance the production and physiology of culture as solid supports mimic natural fungal habits (Bigelis et al. 2006). These strategies have been successfully applied in endophyte strains (Thammajaruk et al. 2011).

4.2.5. Strategies for Genomic Stability

Bioprocess optimization can improve the yield of production with time, genomic instability and silencing of the biosynthetic genes hinders the process in endophytes. Hence, it is very vital to retain the genomic stability of endophytes. This problem can be solved by employing co-cultivation and epigenetic modifiers (Aarthi and Smita 2015).

4.2.6. Co-Cultivation and Mixed Fermentation

Endophytes when growing in an axenic culture no longer have access to specialized microenvironment of the host plant, which may lead to silencing the genes of bioproduct. Maintaining specialized microenvironment conditions like the host plant is not possible which can reduce the production of novel metabolites. The slight change in the conditions affect the metabolites production in the microbes. In the case of endophytes, survival is delicately balanced in relation to the host plants. Paranagama et al. (2007) reported just changing the tap water while making the media with the distilled water activate six novel secondary metabolites. The use of co-cultivation or mixed fermentation will give novel paths like OSMAC strategies that can enable the natural habit of endophytes via mimicking the host microenvironment. Literature studies are required to understand the interaction of various endophytes with the host plant. Co-cultivation process optimization is a very challenging part. Growth balance has to be optimized in such a way that one growth does not overpower the other growth (Aarthi and Smita 2015). In 2.7 and 7.8 fold increase in taxol production reported while co-culturing the *Paraconithyrium* strain with *Alternaria* species and *Alternaria* and *Phmopsis* species (Soliman and Raizada 2013) whereas 38 fold taxol production increase has seen when co-culturing the *Taxus chinensis var. mairei* with *Fusarium mairei* (Li et al. 2009). Hence, it proves that mixed fermentation is one of the powerful strategies to enhance yield of high-value bioactive metabolites.

4.2.7. Use of Epigenetic Modifiers

Chromosomal genes are controlled by epigenetic regulation. Regulation of expression and silencing of the genes are controlled by histone deacetylation and DNA methylation (Pettit 2010). Hence, there is a possibility of using DNA methyltransferase (DNMT) and histone deacetylase (HDAC) inhibitors to unlock the genes. Sodium butyrate is an HDAC inhibitor. In *Leucostoma persooni* a new cytosporones R with known cystosporones B, C, and E was produced on addition of sodium butyrate (Beau et al. 2012).

Conclusion

Endophytes produce a higher yield of metabolites in fewer days of growth in a cost-effective manner, compared to plant cells and tissue culture. Hereafter, transparently we can conclude that the endophytes are a major source of bioactive compounds in the future. The knowledge and strategies to improve and tailor endophytes are obtainable in the pieces of a puzzle. In the future endophyte highlights a very promising tool for the production of high-value natural products. However, there are numerous reports available on production of bio-products but there are no reports on the production in a mass scale and commercialization. The reason for this is hidden in the genes and plant-endophytes interaction and mechanism of the function. Hence, much greater understanding and optimization are required.

Acknowledgments

The authors acknowledge their institutions for encouragement and support.

Conflict of Interests

No conflict of interests was declared.

References

Abreu-Tarazi, Monita F., Acácio A. Navarrete, Fernando D. Andreote, Cristina V. Almeida, Siu M. Tsai, and Marcílio Almeida. "Endophytic bacteria in long-term in vitro cultivated "axenic" pineapple microplants revealed by PCR–DGGE." *World Journal of Microbiology and Biotechnology* 26, no. 3 (2010): 555-560.

Ajikumar, Parayil Kumaran, Wen-Hai Xiao, Keith E. J. Tyo, Yong Wang, Fritz Simeon, Effendi Leonard, Oliver Mucha, Too Heng Phon, Blaine Pfeifer, and Gregory Stephanopoulos. "Isoprenoid pathway optimization for Taxol precursor overproduction in *Escherichia coli*." *Science* 330, no. 6000 (2010): 70-74.

Amna, Touseef, Musarat Amina, P. R. Sharma, S. C. Puri, Hanan M. Al-Youssef, Areej M. Al-Taweel, and G. N. Qazi. "Effect of precursors feeding and media manipulation on production of novel anticancer pro-drug camptothecin from endophytic fungus." *Brazilian Journal of Microbiology* 43 (2012): 1476-1489.

Aravind, R., Dinu Antony, Santosh J. Eapen, A. Kumar, and K. V. Ramana. "Isolation and evaluation of endophytic bacteria against plant parasitic nematodes infesting black pepper (*Piper nigrum* L.)." *Indian Journal of Nematology* 39, no. 2 (2009): 211-217.

Arthur, Ellen L., Pamela J. Rice, Patricia J. Rice, Todd A. Anderson, Sadika M. Baladi, Keri L. D. Henderson, and Joel R. Coats. "Phytoremediation-an overview." *Critical Reviews in Plant Sciences* 24, no. 2 (2005): 109-122.

Atanasov, A. G., Waltenberger, B., Pferschy-Wenzig, E.-M., Linder, T., Wawrosch, C., Uhrin, P., Temml, V., Wang, L., Schwaiger, S., Heiss, E. H., Rollinger, J. M., Schuster, D., Breuss, J. M., Bochkov, V., Mihovilovic, M. D., Kopp, B., Bauer, R., Dirsch, V. M., & Stuppner, H. "Discovery and resupply of pharmacologically active plant-derived natural products: A review." *Biotechnology advances* 33, no. 8 (2015): 1582-1614.

Baldi, A., A. K. Srivastava, and V. S. Bisaria. "Fungal elicitors for enhanced production of secondary metabolites in plant cell suspension cultures." In *Symbiotic fungi*, pp. 373-380. Springer, Berlin, Heidelberg, 2009.

Balunas, Marcy J., and A. Douglas Kinghorn. "Drug discovery from medicinal plants." *Life sciences* 78, no. 5 (2005): 431-441.

Beau, Jeremy, Nida Mahid, Whittney N. Burda, Lacey Harrington, Lindsey N. Shaw, Tina Mutka, Dennis E. Kyle, Betty Barisic, Alberto Van Olphen, and Bill J. Baker. Epigenetic tailoring for the production of anti-infective cytosporones from the marine fungus *Leucostoma persoonii*." *Marine Drugs* 10, no. 4 (2012): 762-774.

Bertalan, M., Albano, R., de Pádua, V., Rouws, L., Rojas, C., Hemerly, A., Teixeira, K., Schwab, S., Araujo, J., Oliveira, A., França, L., Magalhães, V., Alquéres, S., Cardoso, A., Almeida, W., Loureiro, M. M., Nogueira, E., Cidade, D., Oliveira, D., Ferreira, P. C. "Complete genome sequence of the sugarcane nitrogen-fixing endophyte *Gluconacetobacter diazotrophicus Pal5*." *BMC genomics* 10, no. 1 (2009): 1-17.

Bhardwaj, Akanksha, and Pavan Agrawal. "A review fungal endophytes: as a store house of bioactive compound." *World J. Pharm. Pharm. Sci* 3, no. 9 (2014).

Bigelis, Ramunas, Haiyin He, Hui Y. Yang, Li-Ping Chang, and Michael Greenstein. "Production of fungal antibiotics using polymeric solid supports in solid-state and liquid fermentation." *Journal of Industrial Microbiology and Biotechnology* 33, no. 10 (2006): 815-826.

Bittleston, L. S., F. Brockmann, W. Wcislo, and Sunshine A. Van Bael. "Endophytic fungi reduce leaf-cutting ant damage to seedlings." *Biology Letters* 7, no. 1 (2011): 30-32.

Bloemberg, Guido V., and Ben J. J. Lugtenberg. "Molecular basis of plant growth promotion and biocontrol by rhizobacteria." *Current opinion in plant biology* 4, no. 4 (2001): 343-350.

Bode, Helge Björn, Barbara Bethe, Regina Höfs, and Axel Zeeck. "Big effects from small changes: possible ways to explore nature's chemical diversity." *Chem Bio Chem* 3, no. 7 (2002): 619-627.

Bölker, Michael, Heidi U. Böhnert, Karl Heinz Braun, Johannes Görl, and Regine Kahmann. "Tagging pathogenicity genes in Ustilago maydis by restriction enzyme-mediated integration (REMI)." *Molecular and General Genetics MGG* 248, no. 5 (1995): 547-552.

Brower, V. "Back to Nature: Extinction of Medicinal Plants Threatens Drug Discovery." *JNCI Journal of the National Cancer Institute* 100, no. 12 (June 18, 2008): 838–39.

Cao, Ronghua, Xiaoguang Liu, Kexiang Gao, Kurt Mendgen, Zhensheng Kang, Jianfeng Gao, Yang Dai, and Xue Wang. "Mycoparasitism of endophytic fungi isolated from reed on soilborne phytopathogenic fungi and production of cell wall-degrading enzymes in vitro." *Current microbiology* 59, no. 6 (2009): 584-592.

Chen, Jichao, Wenlong Li, Hequan Yao, and Jinyi Xu. "Insights into drug discovery from natural products through structural modification." *Fitoterapia* 103 (2015): 231-241.

Conesa, Héctor M., Michael W. H. Evangelou, Brett H. Robinson, and Rainer Schulin. "A critical view of current state of phytotechnologies to remediate soils: still a promising tool?" *The Scientific World Journal* 2012 (2012).

Cordell, Geoffrey A., and Michael D. Colvard. "Some thoughts on the future of ethnopharmacology." *Journal of ethnopharmacology* 100, no. 1-2 (2005): 5-14.

DeJong, JingHong M., Yule Liu, Arthur P. Bollon, Robert M. Long, Stefan Jennewein, David Williams, and Rodney B. Croteau. "Genetic engineering of taxol biosynthetic genes in Saccharomyces cerevisiae." *Biotechnology and bioengineering* 93, no. 2 (2006): 212-224.

Dudeja, Surjit Singh, and Rupa Giri. "Beneficial properties, colonization, establishment and molecular diversity of endophytic bacteria in legumes and non-legumes." *African journal of microbiology research* 8, no. 15 (2014): 1562-1572.

Eljounaidi, Kaouthar, Seung Kyu Lee, and Hanhong Bae. "Bacterial endophytes as potential biocontrol agents of vascular wilt diseases–review and future prospects." *Biological Control* 103 (2016): 62-68.

Etim, E. E. "Phytoremediation and its mechanisms: a review." *International Journal of Environment and Bioenergy* 2, no. 3 (2012): 120-136.

Fadiji, Ayomide Emmanuel, and Olubukola Oluranti Babalola. "Elucidating mechanisms of endophytes used in plant protection and other bioactivities with multifunctional prospects." *Frontiers in Bioengineering and Biotechnology* 8 (2020): 467.

Fouts, D. E., Tyler, H. L., DeBoy, R. T., Daugherty, S., Ren, Q., Badger, J. H., Durkin, A. S., Huot, H., Shrivastava, S., Kothari, S., Dodson, R. J., Mohamoud, Y., Khouri, H., Roesch, L. F. W., Krogfelt, K. A., Struve, C., Triplett, E. W., & Methé, B. A. "Complete genome sequence of the N2-fixing broad host range endophyte *Klebsiella pneumoniae 342* and virulence predictions verified in mice." *PLoS genetics* 4, no. 7 (2008): e1000141.

Gao, Fu-kang, Chuan-chao Dai, and Xiao-zhen Liu. "Mechanisms of fungal endophytes in plant protection against pathogens." *African Journal of Microbiology Research* 4, no. 13 (2010): 1346-1351.

Gaosheng, H., & Jingming, J. Production of useful secondary metabolites through regulation of biosynthetic pathway in cell and tissue suspension culture of medicinal plants. *Recent advances in plant in vitro culture*, 10 (2012): 53038.

Golinska, Patrycja, Magdalena Wypij, Gauravi Agarkar, Dnyaneshwar Rathod, Hanna Dahm, and Mahendra Rai. "Endophytic actinobacteria of medicinal plants: diversity and bioactivity." *Antonie Van Leeuwenhoek* 108, no. 2 (2015): 267-289.

Gonzalez, J. Barrios, F. J. Fernandez, and A. Tomasini. "Microbial secondary metabolites production and strain improvement." *Indian Journal of Biotechnology,* Vol 2, July 2003, pp 322-333.

González-Menéndez, V., Asensio, F., Moreno, C., de Pedro, N., Monteiro, M. C., de la Cruz, M., Vicente, F., Bills, G. F., Reyes, F., Genilloud, O., and Tormo, J. R. "Assessing the effects of adsorptive polymeric resin additions on fungal secondary metabolite chemical diversity." *Mycology* 5, no. 3 (2014): 179-191.

Gouda, Sushanto, Gitishree Das, Sandeep K. Sen, Han-Seung Shin, and Jayanta Kumar Patra. "Endophytes: a treasure house of bioactive compounds of medicinal importance." *Frontiers in microbiology* 7 (2016): 1538.

Grond, Stephanie, Ina Papastavrou, and Axel Zeeck. "Novel α-L-Rhamnopyranosides from a Single Strain of *Streptomyces* by Supplement-Induced Biosynthetic Steps." *European Journal of Organic Chemistry* 2002, no. 19 (2002): 3237-3242.

Han, J.-I., Choi, H.-K., Lee, S.-W., Orwin, P. M., Kim, J., LaRoe, S. L., Kim, T., O'Neil, J., Leadbetter, J. R., Lee, S. Y., Hur, C.-G., Spain, J. C., Ovchinnikova, G., Goodwin, L., & Han, C. "Complete genome sequence of the metabolically versatile plant growth-promoting endophyte *Variovorax paradoxus S110*." *Journal of bacteriology* 193, no. 5 (2011): 1183-1190.

Hardoim, Pablo R., Leo S. van Overbeek, and Jan Dirk van Elsas. "Properties of bacterial endophytes and their proposed role in plant growth." *Trends in microbiology* 16, no. 10 (2008): 463-471.

Hata, Kunihiko, and Koichi Sone. "Isolation of endophytes from leaves of *Neolitsea sericea* in broadleaf and conifer stands." *Mycoscience* 49, no. 4 (2008): 229-232.

Jalgaonwala, Ruby Erach, Bhavna Vishwas Mohite, and Raghunath Totaram Mahajan. "A review: natural products from plant associated endophytic fungi." *J Microbiol Biotechnol Res* 1, no. 2 (2011): 21-32.

Jia, Min, Ling Chen, Hai-Liang Xin, Cheng-Jian Zheng, Khalid Rahman, Ting Han, and Lu-Ping Qin. "A friendly relationship between endophytic fungi and medicinal plants: a systematic review." *Frontiers in microbiology* 7 (2016): 906.

Joseph, Baby, and R. Mini Priya. "Bioactive Compounds from Endophytes and their Potential in." *Am J Biochem Mol Biol* 1, no. 3 (2011): 291-309.

Kai, Z., Qingshen, S., Yanjun, Z., Wenxiang, P., Tao, J., Dongpo, Z. "Screening and characterization of a high taxol producing fungus by protoplast mutagenesis." *High Technol. Lett.* 15(2009):220–226.

Kai, Zhao, Wang Xuan, Sun Yushi, Wang Ying, Ping Wenxiang, and Zhou Dongpo. "Screening of high taxol producing fungi by NTG combining with UV mutagenesis [J]." *Journal of Natural Science of Heilongjiang University* 1 (2008).

Kaneko, T., Minamisawa, K., Isawa, T., Nakatsukasa, H., Mitsui, H., Kawaharada, Y., Nakamura, Y., Watanabe, A., Kawashima, K., Ono, A., Shimizu, Y., Takahashi, C., Minami, C., Fujishiro, T., Kohara, M., Katoh, M., Nakazaki, N., Nakayama, S., Yamada, M., Sato, S. "Complete genomic structure of the cultivated rice endophyte *Azospirillum sp. B510*." *DNA research* 17, no. 1 (2010): 37-50.

Korkina, L. G. "Phenylpropanoids as naturally occurring antioxidants: from plant defense to human health." *Cellular and molecular biology* 53, no. 1 (2007): 15-25.

Krause, A., Ramakumar, A., Bartels, D., Battistoni, F., Bekel, T., Boch, J., Böhm, M., Friedrich, F., Hurek, T., Krause, L., Linke, B., McHardy, A. C., Sarkar, A., Schneiker, S., Syed, A. A., Thauer, R., Vorhölter, F.-J., Weidner, S., Pühler, A., Goesmann, A. "Complete genome of the mutualistic, N 2-fixing grass endophyte *Azoarcus* sp. *strain BH72*." *Nature biotechnology* 24, no. 11 (2006): 1384-1390.

Krings, Michael, Thomas N. Taylor, Hagen Hass, Hans Kerp, Nora Dotzler, and Elizabeth J. Hermsen. "Fungal endophytes in a 400-million-yr-old land plant: infection pathways, spatial distribution, and host responses." *New Phytologist* 174, no. 3 (2007): 648-657.

Kusari, Souvik, Ján Košuth, Eva Čellárová, and Michael Spiteller. "Survival-strategies of endophytic *Fusarium solani* against indigenous camptothecin biosynthesis." *Fungal Ecology* 4, no. 3 (2011): 219-223.

Kwak, M.-J., Song, J. Y., Kim, S.-Y., Jeong, H., Kang, S. G., Kim, B. K., Kwon, S.-K., Lee, C. H., Yu, D. S., Park, S.-H., & Kim, J. F. "Complete genome sequence of the endophytic bacterium *Burkholderia sp. strain KJ006*." (2012): 4432-4433.

Lam, Kin S. "New aspects of natural products in drug discovery." *Trends in microbiology* 15, no. 6 (2007): 279-289.

Li, Jia-Yao, Rajinder S. Sidhu, A. R. T. Bollon, and Gary A. Strobel. "Stimulation of taxol production in liquid cultures of *Pestalotiopsis microspora*." *Mycological Research* 102, no. 4 (1998)(b): 461-464.

Li, Xuemei, and Lihong Zhang. "Endophytic infection alleviates Pb^{2+} stress effects on photosystem II functioning of *Oryza sativa* leaves." *Journal of hazardous materials* 295 (2015): 79-85.

Li, Yong-Cheng, and Wen-Yi Tao. "Paclitaxel-producing fungal endophyte stimulates the accumulation of taxoids in suspension cultures of *Taxus cuspidate*." *Scientia horticulturae* 121, no. 1 (2009): 97-102.

Linden, James C. "Secondary products from plant tissue culture." *Biotechnology* 6 (2006): 1-9.

Liu, L., Y. M. Wei, X. W. Zhou, J. Lin, X. F. Sun, and K. X. Tang. "*Agrobacterium tumefaciens*-mediated genetic transformation of the Taxol-producing endophytic fungus *Ozonium sp EFY21*." *Genetics and Molecular Research* 12, no. 3 (2013): 2913-2922.

Liu, Yang, Shan Zuo, Liwen Xu, Yuanyuan Zou, and Wei Song. "Study on diversity of endophytic bacterial communities in seeds of hybrid maize and their parental lines." *Archives of microbiology* 194, no. 12 (2012): 1001-1012.

Luo, Haiyu, Hongwei Liu, Yuheng Cao, Dan Xu, Ziling Mao, Yan Mou, Jiajia Meng, Daowan Lai, Yang Liu, and Ligang Zhou. "Enhanced production of botrallin and TMC-264 with in situ macroporous resin adsorption in mycelial liquid culture of the endophytic fungus *Hyalodendriella sp. Ponipodef 12*." *Molecules* 19, no. 9 (2014): 14221-14234.

Luo, S., Chen, L., Chen, J., Xiao, X., Xu, T., Wan, Y., Rao, C., Liu, C., Liu, Y., Lai, C., & Zeng, G. "Analysis and characterization of cultivable heavy metal-resistant bacterial endophytes isolated from Cd-hyperaccumulator *Solanum nigrum* L. and their potential use for phytoremediation." *Chemosphere* 85, no. 7 (2011): 1130-1138.

Ma, Ying, M. N. V. Prasad, M. Rajkumar, and H. J. B. A. Freitas. "Plant growth promoting rhizobacteria and endophytes accelerate phytoremediation of metalliferous soils." *Biotechnology advances* 29, no. 2 (2011): 248-258.

Manivasakam, Palaniyandi, Jiri Aubrecht, Samy Sidhom, and Robert H. Schiestl. "Restriction enzymes increase efficiencies of illegitimate DNA integration but decrease homologous integration in mammalian cells." *Nucleic acids research* 29, no. 23 (2001): 4826-4833.

Meng, Jing-jing, and Xue-li H. E. "Effects of AM fungi on growth and nutritional contents of Salvia miltiorrhiza Bge. under drought stress [J]." *Journal of Agricultural University of Hebei* 1 (2011).

Mu, J. H., A. P. Bollon, and R. S. Sidhu. "Analysis of β-tubulin cDNAs from taxol-resistant *Pestalotiopsis microspora* and taxol-sensitive *Pythium ultimum* and comparison of the taxol-binding properties of their products." *Molecular and General Genetics MGG* 262, no. 4 (1999): 857-868.

Mulabagal, Vanisree, and Hsin-Sheng Tsay. "Plant cell cultures-an alternative and efficient source for the production of biologically important secondary metabolites." *International journal of applied science and engineering* 2, no. 1 (2004): 29-48.

Nair, Dhanya N., and S. Padmavathy. "Impact of endophytic microorganisms on plants, environment and humans." *The Scientific World Journal* 2014 (2014).

Namdeo, A. G. "Plant cell elicitation for production of secondary metabolites: a review." *Pharmacogn Rev* 1, no. 1 (2007): 69-79.

Nejad, Pajand, and Paul A. Johnson. "Endophytic bacteria induce growth promotion and wilt disease suppression in oilseed rape and tomato." *Biological control* 18, no. 3 (2000): 208-215.

Newman, Lee A., and Charles M. Reynolds. "Phytodegradation of organic compounds." *Current opinion in Biotechnology* 15, no. 3 (2004): 225-230.

Omojate Godstime, C., O. Enwa Felix, O. Jewo Augustina, and O. Eze Christopher. "Mechanisms of antimicrobial actions of phytochemicals against enteric pathogens–a review." *J Pharm Chem Biol Sci* 2, no. 2 (2014): 77-85.

Paranagama, Priyani A., E. M. Kithsiri Wijeratne, and A. A. Leslie Gunatilaka. "Uncovering biosynthetic potential of plant-associated fungi: effect of culture conditions on metabolite production by *Paraphaeosphaeria quadriseptata* and *Chaetomium chiversii*." *Journal of natural products* 70, no. 12 (2007): 1939-1945.

Parekh, S., V. A. Vinci, and R. J. Strobel. "Improvement of microbial strains and fermentation processes." *Applied microbiology and biotechnology* 54, no. 3 (2000): 287-301.

Patel, Hardik A., Rajesh K. Patel, Sunil M. Khristi, Kruti Parikh, and Geetha Rajendran. "Isolation and characterization of bacterial endophytes from *Lycopersicon esculentum* plant and their plant growth promoting characteristics." *Nepal Journal of Biotechnology* 2, no. 1 (2012): 37-52.

Pérez-de-Mora, Alfredo, Paula Madejón, Pilar Burgos, Francisco Cabrera, Nicholas W. Lepp, and Engracia Madejón. "Phytostabilization of semiarid soils residually contaminated with trace elements using by-products: sustainability and risks." *Environmental pollution* 159, no. 10 (2011): 3018-3027.

Pettit, Robin K. "Small-molecule elicitation of microbial secondary metabolites." *Microbial biotechnology* 4, no. 4 (2011): 471-478.

Pu, Xiang, Xixing Qu, Fei Chen, Jinku Bao, Guolin Zhang, and Yinggang Luo. "Camptothecin-producing endophytic fungus *Trichoderma atroviride LY357*: isolation, identification, and fermentation conditions optimization for camptothecin production." *Applied microbiology and biotechnology* 97, no. 21 (2013): 9365-9375.

Radman, Romeo, Teresa Saez, Christopher Bucke, and Tajalli Keshavarz. "Elicitation of plants and microbial cell systems." *Biotechnology and applied biochemistry* 37, no. 1 (2003): 91-102.

Roychowdhury, D., S. Mondal, and S. K. Banerjee. "The effect of biofertilizers and the effect of vermicompost on the cultivation and productivity of maize-a review." *Adv. Crop Sci. Technol* 5, no. 01 (2017): 1-4.

Sahoo, Sabuj, Sarmistha Sarangi, and Rout George Kerry. "Bioprospecting of endophytes for agricultural and environmental sustainability." In *Microbial Biotechnology*, pp. 429-458. Springer, Singapore, 2017.

Santoyo, Gustavo, Gabriel Moreno-Hagelsieb, Ma del Carmen Orozco-Mosqueda, and Bernard R. Glick. "Plant growth-promoting bacterial endophytes." *Microbiological research* 183 (2016): 92-99.

Schiavon, Michela, and Elizabeth A. H. Pilon-Smits. "Selenium biofortification and phytoremediation phytotechnologies: a review." *Journal of environmental quality* 46, no. 1 (2017): 10-19.

Schügerl, Karl, and Jürgen Hubbuch. "Integrated bioprocesses." *Current opinion in microbiology* 8, no. 3 (2005): 294-300.

Soliman, Sameh S. M., and Manish N. Raizada. "Interactions between co-habitating fungi elicit synthesis of Taxol from an endophytic fungus in host *Taxus* plants." *Frontiers in microbiology* 4 (2013): 3.

Specian, Vânia, Maria Helena Sarragiotto, João Alencar Pamphile, and Edmar Clemente. "Chemical characterization of bioactive compounds from the endophytic fungus *Diaporthe helianthi* isolated from *Luehea divaricata*." *Brazilian Journal of Microbiology* 43, no. 3 (2012): 1174-1182.

Spiering, Martin J., Christina D. Moon, Heather H. Wilkinson, and Christopher L. Schardl. "Gene clusters for insecticidal loline alkaloids in the grass endophytic fungus *Neotyphodium uncinatum*." *Genetics* 169, no. 3 (2005): 1403-1414.

Stępniewska, Z., and A. Kuźniar. "Endophytic microorganisms-promising applications in bioremediation of greenhouse gases." *Applied microbiology and biotechnology* 97, no. 22 (2013): 9589-9596.

Strobel, Gary, and Bryn Daisy. "Bioprospecting for microbial endophytes and their natural products." *Microbiology and molecular biology reviews* 67, no. 4 (2003): 491-502.

Subbulakshmi, G. K., A. Thalavaipandian, V. Ramesh, and A. Rajendran. "Bioactive endophytic fungal isolates of *Biota orientalis (L) Endl., Pinus excelsa Wall.* and *Thuja occidentalis* L." *International Journal of Advanced Life Sciences (IJALS)* 4 (2012): 9-15.

Taghavi, Safiyh, Craig Garafola, Sébastien Monchy, Lee Newman, Adam Hoffman, Nele Weyens, Tanja Barac, Jaco Vangronsveld, and Daniel van der Lelie. "Genome survey and characterization of endophytic bacteria exhibiting a beneficial effect on growth and development of poplar trees." *Applied and environmental microbiology* 75, no. 3 (2009): 748-757.

Thammajaruk, Narukjaporn, Nongluksna Sriubolmas, Duangnate Israngkul, Vithaya Meevootisom, and Suthep Wiyakrutta. "Optimization of culture conditions for mycoepoxydiene production by *Phomopsis sp. Hant25*." *Journal of Industrial Microbiology and Biotechnology* 38, no. 6 (2011): 679-685.

Tiwari, Pragya, Hanhong Bae. "Horizontal gene transfer and Endophytes: An implication for the acquisition of novel traits." *Plants* 9(2020): 305.

Tiwari, Pragya, Subir Kumar Bose, and Hanhong Bae. "Plant Growth Promoting Soil Microbiomes: beneficial attributes and potential applications." In *Soil Microbiomes for Sustainable Agriculture - Volume 2: Functional Annotation.* edited by Dr. Ajar Nath Yadav, Springer Publication (2021a).

Tiwari, Pragya, Yashdeep Srivastava, Mangalam Bajpai, Abhishek Sharma. "Bioactive metabolites from natural sources: Prospects and significance in drug discovery and research." *Bioingene PSJ* 1, (2021b):1-14.

Tiwari, Pragya, Yashdeep Srivastava, and Hanhong Bae. "Trends of pharmaceutical design of Endophytes as anti-infective." *Current Topics in Medicinal Chemistry* 21, no. 17 (2021c): 1572-1586.

Tiwari, Pragya, Hanhong Bae. "Endophytic fungi: Key insights, emerging prospects, and challenges in natural product drug discovery." *Microorganisms* 10 (2022b): 360. https://doi.org/10.3390/microorganisms10020360.

United States Environmental Protection Agency. "Introduction to phytoremediation: EPA/600/R-99/107." *US Environmental Protection Agency* (2000).

Urban, Mark C. "Accelerating extinction risk from climate change." *Science* 348, no. 6234 (2015): 571-573.

Vasconsuelo, Andrea, and Ricardo Boland. "Molecular aspects of the early stages of elicitation of secondary metabolites in plants." *Plant science* 172, no. 5 (2007): 861-875.

Venugopalan, Aarthi, and Smita Srivastava. "Endophytes as *in vitro* production platforms of high value plant secondary metabolites." *Biotechnology advances* 33, no. 6 (2015): 873-887.

Wang, Miao, Yuwan Xing, Junfang Wang, Yubin Xu, and Gang Wang. "The role of the chi1 gene from the endophytic bacteria *Serratia proteamaculans 336x* in the biological control of wheat take-all." *Canadian journal of microbiology* 60, no. 8 (2014): 533-540.

Wang, Wen-Jing, Dan-Yi Li, Yan-Chun Li, Hui-Ming Hua, En-Long Ma, and Zhan-Lin Li. "Caryophyllene sesquiterpenes from the marine-derived fungus *Ascotricha sp. ZJ-M-5* by the one strain–many compounds strategy." *Journal of natural products* 77, no. 6 (2014-a): 1367-1371.

Weilharter, Alexandra, Birgit Mitter, Maria V. Shin, Patrick SG Chain, Jerzy Nowak, and Angela Sessitsch. "Complete genome sequence of the plant growth-promoting endophyte Burkholderia phytofirmans strain PsJN." (2011). *Journal of Bacteriology,* 193(13), 3383–3384.

White, James F., Kathryn L. Kingsley, Qiuwei Zhang, Rajan Verma, Nkolika Obi, Sofia Dvinskikh, Matthew T. Elmore, Satish K. Verma, Surendra K. Gond, and Kurt P. Kowalski. "Endophytic microbes and their potential applications in crop management." *Pest management science* 75, no. 10 (2019): 2558-2565.

Wisniewski-Dyé, F., Borziak, K., Khalsa-Moyers, G., Alexandre, G., Sukharnikov, L. O., Wuichet, K., Hurst, G. B., McDonald, W. H., Robertson, J. S., Barbe, V., Calteau, A., Rouy, Z., Mangenot, S., Prigent-Combaret, C., Normand, P., Boyer, M., Siguier, P., Dessaux, Y., Elmerich, C., Zhulin, I. B. "*Azospirillum* genomes reveal transition of bacteria from aquatic to terrestrial environments." *PLoS genetics* 7, no. 12 (2011): e1002430.

Xu, Li-Jian, Yuan-Shuai Liu, Li-Gang Zhou, and Jian-Yong Wu. "Enhanced beauvericin production with in situ adsorption in mycelial liquid culture of *Fusarium redolens Dzf2*." *Process Biochemistry* 44, no. 10 (2009): 1063-1067.

Yang, Si Young, Myung Ryeol Park, In Seon Kim, Young Cheol Kim, Jung Wook Yang, and Choong-Min Ryu. "2-Aminobenzoic acid of *Bacillus sp. BS107* as an ISR determinant against *Pectobacterium carotovorum subsp. carotovotrum SCC1* in tobacco." *European Journal of Plant Pathology* 129, no. 3 (2011): 371-378.

Zhao, Jian, Lawrence C. Davis, and Robert Verpoorte. "Elicitor signal transduction leading to production of plant secondary metabolites." *Biotechnology advances* 23, no. 4 (2005)(a): 283-333.

Zhao, Kai, Lixin Sun, Xi Ma, Xiuliang Li, Xin Wang, Wenxiang Ping, and Dongpo Zhou. "Improved taxol production in *Nodulisporium sylviforme* derived from inactivated protoplast fusion." *African Journal of Biotechnology* 10, no. 20 (2011)(a): 4175-4182.

Zhao, Xin-Mei, Zhang-Qian Wang, Shao-Hua Shu, Wen-Juan Wang, Hai-Jie Xu, Young-Joon Ahn, Mo Wang, and Xuebo Hu. "Ethanol and methanol can improve Huperzine A production from endophytic *Colletotrichum gloeosporioides ES026*." *PloS one* 8, no. 4 (2013): e61777.

Chapter 8

Nanoparticles from Bacterial and Fungal Endophytes: Synthesis and Applications

Anjani Devi Chintagunta[1], K. Pallavi[2], N. Satya Vijaya Kumar[3] and N. S. Sampath Kumar[1,*]

[1]Department of Biotechnology, Vignan's Foundation for Science, Technology and Research, Vadlamudi, Andhra Pradesh, India
[2]Vignan Pharmacy College, Vadlamudi, Guntur, Andhra Pradesh, India
[3]Centre of Excellence (CoExAMMPC), Vignan's Foundation for Science, Technology and Research, Vadlamudi, Andhra Pradesh, India

Abstract

Nanotechnology is an interdisciplinary approach developed by the amalgamation of chemical, biological, and engineering concepts. Nanoparticles due to their distinctive functional diversity have been synthesized by physical and chemical means. However, nanoparticle synthesis through biological means is highly appreciated by the research community owing to its unique features such as its non-toxic, eco-friendly, and green approach. Biological entities include plants and other endophyte microbes *viz.*, bacteria, algae, and fungi that maintain a symbiotic relationship with them. This relationship leads to the release of several potential therapeutic secondary metabolites that have the affinity to reduce the metallic ions into nanoparticles. The synthesis of silver, zinc, gold, and copper nanoparticles using endophyte microbes have been reported in several studies, and their applications in the various fields are less focused. Hence, this chapter gives an overview of the synthesis of nanoparticles in presence of bacterial and fungal

[*]Corresponding Author's Email: nssk84@gmail.com.

In: Endophytes
Editor: Pragya Tiwari
ISBN: 979-8-88697-045-6

endosymbionts, parameters that influence the nanoparticle synthesis, and the characterization of nanoparticles with special emphasis on their application in sustainable agriculture and biomedicine.

Keywords: biomedicine, endosymbiont, nanoparticles, nanotechnology, sustainable agriculture

1. Introduction

Nanotechnology is an emerging branch of science that deals with the synthesis of nano-sized particles (<100 nm) with various metals such as silver, zinc, copper, and gold and its application in catalysis, electronics, textiles, biomarker, diagnosis, cytotoxic studies, and for energy (Naik 2020). It is a diverse technology that ranges from extensions of the conventional device to novel approaches applicable in various fields of science *viz.*, plant biology, chemistry, agriculture, and the food industry (Kumar et al. 2021b). In the food industry, nanotechnology improves the flavor, color, taste, texture, and consistency of food besides the absorption of bioactive compounds (Singh et al. 2020). Apart from food processing and quality, nanotechnology is also used in food packaging and safety by eliminating foodborne pathogens (Shaheena et al. 2019). In association with various interdisciplinary approaches, nanotechnology is also involved in settling challenges in food processing and developing sustainable and eco-friendly food products (Tiwari et al. 2020a).

Nanotechnology also finds its application in agriculture as compound fertilizers and nano pesticides deliver the chemical compound to the target molecule of the plant cells (Kumar et al. 2021a). It also diminishes the nutrient loss caused during enrichment and enhances crop yield through nutrient management. Further, nanotechnology-based devices can also play a crucial role in environmental conservation and remediation by solving long-term threatening issues for humans (Chintagunta et al. 2020; Kaki et al. 2022). To summarise, nanotechnology has provided humans with various amenities and made a significant contribution by better material for environmental remedies thereby promising ecological sustainability (Beulah et al. 2021). Thus, nanotechnology has a high aptitude for solving various societal issues in the present as well as in the future.

1.1. Types of Nano-Materials

The materials that have at least one dimension in a three-dimensional space or the materials whose composition has been reduced to nanoscale are referred to as nanomaterials (NMs). On contrary, the size of bulk materials is above 100 nm in all dimensions and physical properties are independent of the size. But, the physical properties of nanomaterials depend upon their size and shape. NMs have exceptional mechanical properties such as strength, and flexibility due to the surface, volume, and quantum effects of nanoparticles. The mechanical properties of the nanomaterials mostly depend upon the process parameters *viz.*, temperature, processing time, method, dosage, and the grain size of nanoparticles

Based on the shape and dimensions of the material, the nanomaterials are classified as zero-dimensional (0D), one-dimensional (1D), two-dimensional (2D), and three-dimensional materials (3D) (Wolcott et al. 2009). Zero dimensional NMs have all their dimensions on the nanoscale. 1D NMs are materials with two dimensions on the nanoscale and one dimension not on the nanoscale. Metallic, ceramic nanotubes, nanorods, nanowires, and nanofibers are included in this category. In 2D NMs, only one dimension will be in the nanoscale, and the rest not be in the nanoscale. Single and multi-layered crystalline/amorphous, nanoplates, nanocoatings, and thin films fall under this category. In 3D NMs, multiple nanocrystals are arranged in different dimensions. Carbon nanobuds, fibers, nanotubes, fullerenes, polycrystals, and pillars are examples of 3D NMs.

Based on the chemical composition, NMs are categorized into four types such as metal-based materials, carbon-based materials, composites, and dendrimers. Carbon-based materials are mostly composed of carbon and are available in various forms like hollow spheres, ellipsoids, and tubes (Kurbanoglu et al. 2018). Ellipsoidal and spherical nanomaterials are referred to as fullerenes whereas, cylindrical nanomaterials are known as nanotubes. Quantum dots, nanosilver, nanogold, and metal oxides like titanium dioxide are examples of metal-based nanomaterials. Besides, dendrimers are nanomaterials built from branched units. The surface of these nanomaterials has several chain ends, which are being tailored in order to perform various chemical functions (Sharma et al. 2015). The three-dimensional drimers have interior cavities, into which molecules such as drugs can be introduced; consequently, the dendrimers can be used in drug delivery. Composite is a combination of various nanomaterials or nanomaterials with other bulk materials. These composites are mostly used as packing materials to increase thermal, mechanical, and flame-retardant properties.

2. Nanoparticles

The size of nanoparticles plays a key role in the physicochemical properties of the nanoparticles. A 20 nm palladium, silver, platinum, and gold nanoparticles show characteristic dark black, black, yellowish-grey, and wine red colors, respectively (Tan et al. 2003). Based on the morphology, size, and chemical properties, NPs can be classified as carbon-based NPs, metal NPs, ceramic NPs, semiconductor NPs, polymeric NPs, and lipid-based NPs. Dendrimers, micelles, ferritin, and liposomes are known as organic nanoparticles whereas metal and metal oxide-based nanoparticles are categorized as inorganic nanoparticles. Fullerenes and carbon nanotubes are commercially important because of their electrical conductivity, structure, electron affinity, strength, and versatility. The carbon in the pentagonal and hexagonal carbon units exhibit sp^2 hybridization. Besides, ceramics NPs are found in amorphous, polycrystalline, or porous forms (Khan et al. 2019). NPs are widely used in the photo-degradation of dyes, and imaging applications. On the other hand, semiconductor NPs possess wide bandgaps which are responsible for alterations in their optical and electric properties. Therefore, they are very important materials in photocatalysis, photo optics, and electronic devices.

Most commonly used metals for nanoparticle synthesis are cadmium (Cd), aluminium (Al), cobalt (Co), gold (Au), copper (Cu), iron (Fe), silver (Ag), lead (Pb), and zinc (Zn) (Ealia et al. 2017). These nanoparticles vary in size, surface charge, shape, structure, color, sensitivity to environmental factors like moisture, heat, air, and sunlight, and reactivity. Among these metal nanoparticles, silver nanoparticles (AgNPs) got more attention because of their distinctive properties like good chemical stability, conductivity, and catalytic properties (Diantoro et al. 2018). AgNPs possess antimicrobial properties and therefore, can be used as nanomedicines, implants, devices, and dressing materials (Manikprabhu and Lingappa 2014; Kumar et al. 2016). AgNPs that are synthesized biologically is more suitable for medical application due to greater biocompatibility than chemically synthesized NPs. Owing to their high surface-to-volume ratio, these NPs exhibit enhanced catalytic activity (Cui and Yu 2013). Besides, metal oxide-based nanoparticles are produced by modifying the properties of metal nanoparticles. For instance, in the presence of oxygen at room temperature, iron nanoparticles rapidly oxidize to iron oxide (Fe_2O_3), the reactivity of which is higher than that of iron (Fe). Similarly, cerium oxide (CeO_2), aluminium oxide (Al_2O_3), Magnetite (Fe_3O_4), titanium oxide (TiO_2), silicon dioxide (SiO_2), and zinc oxide (ZnO) exhibit exceptional properties than their metal counterparts. Upon considering lipid-based NPs, they contain lipid moieties with a solid core with soluble lipophilic molecules stabilized by surfactants. Lipid nanotechnology focus on designing and synthesis of lipid NPs that can be used as drug carriers and for RNA release in cancer therapy (Puri et al. 2009).

2.1. Synthesis of NPS - Pros and Cons of Various Methods

The nanoparticles are synthesized by three methods, *viz.,* physical, chemical, and biological either by top-down or bottom-up approach (Deokar 2017). In the top-down approach, bulk materials are trimmed down to desired NPs whereas, in a bottom-up approach, the NPs are integrated for the formation of nanomaterials. Ball milling, etching, grinding, optical lithography, nanoimprint lithography, e-beam lithography, scanning probe lithography are examples of top-down approach whereas, sedimentation, spinning, green synthesis, reduction, etc. stands as examples of bottom-up approach (Biswas et al. 2012).

Mechanical milling is most prominently used for milling and post-annealing of nanoparticles in an inert environment (Yadav et al. 2012). Nanolithography is a physical technique for printing the desired shape on a light-sensitive material. The advantage of this technique is the production of a single to cluster of nanoparticles with the required size and shape and the disadvantage is the usage of complex equipment and the cost incurred during the process. Laser ablation is another method of nanoparticle synthesis where a high-energy laser beam is used to irradiate a metal submerged in a solution and condense a plasma plume that produces nanoparticles (Zeng et al. 2012). This method provides a stable synthesis of nanoparticles in water and organic solvents. Similarly, sputtering is another process where nanoparticles are deposited and annealed onto a thin surface. The thickness of the layer, duration of annealing, temperature, and type of the substrate determine the size and shape of the nanoparticle.

The bottom-up method is a constructive method from atoms to nanoparticles through clusters (Lu et al. 2016). One such method is the sol-gel method where the precursors are dispersed in the solvent either by stirring, shaking, or sonication and form solid and liquid phases. The phase separation needs to be carried out by filtration, sedimentation, and centrifugation for the recovery of the nanoparticles. Besides, the synthesis of nanoparticles can also be carried out using a spinning disc reactor that contains a rotating disc under controlled temperature conditions. The disc is rotated at various speeds during which the atoms/molecules fuse and precipitate the nanoparticles. The characteristic features of the nanoparticles depend upon flow rate, disc rotation speed, disc surface, location of the feed, etc. Another method for nanoparticle synthesis is chemical vapor deposition during which a thin film of gaseous reactant deposits on the surface of the substrate. In a reaction chamber at the desired temperature, the product is formed when heated substrates react with the combined gas (Morosanu 2016). Through this method, highly pure, hard, and strong nanoparticles will be formed. The disadvantages of this technique include the need for sophisticated equipment and the formation of toxic gaseous by-products. Besides the mentioned methods, pyrolysis is mostly used for the largescale production of nanoparticles. Either liquid or vapor is fed into the furnace through a small hole at high pressure. In place of flame, laser and plasma can be used for easy evaporation. The advantages of this method are simple, efficient, cost-effective, and high yield.

High temperature, pressure, and radiation-associated methods such as thermal decomposition, laser ablations, and mechanical milling are included under physical means of nanoparticle synthesis (Clifford et al. 2017). Whereas, chemical reduction methods are employed for the large-scale homogenous synthesis of nanoparticles in a short span (Tang et al. 2010). However, these are energy-intensive methods that use hazardous chemicals. Likewise, the physicochemical methods are expensive, time-consuming with a low production rate, and show a drastic effect on human health and the environment (Mishra and Singh 2015).

3. Green Approach/Biological Methods

The chemical methods are costly and use chemicals that are toxic and hazardous to the environment. Comparatively, the biosynthetic route is a biocompatible, safe, and eco-friendly green approach to synthesizing nanoparticles. Moreover, the green synthesis can generate nanoparticles with defined morphology and size in a cost-effective, eco-friendly, and less energy-intensive manner (Iravani 2011). Further, nanoparticles synthesized through this approach have wide acceptability in pharmaceutical, agronomical, medical and environmental aspects (Pandey 2018; Tiwari et al. 2022).

The major requirements for the biological method are the selection of suitable solvent, reducing agent, and harmless stabilization materials. Green synthesis encompasses the exploitation of natural resources like plants, bacteria, algae, fungi, and actinomycetes for the reduction of metal ions into metallic nanoparticles (Singh et al. 2018). Because of the ability to accumulate and detoxify heavy metals and higher kinetics, plants are mostly used when compared to other natural resources. Several research groups working in this domain authenticate the accumulation, detoxification, and phytoremediation of toxic metals by some plants *viz., Arabidopsis helleri, Acanthopanax sciadophylloides, Brassica juncea, Clethra barbinervis*, *Maytenus founieri, Sesbania drummondii* and *Thlaspi caerulescens* (Kumar et al. 2016). Plant parts such as leaf, fruit, root, and stem are excellent sources of phytochemicals and are hence suitable for the synthesis of nanoparticles. These plant extracts possess several phytochemicals such as ketones, terpenoids, amides, aldehydes, carboxylic acids, proteins, carbohydrates, and vitamins which are responsible for metal ion reduction. For the green synthesis of the nanoparticles, suitable plant material should be selected, boiled, washed with distilled water, squeezed, filtered, and mixed with a suitable salt solution.

As a result, a change in the color of the solution takes place confirming the formation of the nanoparticles. Predominantly gold, silver, iron, and zinc oxide nanoparticles have been synthesized easily through the green approach (Table 1). The nanoparticles can be characterized by ultraviolet-visible spectroscopy, Fourier transforms infrared spectroscopy, scanning electron microscopy, transmission electron microscopy, Raman spectroscopy, X-ray diffractometer, photoluminescence analysis, energy dispersion analysis of X-ray, X-ray photoelectron microscopy, atomic force microscopy, field emission scanning electron microscopy, thermal-gravimetric differential thermal analysis.

Apart from a plant source, various algal species belonging to *Cyanophyceae*, *Chlorophyceae*, *Rhodophyceae,* and *Phaeophyceae* are being used for the synthesis of nanoparticles because of the following reasons:

1. Ease in handling
2. Capacity to absorb inorganic metallic ions
3. Eco-friendly and cost-effective

Table 1. Green synthesis of nanoparticles using plant extracts (Jadoun et al. 2021)

Nanoparticle	Plant extract	Applications	Shape of nanoparticle	Conditions
Silver (Ag)	*Saccharum officinarum, Helianthus annus, Cinamomum camphora, Aloe vera, Oryza sativa, Capsicum annuum Magnolia Kobus, Medicago sativa*	Pharmaceutical	Spherical	300 K, 30 min
Gold (Au)	*Azadirachta indica, Terminalia catappa, Diopyros kaki, Magnolia Kobus, Coffea Arabica, Croton sparsiflorus, Citrus limonum*	Medical	Triangular, spherical, decahedral, and icosahedral	313 K, 2.5 h
Palladium (Pd) & Platinum (Pt)	*Cinnamomum camphora, Gardenia jasminoides, Pinus resinosa, Anogeissus latifolia, Glycine max, Ocimun sanctum, Curcuma longa, Musa paradisica, Cinnamom zeylanicum, Pulicaria glutinosa, Doipyros kaki*	Medical diagnoses	Spherical	60°C, 1 h
Copper (Cu)	*Aloe vera, Cuscuta reflexa,* henna leaves	Medical	Spherical	50°C, 3 h
Zinc oxide (ZnO)	*Vitex negundo, Azadirachta indica*	wound healing, antifungal, antibacterial, drug delivery, antidiabetic, anticancer	Spherical	60°C, 2 h
Titanium oxide (TiO_2)	*Annona squamosa, Catharanthus roseus, Moringa oleifera, Cucurbita pepo*	wound healing, tissue engineering, disease diagnostics, manufacturing of surgical tools agriculture, and energy production	Spherical	180 W, 1 h

Algae are the autotrophic, photosynthetic eukaryotic organs that are further classified into microalgae and macroalgae based on their morphological features (Uzair et al. 2020). These are non-flowering mostly found in marine and freshwater or over moist rocks. Although algae contain chlorophyll, they differ from plants by lacking structures such as true stems, leaves, roots, and vascular tissue that are prominent in land plants. They play a key role in the aquatic ecosystems and find their application in agriculture, aquaculture, cosmetics, and pharmaceutical industries (Chintagunta et al. 2021).

Two kinds of NPs can be synthesized using algae: organic and inorganic ones. Organic NPs include chitosan (CS), poly-ε-lysine (ε-PL), cationic quaternary polyelectrolytes, and quaternary ammonium compounds (Uzair et al. 2020). At high temperatures, the organic NPs are less stable compared to that inorganic NPs. Algal nanoparticles are synthesized in three steps: During step 1, algal biomass is mixed with water or other solvent and boiled. In step 2, molar ionic metallic solutions are prepared. During step 3, the algal extract is mixed with a molar solution of metal ion precursor under continuous stirring. As a result, a change in color of the mixture is noted down which is due to the formation of stable nanoparticles from adjacent nucleon particles. Based on the location of synthesis, algal nanoparticles can be considered either intracellular or extracellular nanoparticles (Das et al. 2014). In intracellular nanoparticle synthesis, the live algae were collected, washed, and treated with the metallic solution under a specific temperature and pH for a certain period of time. Then algal biomass may be sonicated and centrifuged to harvest stable NPs. In the extracellular synthesis of NPs, the algal biomass is collected, washed, dried under shade, treated with water, and filtered out. Then algal biomass may be sonicated or washed with distilled water to obtain the cell-free extract. Similarly, cyanobacteria-mediated nanoparticles were also synthesized using Au, Ag, Cd, CuO, Fe_3O_4, and Pd. These nanoparticles have applications in biotechnology and medicine along with many other fields (Chintagunta et al. 2021).

4. Endosymbionts

Endosymbionts are organisms that form a symbiotic relationship with another cell or organism, which can be found either intracellular or extracellular (Casadevall 2008). Symbiotic relationships are most successful in nature because both parties get benefits during their association. Based on the mode of transmission, endosymbionts are categorized as horizontal and vertical transmission. Where in is vertical transmission takes place when the parent passes the symbiont to its offspring and in the case of horizontal transmission a new symbiont starts associating with the host directly from the environment (Tiwari and Bae 2020b). Comparatively, high co-dependency is reported in vertical transmission, because of the continuous association between host and symbiont for generations together (Mueller et al. 2015). Another interesting feature of endosymbiosis is that their association becomes obligate as they cannot without each other and any attempt to separate them leads to mortality either for one or both the associates.

Endophytes/endosymbionts include both fungal and bacterial species that colonize tissues of healthy plants without causing any disease (Dutta et al. 2014; Tiwari et al. 2021a). The colonization can be intercellular or intracellular. Owing to their ubiquitous presence in all plants, endophytes are regarded as important and promising organisms for study. Endophytes exhibit the ability to produce large bioactive compounds that promote plant growth and control phytopathogens, as well as enhance plant tolerance toward biotic and abiotic stresses (Alurappa et al. 2018; Tiwari et al. 2021a; Tiwari et al. 2021b). Moreover, endophytes isolated from medicinal plants can synthesize bioactive compounds which are used as raw materials for the production of medicines, cosmetics, fragrances, and foods (Kumar et al. 2020; Tiwari et al. 2022). In particular, due to the complexity of culture conditions and indiscriminate pesticides, the research on fungal/ bacterial endophytes and their environmental interaction in floral communities has been really challenging. But considering their potentiality in enhancing the secondary metabolites such as alkaloids, terpenoids, flavonoids, etc., and their role as reducing agents have drawn the attention of material scientists towards using them in the green synthesis of nanoparticles.

4.1. Fungal Endophytes

The fungal endophytes are found to be living asymptomatically within host plant tissues for at least part of their life cycle, occupying leaves, stems, and branches of plants (Alurappa et al. 2018). Association between fungal endophytes and plants is usually strengthened by the compatible genes in both organisms and encouraged by the environment (Moricca and Ragazzi 2008). Based on the association with the host fungal endophytes are categorized into four classes and parameters such as tissue specificity, host extensiveness, colonization, mode of transmission, etc. are duly given weightage

- Class I endophytes are transmitted from one generation to another and are known to colonize systematically in the shoots. They are phylogenetically related and limited their proliferation within the warm and cool-season grasses. Examples: *Aschersonia* sp., *Balansiopsis* sp., *Blistum* sp., *Cepsiclava* sp. (Rodriguez et al. 2009)
- Class II endophytes consist of a wide range of species capable of infecting tissues above and below the ground. They support the host by conferring tolerance towards adaptation-specific stress. Examples: *Ascomycota* sp., *Basidiomycota* sp. (Mane et al. 2018)
- Class III endophytes are transmitted among a wide range of angiosperms and are diversified within the host tissues, plants, and population. (Sakalidis et al. 2011).
- Class IV endophytes are confined to plant roots and form melanized structures like inter and intracellular hyphae. They are found in the flora of temperate and tropical ecosystems (Rodriguez et al. 2009).

4.2. Bacterial Endophytes

In process of evolution, plants established a symbiotic relationship with bacteria to withstand a wide range of biotic/ abiotic stresses and in addition aid their growth and development. Such non-pathogenic and immune-boosting microorganisms are labeled as bacterial endophytes. More than 300 bacterial endophytes pertaining to *Arthrobacter, Rhodococcus, Kocuria, Pseudonocardia, Microbacterium,* etc. are capable of producing hormones, nitrogen fixation, and, solubilization of phosphates, and enhanced bio-control activity (Jan et al. 2019). Endophyte bacterial association may be obligatory or facultative based on their dependence, controlled by genes, and amended by environmental conditions. Transmission of bacterial endophyte is either through seeds or soil and entry into `plants is achieved via wounds or cracks in root junctions. In some cases, cell-degrading enzymes such as endoglucanases, pectinases, cellulases, chitinases, hemicellulases, amylases, and glucanases also support the invasion process.

Fungal and bacterial endophytes have been considered as possible useful sources for enhancing natural products, which are able to produce substances of biotechnological importance and also protect plants from insect attacks and diseases (Strobel et al. 2004; Borges et al. 2007; Newman and Cragg 2007). Some species of endophytes have been identified as sources of antimicrobial, anticancer, antidiabetic, insecticidal, and immunosuppressive compounds (Strobel and Daisy 2003; Tiwari et al. 2022). Other reports have also shown that certain fungi/ bacterial endophytes produced metabolites similar to those produced by the host plant with therapeutic functions including alkaloids, steroids, terpenoids, flavanoids, quinines, peptides, and phenolic acid (Strobel and Daisy 2003; Alfarano et al. 2005).

5. Synthesis and Characterization of Nanoparticles via Endosymbionts

Owing to their potential enhancement of secondary metabolites especially reducing agents, endosymbionts are successfully employed in synthesizing nanoparticles.

5.1. Synthesis of Nanoparticles by Bacterial Endosymbionts

Burgeoning scientific interest lead to the discovery of various microbes harboring inside the plant tissue referred to as endophytes (Singh et al. 2021). In spite of extensive research work carried out on microorganism-mediated nanoparticle synthesis, only a few reports on the synthesis of nanoparticles from endophytes are available in the scientific domain (Salem et al. 2021). Several researchers believe that interference of nanomaterials and endophytes leads to the eco-friendly synthesis of nanoparticles with the required size, shape, and biological activities (Rana et al. 2020).

Bacterial endophytes from the sterilized explant were isolated and cultured in the media mixed with silver nitrate and incubated at temperatures of 30°C to 80°C until visible growth was observed (Ismail et al. 2021). The bacterial colonies were sub-cultured into the nutrient broth and further incubated for 72 h. Then the broth was subjected to centrifugation at 10000x g at 4°C for 5 min. An aliquot of samples was drawn periodically and the color change was monitored using a UV-Visible spectrophotometer for confirming the synthesis of nanoparticles. During the synthesis of silver nanoparticles, apart from temperature silver nitrate concentration, pH, and the ratio of supernatant with the silver nitrate also play a key role (Raut et al. 2014). In order to arrive at the optimized conditions, the range of pH considered was from 6 to 9 and, the range of silver nitrate concentration was 1.0 to 2.5 mM. The success of nanoparticles synthesis depends on monitoring/maintaining the parameters within the mentioned range. A few studies were reported on the synthesis of nanoparticles using bacterial endophyte isolated from *Coffea arabica* L. and *Euphorbia hirta* L. (Asmari et al. 2020). *C. arabica* has anti-cancerous, anti-inflammatory, anti-bacterial, anti-diabetic, and anti-atherosclerotic properties, whereas, *E. hirtais* is used in treating gastrointestinal disorders (Kumar et al. 2010).

5.2. Synthesis of Nanoparticles by Fungal Endosymbionts

Fungal endophytes are an excellent source of secondary metabolites with bioactive potential to protect against harmful pathogens (Chintagunta et al. 2021; Kumar et al. 2021c). The advantages of fungi over bacteria include easy handling, the requirement of simple nutrients, high yield of the enzyme, high intracellular metal uptake, and wall-binding capabilities. Besides, filtering fungal broth using a filter press is easy and more economical than bacterial broth filtration hence, large-scale fungal nanoparticle synthesis is preferred over another biological system (Kumar et al. 2021d). Various fungal endophytes that are involved in the synthesis of nanoparticles include *Aspergillus conicus, Phomopsis* sp., and *Penicillium janthinellum* isolated from *Avicennia marina, Rhizophora mucronata,* and *Suaedamonica* leaves (Vijayanandan et al. 2018).

Adnan (2020) has isolated fungal endophytes from mature petiole of *Cinnamomum zeylanicum.* The fresh petioles were collected, washed, cut into small pieces, immersed in 70% ethanol for 5 min, washed with 1% sodium hypochlorite for 1 min, further washed with ethanol and distilled water, then inoculated on potato dextrose agar and incubated at 28°C. The aqueous extract of this organism was used to synthesize green silver nanoparticles. The formation of silver nanoparticles was confirmed by a change in color from white to brown color. The nanoparticles are characterized by FTIR, Zeta potential analyzer, Particle size analyzer, FESEM, and EDAX analysis (Anandalakshmi et al. 2016).

5.3. Characterization of Nanoparticles

Due to a constant demand for functional, selective, and high-performance materials rising exponentially, there is a constant need to make those materials and access them. Nanoparticles due to their unique characteristics and properties, though promising materials are relatively hard to characterize compared to their bulk counterparts. It is therefore imperative to have a different route or slightly modify the existing protocols of characterization tools. Hence it is paramount to design techniques and/or protocols to access nanoparticles with precision and reproducibility. As the size of the particle falls below 100 nm, the overall properties vary by a major extent due to their large surface-to-volume ratio. The changes, significantly in surface, chemical, optical, mechanical, and biological properties cannot be properly accessed by traditional methods.

Though different characterization techniques are used to study different parameters and properties of materials, there are few parameters that are unique to nanoparticles (Figure 1). These parameters fundamentally govern the nature of the nanoparticles and are essential in understanding them. They are as follows:

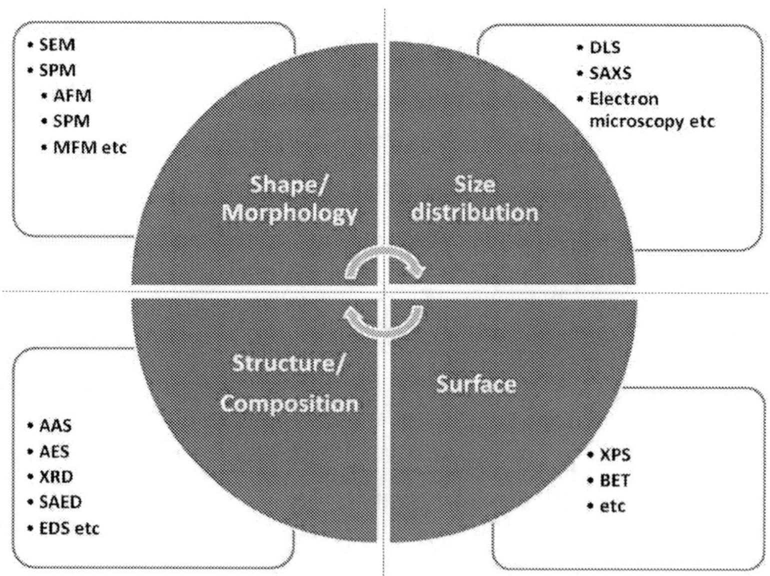

Figure 1. Represents the various techniques used for the characterization of nanoparticles.

5.3.1. Size

It is essential for a particle to have at least one dimension to be less than 100 nm to be identified as a nanoparticle. Hence, it is of utmost importance to confirm the size of the particles and in the case of an asymmetric sample, the size distribution. There are different techniques to confirm the particle size of a given sample, based on the sample's physical state.

Dynamic light scattering (DLS) technique analyzes the particle size in a colloidal solution based on the intensity of scattered monochromatic light caused by particles' diffusive motion as a function of time (Pecora 2000; Stetefeld et al. 2016).

Small-angle X-ray scattering (SAXS), where the particle size can be determined by studying the X-rays that are elastically scattered by the sample over small irradiation angles between 0.1°–10° within the size range of 1 – 100 nm (Narayanan et al. 2018; Gommes et al. 2021).

Electron microscopy is another technique where samples are analyzed under a high-vacuum. Both transmission and scanning use accelerated electrons that are guided into a focused beam. The energy/wavelength of the electron can be tuned by altering the acceleration voltage and the electron beam can be focused onto the sample surface by a series of magnetic condenser lenses. Typically, an electron microscope can give very fine resolutions, 0.05–0.1 nm in the Transmission electron microscope (HRTEM) and 2–3 nm in the Scanning electron microscope (FESEM). However, the intensity of the electron beam or the high vacuum maintained during the analysis may influence the sample. These limitations can be overcome by changing the different protocols or methods of imaging like low-voltage electron microscopy (electron acceleration voltage in the range of 5–25 KV), environmental TEM (maintaining the sample chamber at a low vacuum by differential pumping), cryomicroscopy (covering the sample surface with amorphous ice, by circulating liquid nitrogen or liquid helium), etc. (Crewe et al. 1970; Rose et al. 2008; Rudenberg et al. 2010).

5.3.2. Shape/Morphology

Morphology and topography are the physical parameters of the sample that are related to the shape. Morphology studies mainly comprise techniques that visualize the sample either directly or indirectly.

Scanning electron microscopy (SEM) is a well-versed technique where accelerated electrons scan the sample in a raster pattern to gain information from the atoms present on the sample surface. The topography of the sample can be studied by measuring secondary electrons and backscattered electrons.

The secondary electrons are generated by atoms on the surface due to the excitation by the electron beam. Backscattered electrons are generated by elastic scattering of electrons of the incident beam and the atoms of the sample. However, the major limitation of this technique happens to be the sample conductivity. Non-conducting samples or biological samples may develop surface charging, leading to poor resolution or artifacts. This limitation can be overcome by depositing a thin (~5nm) coating of a good conducting layer of carbon or gold or palladium etc. (McMullan 1995; Rose et al. 2008; Takaku et al. 2015).

Scanning probe microscopy (SPM) is a technique that visualizes a sample surface image by physically 'touching' or 'feeling' it. SPM has many types or variants in visualizing the sample surface based on the interactive or binding forces of the atoms/molecules. The probe usually comprises a sharp tip (radius of curvature being in the nanometer range) on the free end of a cantilever and is sensitive to forces like Casmir forces, capillary forces, electrostatic forces, van der walls forces, solvation forces, magnetic forces, among others. When the tip is in close proximity to the sample surface, the interaction of the forces causes deflections, these cantilever deflections can be monitored and measured, which can be converted into a three-dimensional profile. Unlike SEM, SPM requires less space and relatively very low voltages. However, SPM is limited by its slow scanning rate and the dimensionality confinements (samples should be atomically flat and very small in size) (Binnig et al. 1986; Salapaka et al. 2008).

5.3.3. Structure/Composition

The structure and composition of the sample can be analyzed using a wide variety of characterization techniques. While techniques like atomic spectroscopy and mass spectroscopy are destructive, non-destructive techniques using X-rays are also available.

5.3.3.1. Atomic Spectroscopy

Destructive techniques are used to find elemental composition by studying the absorption or emission of radiation by atoms or ions or molecules in a given sample. Many of the elements in the periodic table when irradiated tend to absorb a certain wavelength and move electrons to a much higher energy level, commonly referred to as an excited state. These exciting elements then turn back to the ground state by emission of energy in the form of photons. Each element has its own characteristic absorption and emission bands of spectra that can give both qualitative and quantitative information about the element.

Atomic spectroscopy usually requires the sample to be vaporized and atomized in order to study the gas phase absorptions or emissions. There are different types of atomization methods- like flame atomization (sample is introduced into a flame with high temperature), graphite furnace atomizer (a flameless technique using a graphite disk housed in a high-temperature furnace), cold vapor atomizer (atomization at room temperature using chemical/electrochemical methods), hydride generation atomization (a method where the sample is converted into a hydride before atomization), inductively coupled plasma atomization (atomization bypassing vaporized sample along with ionized inert gas into a strong magnetic field) (Koirtyohann 1991; Stephansson 2007).

5.3.3.2. Powder Diffraction

Powder diffraction is a non-destructive analytical technique that provides information about crystallinity, lattice parameters, and crystal phase identification by observing the characteristic bending of radiation waves of X-rays or electrons, or neutrons. According to Bragg's law, when radiation of a particular wavelength passes through a crystallite of comparable atomic spacing, the radiation is scattered or bent in a specific pattern. This diffraction pattern is unique like a 'fingerprint' to a crystalline system and can be used for identification/confirmation of the crystallite and also its phase purity based on the peak position as well as peak intensity. Scherrer equation, by taking the width of the peak at half of the maximum peak intensity (FWHM) can also determine the size of crystallites (not necessarily particles) that are below 100 nm. Advanced diffraction studies namely Selected Area Electron Diffraction (SAED) can be performed in transmission electron microscopy. In SAED the diffraction pattern of crystallites in a specific spot of the sample can be measured which provides information corresponding to lattice matching, crystalline defects, and interfaces (Brown et al. 1999; Jung et al. 2017; Guccione et al. 2021).

5.3.3.3. X-ray Spectroscopy

Energy-dispersive X-ray spectroscopy (EDS or EDAX) is a technique that identifies the elemental composition of a specific area of the sample by studying the energy of the characteristic X-rays emitted upon irradiation with a high-energy electron beam. According to Moseley's law, the energy of the characteristic X-rays emitted by an element is distinctive to the atomic number (Z) of the element (Corbari et al. 2008; Kosasih et al. 2021).

5.3.4. Surface

Nanoparticles exhibit unique or enhanced surface properties like surface area, surface charge, porosity, etc.

X-ray Photoelectron Spectroscopy (XPS) is an advanced surface-sensitive technique that can provide information about the chemical or oxidative state of the elements on the sample surface. XPS works on the principle of the photoelectron effect, which states that when radiation of a specific wavelength or energy hits an atom, electrons are emitted from the surface. The energy of the emitted electrons gives information regarding the binding energy of the electron, bonding environment, and chemical state of the elements on the surface (Gumerova et al. 2018; Voiry et al. 2018).

5.3.4.1. Surface Area

Specific surface area is defined as the total surface area of a material per unit mass (m^2/g). Specific surface area significantly gives information about the available active sites for adsorption or catalytic activity of the material. Among the available techniques, Brunauer-Emmett-Teller (BET) adsorption/desorption method is widely used in specific surface area measurement (S_{BET}). The technique derives the surface area of a multi-layered material by studying the adsorption and desorption isotherms of a non-reactive probing gas at different applied pressures (Galarneau et al. 2018; Condon 2019).

6. Applications of Endosymbionts-Mediated Nanoparticles

Endosymbiont microbes are identified as the most potent nano-biotechnological tools for the green synthesis of nanoparticles. The bio-generated nanoparticles (metallic ions) have precise structure, composition, morphology, electrochemical and optical properties resulting in profound applications in terms of biological activities and environmental sustainability (Figure 2). Most of the bio-synthesized nanoparticles are capped with bioactive secondary metabolites on the surface which augmented their efficacy and stability (Nirmala and Sridevi 2021).

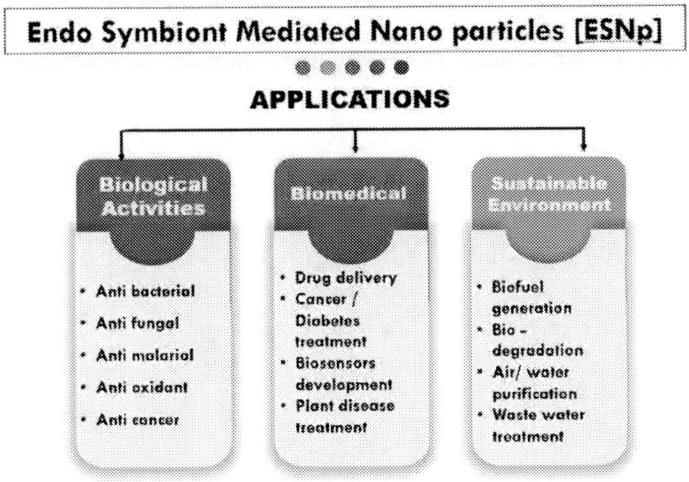

Figure 2. Various applications of endosymbiont mediated nanoparticles.

6.1. Antibacterial Properties of Endosymbiont- Mediated Nanoparticles (ES-NP)

Biosynthesized nanoparticles like copper, cadmium sulfide, zinc sulfide, gold, silver, etc. were observed to exhibit efficient antibacterial properties against bacterial species like *Escherichia coli, Bacillus subtilis,* and *Pseudomonas aeruginosa* (Table 2). The presence of microbial protein-related functional groups like -COOH, -OH, -SH, $-NH_2$, etc. on the surface of the biosynthesized nanoparticles helps to achieve stable metal ion nanoparticles with bacterio-specific binding sites (Lahiri et al. 2021).

Bacterial cell morphologies after treatment with endosymbiont-mediated nanoparticles exhibited discontinuous cell membrane, which may be due to the accumulation of endo-nanoparticles on the bacterial cell wall surface (Xi et al. 2014). Surface morphology studies like STEM scan indicated that the bio-nanoparticles entered into cell cytoplasm resulting in cell inactivation. Secondary implication like leakage of potassium ions was observed with ES-NP, because of change in the permeability of cell wall, which in turn resulted in osmotic imbalance, alteration of intracellular pH, and hindrance of physiological functions of the bacteria (Radchenko et al. 2006). Lipid peroxidation and DNA degradation in the cytoplasm were also found to be the main correlation factors for the effective bacterial inactivation exhibited by endosymbiont-mediated nanoparticles (Harne et al. 2012).

Table 2. Antibacterial properties of endosymbiont mediated nanoparticles

Endosymbiont name	NP	Bacteria affected	Reference
Microbes as Endosymbionts with anti-bacterial properties			
Pantoea anthophila isolated from the stem of *Waltheria indica*	AgNP	*S. epidermis, B. subtilis, S. aureus, E. coli, P. mirabilis, S. typhi, K. pneumonia*	Nirmala 2021
Agaricus bisporus and *Ganoderma lucidum*	AgNP	*S. aureus* and *E. coli*	Sriramulu and Sumathi 2017
Shewanella loihica	CuNp	*Enterococcus* sp., *Acinetobacter* sp.	Lv et al. 2018
Bacillus licheniformis	Cadmium sulfide	*Aspergillus niger*	Shivashankarappa and Sanjay 2015
Serratia nematodiphila	Zinc sulfide	*Bacillus subtilis, Oscillatoria limentica*	Malarkodi and Annadurai 2013
Streptomycetes viridogens HM10	AuNP	*Strptomyces noursei*	Balagurunathan et al. 2011
Actinomycetes sp.	AgNP	*Pencillium corylophilum, Pseudomonas aeruginosa*	Abdeen et al. 2014
Plants as Endosymbionts with anti-bacterial property			
Aqueous extract of corn leaf waste (*Zea mays*)	AgNP	*Bacillus cereus, S. aureus, E. coli,* and *S. typhimurium*	Patra and Baek 2017
Extracts of ginger turmeric and garlic	TiO2 NP	*E. coli, S. mutans, S. aureus, P. aeruginosa,* and *K. pneumonia*	Sharma et al. 2021
Phenolic extract of *Memecylon umbellatum* Burm F. plant	AgNP	*B. subtilis, Acenetobacter baumannii, S. paratyphi* and *S. aureus*	Mohamed et al. 2019
Amaranthus blitum L. extract	AgFeONP	*S. aureus* and *E. coli*	Muthukumar et al. 2020
Aqueous extract of *Centella asiatica*	ZnO NP	*E. coli, P. desmolyticum, P. aeruginosa* and *Aureus* sp.	Pavan et al. 2021
Aqueous extract of *Cassia angustifolia*	CeO2 NP	*K. aerogenes, E. coli, P. desmolyticum* and *S. aureus*	Pavan et al. 2021

6.2. Antifungal Properties of Endosymbiont-Mediated Nanoparticles (ES-NP)

Several bacteria, fungi, and plant species act as endosymbionts for producing nanoparticles with antifungal properties (Table 3).

Table 3. Antifungal properties of endosymbiont-mediated nanoparticles

Endosymbiont name	NP	Fungi affected	Reference
Aspergillus sydowii	AgNP	*C. albican, C. glabrata, C. parapsilosis, C. tropicalis, A. fumigatus, A. favus, A. terreus, Sporothrix schenckii,* and *Cryptococcus neoformans*	Wang et al. 2021
Pseudomonas poae isolated from *Allium sativum*	AgNP	*Fusarium graminearum*	Ibrahim et al. 2020
Aspergillus versicolor isolated from plant *Centella asiatica*	AgNP	*Candida albicans* and *Candida nonalbicans*	Netala et al. 2016
Enterococcus faecilus	AgNP, CuNP, ZnONP	*C.albicans, C. neoformans, A.niger* and *F.oxysporum*	Ashajyothi et al. 2016
Amaranthus retroflexus	AgNP	*Macrophomina phaseolina*	Bahram et al. 2017
Periconium sp. (fungal endophyte)	ZnONP	*Candida albicans*	Ganesan et al. 2020
Bacillus sp.	ZnONP	*Pythium aphanidermatum*	Aswani et. al. 2020
Streptomyces zaomyceticus and *Streptomyces pseudogriseolus* isolated from *Oxalis corniculate* L.	CuONP	*Alternaria alternata, Fusarium oxysporum, Pythium ultimum* and *Aspergillus niger*	Hassan 2019
Pantoea anthophila isolated from the stem of *Waltheria indica*	AgNP	*A. niger, C. albicans* and *P. chrysogenum*	Nirmala et al. 2021

Antifungal activity of ES-NP measured in terms of inhibitory effect on mycelial growth exhibited higher-level activity compared to NP synthesized under *in vitro* conditions. Experimental results from the use of SEM and TEM showed that the ES-NPs could efficiently damage the complex fungal cell wall thus making the fungi susceptible to temperature change, pH variation, and other environmental stress. Damaged cell walls of the affected fungi, also resulted in leakage of DNA and proteins from spore cells. Indeed, the ES-NP's could effectively suppress the spore germination and germ tube growth rate. This anti-sporadic effect finds its prominent application in reducing the fungal infection risk in wheat plants. ELISA studies performed on various ES-NP confirmed the reduction of mycotoxin (deoxynivalenol) concentration thereby reducing damage to human and animal health.

6.3. Antioxidant Properties of Endosymbiont Mediated Nanoparticles (ES-NP)

ES-NP's exhibited strong antioxidant properties by scavenging various free radicals like 1, 1-Diphennyl-2-picryl Hydrazyl (DPPH), nitric oxide, hydrogen peroxide, 2,2-azino-bis-3-ethyl benzothiazole-6-sulfonic acid (ABTS). Superior antioxidant potential of EP-NP's was confirmed using various assay techniques like total antioxidant capacity assay and reducing power assay. Silver nanoparticles synthesized from extracts of edible mushroom (*Agaricus bisporus*) and forest mushroom (*Ganoderma lucidum*) have exhibited excellent antioxidant properties (Sriramulu and Sumathi 2017). This activity is due to the transfer of phenolic and flavonoid compounds present in the extract and their ability to donate hydrogen to free radicals and break the macromolecule's damage. Similarly, Nirmala and Sridevi (2021) reported antioxidant properties of Ag nanoparticles synthesized from the extracellular extract of bacterial endophyte *Pantoea anthophila* isolated from the stem of *Waltheria indica.*

6.4. Anti-Cancer Properties of Endosymbiont-Mediated Nanoparticles (ES-NP)

ES-NP's have diverse cellular responses imparting anticancer properties. Reactive oxygen species (ROS) generated by the ES-NP alter mitochondrial signaling pathways, creates oxidative stress in the tumor cells, and injure DNA which gradually leads to cell death. TiO_2 NP prepared using extracts of garlic, turmeric and ginger reported anticancer activity when investigated with KB oral cancer cell lines (Sharma et al. 2021). Whereas, the anti-tumor activity of Ag NP prepared using phenolic extract of *Memecylon umbellatum* investigated on human breast cancer cell lines (MCF-7) reported efficient anti-proliferative activity with an IC_{50}value of 42.19 mg/ml (Mohamed et al. 2019). Similarly, the anti-angiogenic properties of ES-Ag NP and Au NP were successful in inhibiting the proliferation of HUVEC cells (Mukherjee et al. 2005 & 2016). ZnO nanoparticles synthesized from *C. asiatica* leaf recorded higher cytotoxic effects on the HeLa cell line with an IC_{50} value of 232 μg/ml (Pavan et al. 2021). The cytotoxic effects of CeO_2 NPs as studied by Pavan et al. (2021) reported dose-dependent cytotoxic response on human breast cancer cell lines MCF7 with an IC_{50} value of 259 μg/ml.

6.5. Biomedical and Pharmaceutical Applications of ES-NP

As mentioned earlier numerous ES-NP have proven their potential as antimicrobial agents, which has extended their scope for various other biomedical applications like drug delivery, development of anticancer agents, drug targeting agents, anti-biofilm agents, diagnostic and imaging agents, and tissue engineering materials (Chidambaram et al. 2021; Ghosh et al. 2021). Green synthesized AgNP was used to develop 3D hybrid structures based on biomimetic membranes with improved performance and enhanced biomedical activities (Barbinta et al. 2019). Similarly, Zinc oxide NPs are also being considered for use in next-generation biological applications like bio-imaging probes, biosensors, etc. (Jayaseelan et al. 2012). Especially, Ag NP coated fibers were reported to be used in the preparation of optical sensors, biosensors, and other electrical applications (Rout et al. 2017).

6.6. Applications of ES-NP's in Creating a Sustainable Environment

ES-NP's have profound applications for creating a sustainable environment in terms of biofuel generation, dye degradation, water/air purification, bio-fertilizers development, and wastewater treatment. Bio nano CaO NP produced from chicken feather meal oil using chicken eggshell as a precursor as reported by Minglong et al. (2021) was efficiently used for the generation of biodiesel. The biofuel produced was found to be chemical-free owing to its green methodology employed and environmentally beneficial. The work concluded the potential of the produced biodiesel as an economical alternative for the existing fuel resources. Microbes like *Micrococcus aloeverae*, *M. lylae*, *Chlorella pyrenoidosa*, and *Cellolosi microbium* species produce TiO_2 NP which can show dye degradation ability (Fulekar et al. 2018). Likewise, aqueous extract of *Cassia angustifolia* mediated CeO_2 NPs reported efficient photocatalytic degradation of cancer-causing dyes like Methylene Blue and Malachite Green within a span of 3 hrs (Pavan et al. 2021). *Streptomyces griseoruber* induce Au NP, *Cordyceps militaris* fabricates ZnONP, *Ganoderma lucidum* (forest mushroom) produces AgNP which are proven to exhibit catalytic activity by degradation of methylene blue dye. This dye degradation activity can be used as remediation for the purification of wastewater (Li et al. 2019). Green alga *Chlorella pyrenoidosa*, *Aspergillus* sp. was used to synthesize TiO_2NP and AuNP which have a promising role in the removal of environmentally toxic substances like 4-nitro phenol thus supporting bioremediation (Shen et al. 2017; Ghosh et al. 2021).

Conclusion: Applications of ES-NP's in Creating Sustainable Agriculture

Nano biopesticides developed using metals such as Cu, Ag, ZnO, and SiO_2 were reported to possess broad-spectrum efficiency on plant pests and can serve as the best crop protection systems as reviewed by Camelia (2022). ES-NP's have the potential to deliver nutrition, inert ingredients, and desirable genes to specific plant parts and thus aid in improving productivity (Marchiol et al. 2014). Further, research evidence proved the potency of Ag and Au ESNP's in improving seed germination and root elongation (Ribeiro et al. 2019). Nano fertilizers developed using Au and Ag ESNP's provided reduced oxidative stress, toxicity, and enhanced cellular uptake (Chaudhry et al. 2018; De et al. 2021). Above all, Au and Ag-based ESNP's were proved to have potential in water treatment, and agricultural wastewater management which is attributed to their improved photocatalytic degradation abilities imbibed due to the presence of endosymbiont-mediated secondary metabolites on the surface (Palit et al. 2020; Graily-Moradi et al. 2021).

Acknowledgements

The authors acknowledge their respective institutions for encouragement and support.

Conflict of Interests

None to declare.

References

Adnan, Mohd. "Synthesis, characterization, and applications of endophytic fungal nanoparticles." *Inorganic and Nano-Metal Chemistry* 51, no. 2 (2021): 280-287.

AL Asmari, Khalid Mushabbab, Isam M. Abu Zeid, and Atef M. Al-Attar. "Medicinal properties of *Arabica* coffee (*Coffea arabica*) oil: An overview." *Advancements in Life Sciences* 8, no. 1 (2020): 20-29.

Al Salhi, Mohamad S., Kannan Elangovan, Amirtham Jacob A. Ranjit singh, P. Murali, and Sandhanasamy Devanesan. "Synthesis of silver nanoparticles using plant-derived 4-N-methyl benzoic acid and evaluation of the antimicrobial, antioxidant and antitumor activity." *Saudi Journal of Biological Sciences* 26, no. 5 (2019): 970-978.

Alfarano, C., C. E. Andrade, K. Anthony, N. Bahroos, M. Bajec, K. Bantoft, D. Betel, B. Bobechko, K. Boutilier, E. Burgess, K. Buzadzija, R. Cavero, C. D'Abreo, I. Donaldson, D. Dorairajoo, M. J. Dumontier, M. R. Dumontier, V. Earles, R. Farrall, H. Feldman, E. Garderman, Y. Gong, R. Gonzaga, V. Grytsan, E. Gryz, V. Gu, E. Haldorsen, A. Halupa, R. Haw, A. Hrvojic, L. Hurrell, R. Isserlin, F. Jack, F. Juma, A. Khan, T. Kon, S. Konopinsky, V. Le, E. Lee, S. Ling, M. Magidin, J. Moniakis, J. Montojo, S. Moore, B. Muskat, I. Ng, J. P. Paraiso, B. Parker, G. Pintilie, R. Pirone, J. J. Salama, S. Sgro, T. Shan, Y. Shu, J. Siew, D. Skinner, K. Snyder, R. Stasiuk, D. Strumpf, B. Tuekam, S. Tao, Z. Wang, M. White, R. Willis, C. Wolting, S. Wong, A. Wrong, C. Xin, R. Yao, B. Yates, S. Zhang, K. Zheng, T. Pawson, B. F. F. Ouellette, C. W. V. Hogue. "The biomolecular interaction network database and related tools 2005 update." *Nucleic acids Research* 33, no. suppl_1 (2005): D418-D424.

Alurappa, Ramesha, Srinivas Chowdappa, Radhakrishnan Narayanaswamy, Uma Rani Sinniah, Sudipta Kumar Mohanty, and Mallappa Kumara Swamy. "Endophytic fungi and bioactive metabolites production: an update." In *Microbial biotechnology*, pp. 455-482. Springer, Singapore, 2018.

Anandalakshmi, K., J. Venugobal, and V. Ramasamy. "Characterization of silver nanoparticles by green synthesis method using *Pedalium murex* leaf extract and their antibacterial activity." *Applied Nanoscience* 6, no. 3 (2016): 399-408.

Ashajyothi, C., C. Prabhurajeshwar, and Harish K. Handral. "Investigation of antifungal and anti-mycelium activities using biogenic nanoparticles: an eco-friendly approach." *Environmental Nanotechnology, Monitoring & Management* 5 (2016): 81-87.

Bahrami-Teimoori, Bahram, Yaser Nikparast, Mostafa Hojatianfar, Mahdi Akhlaghi, Reza Ghorbani, and Hamid Reza Pourianfar. "Characterisation and antifungal activity of silver nanoparticles biological synthesized by *Amaranthus retroflexus* leaf extract." *Journal of Experimental Nanoscience* 12, no. 1 (2017): 129-139.

Barbinta-Patrascu, Marcela Elisabeta; Badea, Nicoleta; Bacalum, Mihaela; Ungureanu, Camelia; Suica-Bunghez, Ioana Raluca; Iordache, Stefan Marian; Pirvu, Cristian; Zgura, Irina; Maraloiu, Valentin Adrian. "3D hybrid structures based on biomimetic membranes and *Caryophyllus aromaticus* - green synthesized nano-silver with improved bio-performances." *Materials Science and Engineering: C.* (2019): 120–137. doi:10.1016/j.msec.2019.03.069.

Beulah, G., D. Divya, N. S. Kumar, M. V. N. Sravya, K. Govinda Rao, Anjani Devi Chintagunta, G. Divya, S. Hari Chandana, B. D. Blessy, and G. Simhachalam. "Purification and characterization of bioactive compounds extracted from *Suaeda maritima* leaf and its impact on pathogenicity of *Pseudomonas aeruginosa* in *Catla catla* fingerlings." *AMB Express* 11, no. 1 (2021): 1-10.

Bhattacharya, R., P. Mukherjee, Z. Xiong, A. Atala, S. Soker, and D. Mukhopadhyay. "Gold nanoparticles inhibit VEGF165-induced proliferation of HUVEC cells." *Nano Letters* 4, no. 12 (2004): 2479-2481.

Binnig, Gerd, Calvin F. Quate, and Ch Gerber. "Atomic force microscope." *Physical review letters* 56, no. 9 (1986): 930.

Biswas, Abhijit, Ilker S. Bayer, Alexandru S. Biris, Tao Wang, Enkeleda Dervishi, and Franz Faupel. "Advances in top-down and bottom-up surface nanofabrication: Techniques, applications & future prospects." *Advances in colloid and interface science* 170, no. 1-2 (2012): 2-27.

Borges, Keyller Bastos, Warley De Souza Borges, Mônica Tallarico Pupo, and Pierina Sueli Bonato. "Endophytic fungi as models for the stereoselective biotransformation of thioridazine." *Applied Microbiology and Biotechnology* 77, no. 3 (2007): 669-674.

Brown, Paul D. "Transmission Electron Microscopy-A Textbook for Materials Science, by David B. Williams and C. Barry Carter." *Microscopy and Microanalysis* 5, no. 6 (1999): 452-453.

Casadevall, Arturo. "Evolution of intracellular pathogens." *Annual Reviews in Microbiology* 62 (2008): 19-33.

Chaudhry, Niladri, Sourabh Dwivedi, Vasvi Chaudhry, Anuradha Singh, Quaiser Saquib, Ameer Azam, and Javed Musarrat. "Bio-inspired nanomaterials in agriculture and food: Current status, foreseen applications, and challenges." *Microbial pathogenesis* 123 (2018): 196-200.

Chintagunta, Anjani Devi, Ashutosh Kumar, SP Jeevan Kumar, and Madan L. Verma. "Contribution of metallic nanomaterials in algal biofuel production." *In Metal and metal oxides for energy and electronics*, pp. 331-353. Springer, Cham, 2021.

Chintagunta, Anjani Devi, Gaetano Zuccaro, Mahesh Kumar, SP Jeevan Kumar, Vijay Kumar Garlapati, Pablo D. Postemsky, NS Sampath Kumar, Anuj K. Chandel, and Jesus Simal-Gandara. "Biodiesel production from Lignocellulosic biomass using Oleaginous microbes: Prospects for integrated biofuel production." *Frontiers in Microbiology* 12 (2021).

Chintagunta, Anjani Devi, Mahesh Kumar, NS Sampath Kumar, and SP Jeevan Kumar. "Differential diagnosis and possible therapeutics for coronavirus disease 2019." *In Diagnostic Strategies for COVID-19 and other Coronaviruses*, pp. 51-71. Springer, Singapore, 2020.

Chintagunta, Anjani Devi, Sanjana Nalluru, and Sampath Kumar NS. "Nanotechnology: an emerging approach to combat COVID-19." *Emergent Materials* (2021): 1-12.

Clifford, D. M., C. E. Castano, and J. V. Rojas. "Supported transition metal nanomaterials: nanocomposites synthesized by ionizing radiation." *Radiation Physics and Chemistry* 132 (2017): 52-64.

Condon, James B. *Surface area and porosity determinations by physisorption: measurement, classical theories, and quantum theory*. Elsevier, 2019.

Corbari, Laure, M-A. Cambon-Bonavita, Gary J. Long, Fernande Grandjean, Magali Zbinden, Françoise Gaill, and Philippe Compère. "Iron oxide deposits associated with the ectosymbiotic bacteria in the hydrothermal vent shrimp *Rimicaris exoculata*." *Biogeosciences* 5, no. 5 (2008): 1295-1310.

Crewe, Albert V., J. Wall, and J. Langmore. "Visibility of single atoms." *Science* 168, no. 3937 (1970): 1338-1340.

Cui, Chun-Hua, and Shu-Hong Yu. "Engineering interface and the surface of noble metal nanoparticle nanotubes toward enhanced catalytic activity for fuel cell applications." *Accounts of chemical research* 46, no. 7 (2013): 1427-1437.

Das, Vidhya Lakshmi, Roshmi Thomas, Rintu T. Varghese, E. V. Soniya, Jyothis Mathew, and E. K. Radhakrishnan. "Extracellular synthesis of silver nanoparticles by the *Bacillus strain CS 11* isolated from the industrialized area." *3 Biotech* 4, no. 2 (2014): 121-126.

De, Anindita, and Dristie Kalita. "Bio-Fabricated Gold and Silver Nanoparticle-Based Plasmonic Sensors for Detection of Environmental Pollutants: An Overview." *Critical Reviews in Analytical Chemistry* (2021): 1-17.

Deokar, Archana R. "Nanomaterials: Synthesis, properties, and applications." *Dnyanamay Journal* 3, no. 1 (2017): 16-19.

Dutta, Devanushi, Keshab Chandra Puzari, Robin Gogoi, and Pranab Dutta. "Endophytes: exploitation as a tool in plant protection." *Brazilian Archives of Biology and Technology* 57 (2014): 621-629.

Ealia, S. Anu Mary, and M. P. Saravanakumar. "A review on the classification, characterization, synthesis of nanoparticles and their application." *In IOP Conference Series: Materials Science and Engineering*, vol. 263, no. 3, p. 032019. IOP Publishing, 2017.

Fariq, Anila, Tabeer Khan, and Azra Yasmin. "Microbial synthesis of nanoparticles and their potential applications in biomedicine." *Journal of Applied Biomedicine* 15, no. 4 (2017): 241-248.

Galarneau, Anne, Dirk Mehlhorn, Flavien Guenneau, Benoit Coasne, François Villemot, Delphine Minoux, Cindy Aquino, and Jean-Pierre Dath. "Specific surface area determination for microporous/mesoporous materials: The case of mesoporous FAU-Y zeolites." *Langmuir* 34, no. 47 (2018): 14134-14142.

Ganesan, V., M. Hariram, S. Vivekanandhan, and S. Muthuramkumar. "*Periconium* sp. (endophytic fungi) extract mediated sol-gel synthesis of ZnO nanoparticles for antimicrobial and antioxidant applications." *Materials Science in Semiconductor Processing* 105 (2020): 104739.

Ghosh, Shubhrima, Razi Ahmad, Md Zeyaullah, and Sunil Kumar Khare. "Microbial nano-factories: synthesis and biomedical applications." *Frontiers in Chemistry* 9 (2021): 194.

Gommes, Cedric J., Sebastian Jaksch, and Henrich Frielinghaus. "Small-angle scattering for beginners." *Journal of Applied Crystallography* 54, no. 6 (2021).

Graily-Moradi, Fatemeh, and Behnam Asgari Lajayer. "Nanoinsecticides: Preparation, Application, and Mode of Action." *Nanotechnology Applications in Health and Environmental Sciences* (2021): 385-404.

Guccione, Pietro, Mattia Lopresti, Marco Milanesio, and Rocco Caliandro. "Multivariate Analysis Applications in X-ray Diffraction." *Crystals* 11, no. 1 (2021): 12.

Gumerova, Nadiia I., and Annette Rompel. "Synthesis, structures, and applications of electron-rich polyoxometalates." *Nature Reviews Chemistry* 2, no. 2 (2018): 1-20.

Hafez, Yaser M., Kotb A. Attia, Said Kamel, Salman F. Alamery, Safaa El-Gendy, Abdullah A. Al-Doss, Farid Mehiar, Abdelhalim I. Ghazy, Eid I. Ibrahim, and Khaled AA Abdelaal. "*Bacillus subtilis* as a bio-agent combined with nano molecules can control powdery mildew disease through histochemical and physio biochemical changes in cucumber plants." *Physiological and Molecular Plant Pathology* 111 (2020): 101489.

Harne, Shrikant, Ashwini kumar Sharma, Mayur Dhaygude, Shreeram Joglekar, Kisan Kodam, and Manish Hudlikar. "Novel route for rapid biosynthesis of copper nanoparticles using aqueous extract of *Calotropis procera* L. latex and their cytotoxicity on tumor cells." *Colloids and Surfaces B: Biointerfaces* 95 (2012): 284-288.

Hassan, Saad El-Din, Amr Fouda, Ahmed A. Radwan, Salem S. Salem, Mohammed G. Barghoth, Mohamed A. Awad, Abdullah M. Abdo, and Mamdouh S. El-Gamal. "Endophytic actinomycetes *Streptomyces* spp. mediated biosynthesis of copper oxide nanoparticles as a promising tool for biotechnological applications." *Journal of Biological Inorganic Chemistry* 24, no. 3 (2019): 377-393.

Ibrahim, Ezzeldin, Muchen Zhang, Yang Zhang, Afsana Hossain, Wen Qiu, Yun Chen, Yanli Wang, Wenge Wu, Guochang Sun, and Bin Li. "Green synthesis of silver nanoparticles using endophytic bacteria isolated from garlic and its antifungal activity against wheat *Fusarium* head blight pathogen *Fusarium graminearum*." *Nanomaterials* 10, no. 2 (2020): 219.

Iravani, Siavash. "Green synthesis of metal nanoparticles using plants." *Green Chemistry* 13, no. 10 (2011): 2638-2650.

Ismail, M. A., Amin, M. A., Eid, A. M., Hassan, S. E.-D., Mahgoub, H. A. M., Lashin, I., Abdelwahab, A. T., Azab, E., Gobouri, A. A., Elkelish, A., & Fouda, A. (2021). "Comparative study between exogenously applied plant growth hormones versus metabolites of microbial Endophytes as plant growth-promoting for *Phaseolus vulgaris* L." Cells 10, no. 5 (2021): 1059.

Jadoun, Sapana, Rizwan Arif, Nirmala Kumari Jangid, and Rajesh Kumar Meena. "Green synthesis of nanoparticles using plant extracts: A review." *Environmental Chemistry Letters* 19, no. 1 (2021): 355-374.

Jan, Rahmatullah, Muhammad Aaqil Khan, Sajjad Asaf, In-Jung Lee, and Kyung Min Kim. "Metal resistant endophytic bacteria reduce cadmium, nickel toxicity, and enhance expression of metal stress-related genes with improved growth of *Oryza sativa*, via regulating its antioxidant machinery and endogenous hormones." *Plants* 8, no. 10 (2019): 363.

Jayaseelan, C., A. Abdul Rahuman, A. Vishnu Kirthi, S. Marimuthu, T. Santhoshkumar, A. Bagavan, K. Gaurav, L. Karthik, and K. V. Bhaskara Rao. "Novel microbial route to synthesize ZnO nanoparticles using *Aeromonas hydrophila* and their activity against pathogenic bacteria and fungi." *Spectrochimica Acta Part A: Molecular and Biomolecular Spectroscopy* 90 (2012): 78-84.

Jung, Dubravka Šišak, Tilman Donath, Oxana Magdysyuk, and Jozef Bednarcik. "High-energy X-ray applications: current status and new opportunities." *Powder Diffraction* 32, no. S2 (2017): S22-S27.

Kaki, Satish Babu, Anjani Devi Chintagunta, A. Naga Prasad, N. S. Kumar, Vijaya R. Dirisala, M. Sai Krishna, S. J. K. Naidu, and B. Ramesh. "Production and purification of recombinant glargine insulin from *Escherichia coli BL-21* strain." *Emergent Materials* (2021): 1-12.

Karthik, L., Gaurav Kumar, Tarun Keswani, Arindam Bhattacharyya, B. Palakshi Reddy, and K. V. Bhaskara Rao. "Marine actinobacterial mediated gold nanoparticles synthesis and their antimalarial activity." *Nanomedicine: Nanotechnology, Biology and Medicine* 9, no. 7 (2013): 951-960.

Khalil, Ali Talha, Muhammad Ovais, Javed Iqbal, Arbab Ali, Muhammad Ayaz, Manzar Abbas, Irshad Ahmad, and Hari Parsad Devkota. "Microbes-mediated synthesis strategies of metal nanoparticles and their potential role in cancer therapeutics." In *Seminars in Cancer Biology*. Academic Press, 2021.

Khan, Ibrahim, Khalid Saeed, and Idrees Khan. "Nanoparticles: Properties, applications, and toxicities." *Arabian journal of chemistry* 12, no. 7 (2019): 908-931.

Koirtyohann, S. R. "A history of atomic absorption spectroscopy from an academic perspective." *Analytical Chemistry* 63, no. 21 (1991): 1024A-1031A.

Kosasih, Felix Utama, Stefania Cacovich, Giorgio Divitini, and Caterina Ducati. "Nanometric chemical analysis of beam-sensitive materials: A case study of STEM-EDX on Perovskite Solar Cells." *Small Methods* 5, no. 2 (2021): 2000835.

Kumar, Ashutosh, SP Jeevan Kumar, Anjani Devi Chintagunta, Dinesh K. Agarwal, Govind Pal, Arvind Nath Singh, and Jesus Simal-Gandara. "Biocontrol potential of *Pseudomonas stutzeri* endophyte from *Withania somnifera* (Ashwagandha) seed extract against pathogenic *Fusarium oxysporum* and *Rhizoctonia solani*." *Archives of Phytopathology and Plant Protection* (2021a): 1-18.

Kumar, NS Sampath, Norizah Mhd Sarbon, Sandeep Singh Rana, Anjani Devi Chintagunta, S. Prathibha, Satheesh Kumar Ingilala, SP Jeevan Kumar, B. Sai Anvesh, and Vijaya Ramu Dirisala. "Extraction of bioactive compounds from *Psidium guajava* leaves and its utilization in preparation of jellies." *AMB Express* 11, no. 1 (2021b): 1-9.

Kumar, Rajesh, Rohit Kumar Mishra, Vani Mishra, Afifa Qidwai, Anand Pandey, Shashi Kant Shukla, Madhu Pandey, Ashutosh Pathak, and Anupam Dikshit. "Detoxification and tolerance of heavy metals in plants." In *Plant Metal Interaction*, pp. 335-359. Elsevier, 2016.

Kumar, SP Jeevan, Anjani Devi Chintagunta, Y. Mohan Reddy, Ashutosh Kumar, Dinesh K. Agarwal, Govind Pal, and Jesus Simal-Gandara. "Application of phenolic extraction strategies and Evaluation of the antioxidant activity of Peanut skins as an agricultural by-product for food industry." *Food Analytical Methods* (2021c): 1-12.

Kumar, SP Jeevan, Anjani Devi Chintagunta, Y. Mohan Reddy, Loïc Rajjou, Garlapati Vijay Kumar, Dinesh K. Agarwal, S. Rajendra Prasad, and Jesus Simal-Gandara. "Implications of reactive oxygen and nitrogen species in seed physiology for sustainable crop productivity under changing climate conditions." *Current Plant Biology* (2021d): 100197.

Kumar, SP Jeevan, NS Sampath Kumar, and Anjani Devi Chintagunta. "Bioethanol production from cereal crops and lignocelluloses rich agro-residues: prospects and challenges." *SN Applied Sciences* 2, no. 10 (2020): 1-11.

Kumar, Sunil, Rashmi Malhotra, and Dinesh Kumar. "*Euphorbia hirta*: Its chemistry, traditional and medicinal uses, and pharmacological activities." *Pharmacognosy reviews* 4, no. 7 (2010): 58.

Kurbanoglu, Sevinc, and Sibel A. Ozkan. "Electrochemical carbon-based nanosensors: A promising tool in pharmaceutical and biomedical analysis." *Journal of pharmaceutical and biomedical analysis* 147 (2018): 439-457.

Lahiri, Dibyajit, Moupriya Nag, Hassan I. Sheikh, Tanmay Sarkar, Hisham Atan Edinur, Siddhartha Pati, and Rina Rani Ray. "Microbiologically-synthesized nanoparticles and their role in silencing the biofilm signaling cascade." *Frontiers in Microbiology* 12 (2021): 180.

Lu, Junling, Jeffrey W. Elam, and Peter C. Stair. "Atomic layer deposition—Sequential self-limiting surface reactions for advanced catalyst "bottom-up" synthesis." *Surface Science Reports* 71, no. 2 (2016): 410-472.

Mane, R., and A. Vedamurthy. "The fungal endophytes: Sources and future prospects." *Journal of Medicinal Plants Studies* 6 (2018): 121-126.

Mishra, Sandhya, and H. B. Singh. "Biosynthesized silver nanoparticles as a nano weapon against phytopathogens: exploring their scope and potential in agriculture." *Applied microbiology and biotechnology* 99, no. 3 (2015): 1097-1107.

Morosanu, Constantin E. *Thin films by chemical vapor deposition*. Vol. 7. Elsevier, 2016.

Mueller, Ulrich G., and Joel L. Sachs. "Engineering microbiomes to improve plant and animal health." *Trends in Microbiology* 23, no. 10 (2015): 606-617.

Mukherjee, Priyabrata, Resham Bhattacharya, Ping Wang, Ling Wang, Sujit Basu, Janice A. Nagy, Anthony Atala, Debabrata Mukhopadhyay, and Shay Soker. "Antiangiogenic properties of gold nanoparticles." *Clinical cancer research* 11, no. 9 (2005): 3530-3534.

Mukherjee, Sudip, and Chitta Ranjan Patra. "Therapeutic application of anti-angiogenic nanomaterials in cancers." *Nanoscale* 8, no. 25 (2016): 12444-12470.

Mullan, D. "Scanning electron microscopy 1928–1965." *Scanning* 17, no. 3 (1995): 175-185.

Muthukumar, Harshiny, Santosh Kumar Palanirajan, Manoj Kumar Shanmugam, and Sathyanarayana N. Gummadi. "Plant extract mediated synthesis enhanced the functional properties of silver ferrite nanoparticles over chemical mediated synthesis." *Biotechnology Reports* 26 (2020): e00469.

Nagajyothi, P. C., T. V. M. Sreekanth, Jae-il Lee, and Kap Duk Lee. "Mycosynthesis: antibacterial, antioxidant and antiproliferative activities of silver nanoparticles synthesized from *Inonotus obliquus* (Chaga mushroom) extract." *Journal of Photochemistry and Photobiology B: Biology* 130 (2014): 299-304.

Naik, B. Shankar. "Biosynthesis of silver nanoparticles from endophytic fungi and their role in plant disease management." *Microbial Endophytes* (2020): 307-321.

Narayanan, Theyencheri, Michael Sztucki, Pierre Van Vaerenbergh, Joachim Léonardon, Jacques Gorini, Laurent Claustre, Franc Sever, John Morse, and Peter Boesecke. "A multipurpose instrument for time-resolved ultra-small-angle and coherent X-ray scattering." *Journal of applied crystallography* 51, no. 6 (2018): 1511-1524.

Netala, Vasudeva Reddy, Venkata Subbaiah Kotakadi, Pushpalatha Bobbu, Susmila Aparna Gaddam, and Vijaya Tartte. "Endophytic fungal isolate mediated biosynthesis of silver nanoparticles and their free radical scavenging activity and antimicrobial studies." *3 Biotech* 6, no. 2 (2016): 1-9.

Newman, David J., and Gordon M. Cragg. "Natural products as sources of new drugs over the last 25 years." *Journal of natural products* 70, no. 3 (2007): 461-477.

Nirmala, C., Sridevi, M. "Characterization, antimicrobial and antioxidant evaluation of bio-fabricated silver nanoparticles from endophytic *Pantoea anthophila*." *J Inorg Organommet Polym.* (2021): 31, 3711–3725.

Palit, Sukanchan, and Chaudhery Mustansar Hussain. "Functionalization of nanomaterials for industrial applications: recent and future perspectives." In *Handbook of Functionalized Nanomaterials for Industrial Applications*, pp. 3-14. Elsevier, 2020.

Pandey, Garima. "Challenges and future prospects of agri-nanotechnology for sustainable agriculture in India." *Environmental Technology & Innovation* 11 (2018): 299-307.

Patra, Jayanta Kumar, and Kwang-Hyun Baek. "Antibacterial activity and synergistic antibacterial potential of biosynthesized silver nanoparticles against foodborne pathogenic bacteria along with its anticandidal and antioxidant effects." *Frontiers in Microbiology* 8 (2017): 167.

Pecora, R. "Dynamic light scattering measurement of nanometer particles in liquids." *Journal of nanoparticle research* 2, no. 2 (2000): 123-131.

Prasad, Kumar Suranjit, Pooja Gandhi, and Kaliaperumal Selvaraj. "Synthesis of green nano iron particles (GnIP) and their application in adsorptive removal of As (III) and As (V) from aqueous solution." *Applied Surface Science* 317 (2014): 1052-1059.

Puri, Anu, Kristin Loomis, Brandon Smith, Jae-Ho Lee, Amichai Yavlovich, Eliahu Heldman, and Robert Blumenthal. "Lipid-based nanoparticles as pharmaceutical drug carriers: from concepts to clinic." *Critical Reviews in Therapeutic Drug Carrier Systems* 26, no. 6 (2009).

Radchenko, Martha V., Kimihiro Tanaka, Rungaroon Waditee, Sawako Oshimi, Yasutomo Matsuzaki, Masahiro Fukuhara, Hiroshi Kobayashi, Teruhiro Takabe, and Tatsunosuke Nakamura. "Potassium/proton antiport system of *Escherichia coli.*" *Journal of Biological Chemistry* 281, no. 29 (2006): 19822-19829.

Rana, Kusam Lata, Divjot Kour, Neelam Yadav, and Ajar Nath Yadav. "Endophytic microbes in nanotechnology: current development, and potential biotechnology applications." In *Microbial Endophytes*, pp. 231-262. Woodhead Publishing, 2020.

Raut, Rajesh Warluji, Vijay Damodhar Mendhulkar, and Sahebrao Balaso Kashid. "Photosensitized synthesis of silver nanoparticles using *Withania somnifera* leaf powder and silver nitrate." *Journal of Photochemistry and Photobiology B: Biology* 132 (2014): 45-55.

Ravi, Aswani, Vipina Vinod Thazhe Nandayipurath, Sukanya Rajan, Simi Asma Salim, Nejumal Kannankeril Khalid, Charuvila T. Aravindakumar, and Radhakrishnan Edayileveettil Krishnankutty. "Effect of zinc oxide nanoparticle supplementation on the enhanced production of surfactin and iturin lipopeptides of endophytic *Bacillus sp. Fcl1* and its ameliorated antifungal activity." *Pest Management Science* 77, no. 2 (2021): 1035-1041.

Ribeiro, Caroline A. S., Lindomar J. C. Albuquerque, Carlos E. de Castro, Bruno L. Batista, André LM de Souza, Brunno L. Albuquerque, Morgana S. Zilse, Ismael C. Bellettini, and Fernando C. Giacomelli. "One-pot synthesis of sugar-decorated gold nanoparticles with reduced cytotoxicity and enhanced cellular uptake." *Colloids and Surfaces A: Physicochemical and Engineering Aspects* 580 (2019): 123690.

Rodriguez, R. J., J. F. White Jr, A. E. Arnold, and A. RS and Redman. "Fungal endophytes: diversity and functional roles." *New Phytologist* 182, no. 2 (2009): 314-330.

Rose, Harald H. "Optics of high-performance electron microscopes." *Science and Technology of Advanced Materials* 9, no. 1 (2008): 014107.

Rout, Sanjay Kumar, Bankim Chandra Tripathy, Payodhar Padhi, Bikash Ranjan Kar, and Krushna Gopal Mishra. "A green approach to produce silver nanoparticles coated agro waste fibers for special applications." *Surfaces and Interfaces* 7 (2017): 87-98.

Rudenberg, H. Gunther, and Paul G. Rudenberg. "Origin and background of the invention of the electron microscope: Commentary and expanded notes on the memoir of Reinhold Rüdenberg." *Advances in Imaging and Electron Physics* 160 (2010): 207-286.

Sakalidis, Monique L., Giles E. StJ Hardy, and Treena I. Burgess. "Class III endophytes, clandestine movement amongst hosts and habitats and their potential for disease; a focus on *Neofusicoccum australe*." *Australasian Plant Pathology* 40, no. 5 (2011): 510-521.

Salapaka, Srinivasa M., and Murti V. Salapaka. "Scanning probe microscopy." *IEEE Control Systems Magazine* 28, no. 2 (2008): 65-83.

Salem, Salem S., and Amr Fouda. "Green synthesis of metallic nanoparticles and their prospective biotechnological applications: an overview." *Biological trace element research* 199, no. 1 (2021): 344-370.

Shaheena, Shaik, Anjani Devi Chintagunta, Vijaya Ramu Dirisala, and NS Sampath Kumar. "Extraction of bioactive compounds from *Psidium guajava* and their application in dentistry." *AMB express* 9, no. 1 (2019): 1-9.

Sharma, Anjali, and Ashok Kakkar. "Designing dendrimer and miktoarm polymer-based multi-tasking nanocarriers for efficient medical therapy." *Molecules* 20, no. 9 (2015): 16987-17015.

Sharma, D., P. Shandilya, N. K. Saini, P. Singh, V. K. Thakur, R. V. Saini, D. Mittal, G. Chandan, V. Saini, and A. K. Saini. "Insights into the synthesis and mechanism of green synthesized antimicrobial nanoparticles, answer to the multidrug resistance." *Materials Today Chemistry* 19 (2021): 100391.

Singh, Archana, Dheeraj K. Singh, Ravindra N. Kharwar, James F. White, and Surendra K. Gond. "Fungal endophytes as efficient sources of plant-derived bioactive compounds and their prospective applications in natural product drug discovery: Insights, avenues, and challenges." *Microorganisms* 9, no. 1 (2021): 197.

Singh, Jagpreet, Tanushree Dutta, Ki-Hyun Kim, Mohit Rawat, Pallabi Samddar, and Pawan Kumar. "'Green' synthesis of metals and their oxide nanoparticles: applications for environmental remediation." *Journal of nanobiotechnology* 16, no. 1 (2018): 1-24.

Singh, R. P., Anjani Devi Chintagunta, Dinesh K. Agarwal, R. S. Kureel, and S. P. Jeevan Kumar. "Varietal replacement rate: Prospects and challenges for global food security." *Global Food Security* 25 (2020): 100324.

Song, Hongyuan, Wenbo Wang, Ping Zhao, Zhongtian Qi, and Shihong Zhao. "Cuprous oxide nanoparticles inhibit angiogenesis via down-regulation of VEGFR2 expression." *Nanoscale* 6, no. 6 (2014): 3206-3216.

Sriramulu, Mohana, and Shanmugam Sumathi. "Photocatalytic, antioxidant, antibacterial and anti-inflammatory activity of silver nanoparticles synthesized using forest and edible mushroom." *Advances in Natural Sciences: Nanoscience and Nanotechnology* 8, no. 4 (2017): 045012.

Stephánsson, A. "New methods for the direct determination of dissolved inorganic, organic, and total carbon in natural waters by Reagent-Free? Ion Chromatography and inductively coupled plasma atomic emission, spectrometry." *Anal Chim Acta* 582 (2007): 69-74.

Stetefeld, Jörg, Sean A. McKenna, and Trushar R. Patel. "Dynamic light scattering: a practical guide and applications in biomedical sciences." *Biophysical reviews* 8, no. 4 (2016): 409-427.

Strobel, Gary, and Bryn Daisy. "Bioprospecting for microbial endophytes and their natural products." *Microbiology and molecular biology reviews* 67, no. 4 (2003): 491-502.

Strobel, Gary, Bryn Daisy, Uvidelio Castillo, and James Harper. "Natural products from endophytic microorganisms." *Journal of Natural products* 67, no. 2 (2004): 257-268.

Takaku, Yasuharu, Hiroshi Suzuki, Isao Ohta, Takami Tsutsui, Haruko Matsumoto, Masatsugu Shimomura, and Takahiko Hariyama. "A 'NanoSuit' surface shield successfully protects organisms in a high vacuum: observations on living organisms in an FE-SEM." *Proceedings of the Royal Society B: Biological Sciences* 282, no. 1802 (2015): 20142857.

Tan, Yiwei, Xinhua Dai, Yongfang Li, and Daoben Zhu. "Preparation of gold, platinum, palladium and silver nanoparticles by the reduction of their salts with a weak reductant–potassium bitartrate." *Journal of Materials Chemistry* 13, no. 5 (2003): 1069-1075.

Tang, Xiao-Feng, Zhen-Guo Yang, and Wei-Jiang Wang. "A simple way of preparing high-concentration and high-purity nano copper colloid for conductive ink in inkjet printing technology." *Colloids and Surfaces A: Physicochemical and Engineering Aspects* 360, no. 1-3 (2010): 99-104.

Tiwari, Pragya, Anjani Devi Chintagunta, Vijaya R. Dirisala, and NS Sampath Kumar. "Legume Derived Bioactive Peptides." In *Sustainable Agriculture Reviews* 45, pp. 29-52. Springer, Cham, 2020.

Tiwari, Pragya, Hanhong Bae. "Horizontal gene transfer and Endophytes: an implication for the acquisition of novel traits." *Plants* 2020, 9, 305.

Tiwari, Pragya, Hanhong Bae. 2022. "Endophytic Fungi: Key insights, emerging prospects, and challenges in natural product drug discovery." *Microorganisms* 10, 360 (2022). https://doi.org/10.3390/microorganisms10020360.

Tiwari, Pragya, Yashdeep Srivastava, and Hanhong Bae. "Trends of pharmaceutical design of Endophytes as anti-infective." *Current Topics in Medicinal Chemistry*, 21, no. 17 (2021a): 1572-1586.

Tiwari, Pragya, Yashdeep Srivastava, Y, Mangalam Bajpai, M, Abhishek Sharma. "Bioactive metabolites from natural sources: Prospects and significance in drug discovery and research." *Bioingene PSJ*, no. 1, (2021b): 1–14.

Tiwari, Pragya. "Sustainable agriculture and nanotechnologies for food and nutraceutical production- An Update". In Chen J. (edited), *Plant and Nanoparticles*, Springer book series, (2022), ISBN 978-981-19-2502-3.

Ungureanu, Camelia. "Nano biopesticide: today and future perspectives." In *Biopesticides*, pp. 201-206. Woodhead Publishing, 2022.

Uzair, Bushra, Ayesha Liaqat, Haroon Iqbal, Bouzid Menaa, Anam Razzaq, Gobika Thiripuranathar, Nosheen Fatima Rana, and Farid Menaa. "Green and cost-effective synthesis of metallic nanoparticles by algae: Safe methods for translational medicine." *Bioengineering* 7, no. 4 (2020): 129.

Vijayanandan, Ajuy Sundar, and Raj Mohan Balakrishnan. "Biosynthesis of cobalt oxide nanoparticles using endophytic fungus *Aspergillus nidulans*." *Journal of environmental management* 218 (2018): 442-450.

Voiry, Damien, Hyeon Suk Shin, Kian Ping Loh, and Manish Chhowalla. "Low-dimensional catalysts for hydrogen evolution and CO_2 reduction." *Nature Reviews Chemistry* 2, no. 1 (2018): 1-17.

Wang, Dongyang, Baiji Xue, Lin Wang, Yidi Zhang, Lijun Liu, and Yanmin Zhou. "Fungus-mediated green synthesis of nano-silver using *Aspergillus sydowii* and its antifungal/antiproliferative activities." *Scientific Reports* 11, no. 1 (2021): 1-9.

Willner, Itamar, Ronan Baron, and Bilha Willner. "Integrated nanoparticle–biomolecule systems for biosensing and bioelectronics." *Biosensors and Bioelectronics* 22, no. 9-10 (2007): 1841-1852.

Wolcott, Abraham. Zero-dimensional (0D), one-dimensional (1D), and two-dimensional (2D) nanostructures for sustainable energy production and biolabeling applications. University of California, Santa Cruz, 2009.

Xi, Baojuan, Xiaona Chu, Jiangyong Hu, Charanjit Singh Bhatia, Aaron James Danner, and Hyunsoo Yang. "Preparation of $Ag/TiO_2/SiO_2$ films via photo-assisted deposition and adsorptive self-assembly for catalytic bactericidal application." *Applied surface science* 311 (2014): 582-592.

Yadav, Thakur Prasad, R. Manohar Yadav, and D. Pratap Singh. "Mechanical milling: a top-down approach for the synthesis of nanomaterials and nanocomposites." *Nanoscience and Nanotechnology* 2, no. 3 (2012): 22-48.

Zeng, Haibo, Xi-Wen Du, Subhash C. Singh, Sergei A. Kulinich, Shikuan Yang, Jianping He, and Weiping Cai. "Nanomaterials via laser ablation/irradiation in liquid: a review." *Advanced Functional Materials* 22, no. 7 (2012): 1333-1353.

Zhang, Minglong, Ganesan Ramya, Kathirvel Brindhadevi, Mishal Alsehli, Ashraf Elfasakhany, Changlei Xia, Nguyen Thuy Lan Chi, and Arivalagan Pugazhendhi. "Microwave-assisted biodiesel production from chicken feather meal oil using Bio-Nano Calcium oxide derived from the chicken eggshell." *Environmental Research* (2021): 112509.

Chapter 9

Contributions of Computational Biology Resources in Endophyte Biology and Research

Ankit Singh[1,*], Abhay Tiwari[1,*], Gourav Choudhir[1,*], Anu Dalal[2,*], Anuj Rana[3], Pravin Kumar[4], Simranjeet Kaur[5] and Avneesh Kumar[5,†]

[1]Centre for Rural Development and Technology, Indian Institute of Technology Delhi, Hauz Khas, New Delhi, India
[2]Department of Chemistry, Indian Institute of Technology Delhi, Hauz Khas, New Delhi, India
[3]Department of Microbiology, CCS Haryana Agriculture University, Hisar, Haryana, India
[4]Department of Computer Science and Engineering, Akal University, Talwandi Sabo, Bathinda, Punjab, India
[5]Department of Botany, Akal University, Talwandi Sabo, Bathinda, Punjab, India

Abstract

Endophytes are endosymbionts that live inside plants. Endophytes assist in enhancing the plants' fitness throughout their life span via modulation of various activities such as enzymatic activity, immune responses, bioactive metabolites accumulations, genetic regulation, etc. The enzymes of endophytic origins help in the effective coordination of various processes associated with metabolism, catabolism, and the host's defense system. Endophytes trigger the host plant by the production of

---

*Author contributed equally.
†Corresponding Author's Email: avnishenator@gmail.com.

In: Endophytes
Editor: Pragya Tiwari
ISBN: 979-8-88697-045-6

macromolecules or micromolecules and influence the upregulation and downregulation process of genes. The endophytes stimulated plant development leads by the impact of endophytes macromolecules/small molecules on the host plants at various levels such as gene expression, gene regulation, and functionality of the macromolecules/small molecules. Computational biology helps in the understanding of the central dogma at various levels, from structurality to functionality. Computational genomics assists in interpreting gene expression profiling, gene network pathway exploration, and interaction of the gene. Computational proteomics help in the understanding of the protein structure and functionality. The interaction of the protein with the components from the endophytes helps to unveil the endophyte's roles in the plants' growth and development.

Keywords: computational tools, endophytes identification, EMBOSS seqret, host gene regulation, phenotypic mapping

1. Introduction

The term endophyte refers to "in the plant" where "endon" means within and "phyton" means plant. The term is used for broad referencing to an array of microorganisms that have the ability to colonize their potential hosts and includes groups like fungi (Stone et al. 2000), plants (Marler et al. 1999), bacteria (Kobayashi and Palumbo 2000), and insects in plants (Feller 1995). It can also be used to define the relationship where algae live within another algae (Peters 1996). The interaction of mycorrhizal fungi with the roots of the host plant is considered an endophytic relationship by some authors (reviewed by Sieber 2007). However, due to the fact that plants and fungi involved in mycorrhizal interactions are existentially dependent on each other as specialized surfaces help in nutrient transfer, sometimes mycorrhizal relationships are considered different from endophytic interactions. There have been incidences where mycorrhizal fungi have become pathogenic and even dark septate endophytes (DSE) have assumed the role of mycorrhizal fungi (Jumpponen and Trappe 1998). Some examples of saprophytic behavior of fungal root endophytes are *Oidiodendron maius* (Rice and Currah) and *Phialocephala fortinii* (Jumpponen and Trappe 1998; Jumpponen et al. 1998). There are two major groups of bacteria that have the ability to behave as plant growth-promoting bacteria (PGPB) (i) Rhizospheric bacteria are typically found around the roots of the plants (ii) bacterial endophytes which can be found living inside tissues of the plants. However, they can also be found living independently in soil. The plant-growth promotion mechanism for both the rhizospheric and endophytic PGPB is similar. There exist multiple definitions of the word endophytes but the most commonly accepted is that those organisms which cause infections that are inconspicuous, the infected host tissues are at least transiently symptomless, and the microbial colonization can be demonstrated to be internal (Stone et al. 2000; Tiwari and Bae 2020) The colonization approach followed by bacteria is generally inter-cellular, but it can be intra-cellular as well such as *Azoarcus* spp. (Hurek et al. 1994; Hinton and Bacon 1995; Quadt-Hallmann et al. 1997). The usual location for these microorganisms is the vascular tissues of host plants (Kobayashi and Palumbo 2000) which proves to be beneficial for their distribution. Alternatively, in the case of fungi, the asymptomatic colonization can be inter- or intra-cellular by means of a root network. Sometimes, the asymptomatic interactions by dark septate endophytes lead to the colonization of vascular cylinders, and these types of interactions are frequently associated with pathogenicity (Bacon and Hinton 1996; Barrow 2003; Schulz and Boyle 2005). It should be kept in mind that all the definitions are momentary in nature. Therefore, the definition broadly includes an array of microorganisms that vary in terms of their reproductive life cycle and survival strategies. These can include organisms with the saprophytic life cycle of feeding on dead or decaying tissues following the growth phase of endophytic organisms, non-virulent microorganisms along with dormant pathogens as well as virulent pathogens in early infectious stages (Sinclair and Cerkauskas 1996; Kobayashi and Palumbo 2000). However, if we go for the literal meaning of the definition then endophytes can include all pathogens at some specific stage of their developmental cycle. Petrini's characterization of endophytic interactions not causing harm seems to be holding true due to the absence of any visible symptoms as some host plants respond to some infections by mechanical defense reactions (Petrini 1991; Narisawa et al. 2004). Overall, the generally accepted definition of the term "endophytes" is a group of those bacteria and fungi whose existence has been confirmed inside a tissue at any particular moment within a host plant that is in ambient health condition (Schulz and Boyle 2005). The use of the term endophytes is common for organisms found in asymptomatic plant tissues which may be at different stages of their life history. The term endophytes can represent a variety of relationships between host plants and microorganisms including some of the most commonly known ones such as commensalism, and mutualism to sparse ones such as latent pathogenicity to plain simple exploitation. The phenotypic expression of these interactions is varied and depends on the genetic constitution of the interacting partners, their nutritional status, and developmental life stage along with external factors such as environment (Girlanda et al. 2006). The importance of genetic constitution was confirmed when a single mutation related to the gene responsible for virulence factor in pathogenic fungi *Colletotrichum magna* converted it into the endophytic organism (Freeman and Rodriguez 1993). Similar cases of the dysfunctional mechanism of pathogenicity, its suppression, or avirulence gene can also be found in bacterial endophytes (Kobayashi and Palumbo 2000). In keeping with the phenomenon of fungi growing as ectomycorrhiza in some host plants while are ericoid mycorrhiza in others; even mycorrhizal fungus can have endophytic relationships with non-host plants (Villarreal Ruiz et al. 2004; Girlanda et al. 2006). The influence of the plant and associated microorganism on the endophytic l relationship and its reciprocity becomes clear when a non-host plant is infected with a fungal or bacterial pathogen and the pathogen loses its virulence and behaves like an asymptomatic endophyte during colonization (Kobayashi and Palumbo 2000; Schulz and Boyle 2005). The main factor in the determination of the mycorrhizal, endophytic, or pathogenic character of dark septate endophyte is the influence of the host plant. The interactions between plants in any environmental niche are favored by several fungal species with mutualistic relationships that can form hyphal links or act as inoculum reservoirs. The bacterial and fungal endophytes have the ability to interact with mycorrhizal fungi and metazoan along with host plants. An example of this is nematophagous fungi that are commonly found in soil and have life cycles that vary from saprophytic to parasitic. They are capable of killing and digesting live nematodes while growing endophytically on the roots of host plants. The fungi or bacteria-based mutualistic relationships for endophytic colonization of plant roots by microorganisms are beneficial in the protection of plants from environmental stresses and ensuring a regular supply of nutrients.

Various mechanisms employed by endophytes to promote plant growth are well-studied and understood (Glick 1995; Gamalero and Glick 2011; Glick 2012). The endophytes might function directly or indirectly. Plant growth is promoted directly in two cases (i) The resources such as nitrogen, phosphorus, and iron are acquired from the environment through bacterial or fungal endophytes (ii) Various plant growth hormones such as auxin, cytokinin, and ethylene are either produced or regulated by endophytes to stimulate plant growth. The endophytes can be indirectly involved in plant growth promotion by limiting or preventing the damages that plants suffer due to various pathogenic agents such as bacteria, fungi, or nematodes. There is an array of mechanisms, which endophytes can use to indirectly stimulate plant growth including pathways for antibiotics production, causative enzymes for cell wall degradation, reducing the levels of ethylene in plants, induced systemic resistance, lowering the availability of iron for pathogens, and pathways for the production of volatile compounds, which might have pathogen inhibiting effects (Glick 2015).

Depending on the genotype of the host plant, the relationship studies focusing on host genotype and symbiont/endophyte suggest that it can have either a parasitic or mutualistic type of lifestyle as revealed by relationship studies between the host genotype and endophyte (Redman et al. 2001; Unterseher and Schnittler 2010). The relationship between the endophyte and host plant is flexible and the direction of which is determined by small differences in the gene expression pattern of fungal species in response to reactions of the host plant or alternatively by host recognition response of the fungi/bacteria. Therefore, it is concluded that small genetic variations govern the outcomes for both partners in a symbiotic interaction (Moricca and Ragazzi 2008). The genetic background of the host plant can significantly affect the population structure of endophytes. The taxonomy-based analysis of all medicinally significant plants with endophytic partners in the last 30 years indicates that the number of endophytic species related to Taxaceae and Orchidaceae is highest among all families which currently stands at 11 and 8 respectively. All the members of the family Taxaceae are endosymbiont for fungal species which can produce taxol that has anti-tumorous effects. The first evidence of the anti-cancerous property of fungal endophytes *Taxomyces andreanae* was studied in 1993 when fungi were isolated from the bark of *Taxus brevifolia* and were cultured *in vitro* under axenic conditions to produce Taxol (Stierle et al. 1993). Under batch conditions, various fungal endophyte isolates were studied to display variable Taxol production efficiency (Gangadevi and Muthumary 2009; Tiwari et al. 2022). In most cases, the fungi which are endophytic on orchid species belong to the genus *Rhizoctonia* and are considered obligatory symbionts for the germination of seeds as well as the development of new heterotrophic plantlets. Generally, orchids tend to become self-reliable by developing abilities of photosynthesis but some of them remain heterotropic throughout their life in order to meet their nutritional needs. The fungal endosymbionts that are present in adult orchid plants are considered true symbionts for seeds and protocorms of the orchid plants. The behavioral pattern studies based on the isolates from root culturing of host species seeds indicate that there exists an obligatory relationship that is highly specific between hosts and endophytes. Therefore, it becomes necessary to scientifically analyze the importance of the relationship between plant species and the bacterial/fungal species which are endophytes on them along with members from their respective families.

1.1. Impact of Endophytes on the Host Gene Regulation

The term endophyte is used quite liberally to define an array of potential plants and their inhabitants. The organisms which can commonly be identified as endophytes cause infections that are internal and inconspicuous, and the infected tissues of the host plant are symptom-less, albeit temporarily. The term can also broadly include bacterial prokaryotes and fungal eukaryotes. The life history strategies followed by endophytic microorganisms can include saprophytic growth on dead and decaying tissues, exponential growth by endophytic mechanisms, behaving as avirulent microorganisms, incidental or latent pathogenicity as well as virulent pathogenicity in early infection stages. Literature suggests that the fitness of fungal endophytes is largely dependent on the fitness of the host plant which is indicative of the fact that colonization and distribution of fungal endophytes in the host plant are governed mainly by the host plant (Saikkonen et al. 2004). Moreover, the phase deposition (age) of host plants and tissues might influence the variety of species that are found in an endophytic community (Sieber 2007). An example of it can be variation in endophytic species which are found in differentially-aged host plants and within different tissues such as parenchyma, vascular ducts, and dermis. The differential distribution of endophytic species can be indicative of the fact that they use specific substrates for their growth (Rodrigues 1994). Once the endophytic species have successfully colonized the internal tissues of the plants and have established themselves successfully, the vagaries of changing soil conditions cannot affect them. However, the functioning and proliferation of rhizospheric PGPB are significantly hampered by variation in soil pH, temperature, moisture content, and the presence of competitive bacterial species in the niche whose binding sites are present on root surfaces of the host plant (Rashid et al. 2012).

The commonly observed ecological phenomenon of resource distribution for organisms that live near each other within the same environmental niche also exists in endophytes and helps to reduce competition among them (Carroll and Petrini 1983). The colonization of endophytic species depends on various types of substrates that are produced by different plant tissues. The beneficial effects of endophytic species are mediated through three different types of interactions based on different approaches. Firstly, various types of plant hormones are secreted by endophytic species that enhance the growth of their host plants (Waqas et al. 2012). Secondly, an array of bioactive compounds such as alkaloids, diterpenes, flavonoids, and isoflavonoids are produced by host plants that increase the resistance of plants to biotic and abiotic stresses (Firáková et al. 2007; Rodriguez et al. 2009). Thirdly, many endophytic species help in the assimilation and accumulation of secondary metabolites by their host plant (Tiwari et al. 2021). These metabolites can be produced by either host plant and/or fungal/bacterial endophytes. In the 1960s, American scientists were able to isolate the diterpenoids called Paclitaxel (taxol) from the Taxaceae family plant Pacific Yew whose extract contained naturally occurring secondary metabolites. It has been proven to possess anti-tumor activity against ovarian cancer, uterine cancer, and breast cancer with their high incidence rates. Another family which is significant for its endosymbiotic relationship is Orchidaceae as at least eight members from this family have the potential for further research and development. Most of these species are reliant on endophytes for their growth (Guo and Wang 2001; Zhang et al. 2009).

1.2. Impact of Endophytes on the Host Transcription

Previous studies have revealed that fungal endophytes can help to boost the fitness and growth of host plants by enhancing the production of hormones such as indole-3-acetonitrile, cytokinin, and indole-3-acetic acid. The host plant growth is promoted by fungal endophyte due to their ability to absorb nutritional elements that are useful for plants (Zhang et al. 2006; Hartley and Gange 2009). One example can be *Mycena dendrobii* that helps in seed germination and promotes healthy growth of the host plant due to its ability to secrete indole acetic acid (Guo and Wang 2001). When exudate from roots of rice (*Oryza sativa* cv. Nipponbare) plants was treated with bacterial endophyte *Azoarcus sp.* strain BH72 followed by gene expression studies, it was observed that many genes that are associated with rhizospheric competence were upregulated and some whose exact functions are not known were also modulated. Mutations in many genes confirmed their involvement in endophytic colonization such as genes encoding for minor pilin PilX, a serine-threonine kinase, and signal transduction proteins as a putative element of the type IV secretion system (T6SS) (Shidore et al. 2012). The hyphae of

Metarhizium robertsii help in the translocation of nitrogen from arthropodal species to host plants (Behie et al. 2012). Evidential studies indicate that most of the hormones produced by fungal endophytes were isolated from the roots of the host plants. In most cases, the hormones produced by fungal endophytes promote the secretion of certain enzymes and the expression of specific genes (Chen et al. 2005). The stimulated expression of starch degrading enzymes (i.e., glucan-water dikinase) and nitrate reductase by *Piriformospora indica* promotes growth in tobacco roots (Sherameti et al. 2005). The production of bioactive compounds/chemicals by certain fungal endophytes promote resistance to biotic and abiotic stresses within the host plant (Nejad and Johnson 2000; Cavaglieri et al. 2004).

1.3. Impact of Endophytes at the Host Translational Level

In cases where stress tolerance is attained through symbiotic relationships, the fungal endophytes function as the biological triggers that activates the defense mechanism of the host plant (Rodriguez and Redman 2008). When fungal endophytes were inoculated to crop plants, they helped to improve the resistance and yield of the crop (Kozyrovska et al. 1996). There exists a direct competition between host plants and pathogens in cases where plants develop resistance to pathogens due to endophytic relationships. The fungal endophyte is known to promote tolerance to biotic stresses by producing compounds that have antimicrobial activity against pathogens (Zou and Tan 2001; Saikkonen et al. 2004; Gunatilaka 2006; Zhang et al. 2006). The toxic or distasteful nature of chemicals produced by fungal endophytes acts as a repellent for insects that helps to protect the host plants against the attacks of insects. The fungal endophytes from the genus *Neotyphodium* produce alkaloids that impart deterrence to their respective host plants. Thereby, increasing the chances of survival during an attack by insects.

More than 40 genes are involved in the governance of endophytic behavior that involves genes that encode transporter proteins, control secretion, and delivery systems, involved in the degradation and modification of plant polymers, regulate the transcriptional process, maintenance of redox potential, detoxification, functioning of enzymes such as 2-isopropyl-malate synthase and diaminopimelate decarboxylase. They are also correlated with family efflux transporter MFP (membrane fusion proteins), branched-chain amino acid ABC transporter ATPase, RND (resistance nodulation cell division), and other ABC transporter inner membrane proteins. The process of colonization by endophytes is complicated and is reliant on the capacity of endophytic species to compete in the rhizospheric niche in order to establish themselves, set up communication channels, and interact with the roots of the plant. The secretions coming from roots i.e., root exudates consist of an array of macro and micro molecules, which help in attracting microorganisms to the root and help endophytes to colonize the internal plant tissues.

1.4. Impact of Endophytes at the Post-Translation Level of Host

Some studies have indicated that the presence of endophytes imparts a competitive advantage to host plants and converts it into an invasive species in cases where native plants are not fit for survival under stressed conditions (Tofern et al. 1999; Clement et al. 2005). An array of bioactive compounds having antioxidant potential produced by fungal endophytes helps to protect their host plant by enhancing tolerance to abiotic stresses (Herrera-Carrillo et al. 2009; Torres et al. 2009). Several studies have indicated that the production of compounds such as flavonoids and other phenolics which have antioxidant properties are ramped up by endophytic infections in plants (Richardson et al. 1992; Harper et al. 2003; Huang et al. 2007a,b). The metal sequestration or chelation systems that are present in fungal endophytes help in the survival of host plants in environs where the soil is contaminated with heavy metals and therefore can be useful in pollution management (Weyens et al. 2009). Certain groups of endophytes have the capability to enhance the accretion of secondary metabolites within the host plant and can be useful for quantitative as well as qualitative improvements of drugs that are sourced from plants (Chen et al. 2016). Some endophytes are responsible for the production of various phytochemicals through mediation within the host plant for the accumulation of secondary metabolites which includes compounds such as podophyllotoxin (Eyberger et al. 2006; Puri et al. 2006), azadirachtin (Kusari et al. 2012), deoxypodophyllotoxin (Kusari et al. 2009a), paclitaxel/taxol (Stierle et al. 1993), camptothecin as well as its structural analogs (Puri et al. 2005; Kusari et al. 2009c, 2011; Shweta et al. 2010), hypericin and emodin (Kusari et al. 2008, 2009b). The most commonly known anti-cancer activity containing a drug named taxol was isolated from a taxol-producing fungal endophyte named *Taxomyces andreanae* which was symbiotic with *Taxus brevifolia* (Stierle et al. 1993). Several other endophytes that are symbiotic in other plants host species such as *Seimatoantlerium tepuiense*, *Seimatoantlerium nepalense* (Bashyal 1999), *Tubercularia sp. strain TF5*

(Wang et al. 2006), and *Metarhizium anisopliae* (Liu et al. 2009) are also known producers of taxol. Some fungal endophytes are known to synthesize and assimilate secondary metabolites that may possibly only be produced by host plants. One example of it could be *Coetotrichum gloesporioides* which could stimulate the production of Artemisinin from hairy root cultures of *Artemisia annua* (Wang et al. 2006). These chemicals are useful in the treatment and are found to act as antitumor, anti-inflammatory, analgesic, antimalarial or antipyretic agents.

2. Computational Tools for the Understanding of Endophytes and Host Interactions

Endophytes modulate host activity by binding their macromolecules such as nucleic acid and protein at a distinct level. Several studies showed that the endophytes positively impact the host by modulating the host activity at various levels. Many computational tools are available to develop an understanding of the interaction among the protein-protein, protein-DNA, protein-RNA, etc. (Table 1). Additionally, tools are also available to understand the interaction of endophytes' primary/ secondary metabolites with the host. Computational tools are appealing due to the fact that they can predict interactions of host and endophytes at the different levels of the central dogma. The interactions of the endophytes with the host endorse structural and functional perturbation of the macromolecules. *In silico* tools ease the identification of endophytes species by implementing available data. Mahdavi et al. (2013) conceive that the interaction of the protein-protein and protein-nucleic acid assists in the impact of endophytes on the host or vice-versa.

2.1. Computational Tools for the Understanding of Nucleic Acid

Nucleic acid is an essential asset of living organisms that assists in completing all fundamental activities. Understanding the impact of endophytes and other organisms on the host is essential to understanding the impact that different forms of nucleic acid have in assisting the fundamental process of life. The reciprocity of host and endophytes leads to the regulation of macromolecules at the distinct level of the central dogma. Computational methods assist in the identification of endophytes species at the taxonomic level.

Table 1. Tools and techniques for examining endophytes (DNA, RNA, and interactions with plants)

Tool/Technique Name	Application	References/ Link
Egassembler	Used in aligning and merging gene transcripts or shotgun sequences for reconstructing gene or segment	genome.jp/tools/egassembler/
FastQC	Used for scrutinizing the quality of the sequence	https://www.bioinformatics. babraham.ac.uk/projects/fastqc/
STRING	Database for protein-protein interactions network	https://string-db.org/
EMBOSS seqret	Use for the sequence: extraction, reformatting, producing reverse complement, etc.	https://www.ebi.ac.uk/Tools/ sfc/emboss_seqret/
ExaBaye	Phylogenetic studies	https://cme.h-its.org/exelixis/ web/software/exabayes/index.html
NCBI Tree Viewer (TV)	Use for graphically representing Phylogenetic trees	https://www.ncbi.nlm.nih. gov/projects/treeview/
FUNGuild	Used for parsing fungal community datasets	http://www.funguild.org/
MEGAN (MEtaGenome ANalyzer)	Used for analyzing large metagenomes datasets	https://bio.tools/megan
Rclick	Used for comparing RNA 3D structures	http://mspc.bii.a-star.edu.sg/minhn/rclick.html
metAMOS	Used for metagenomic data (gene finding, taxonomic classification etc.)	https://www.cbcb.umd.edu/ software/metamos
CRISPR-Cas9	Use for gene editing from DNA	Xu, et al. 2021

2.1.1. Computational Tools for the Understanding of DNA

DNA is a vital macromolecule that contains the genetic information of the living organism. Exploring the veritable genome of host and endophytes assists in the realization of upregulation of growth and development (Orozco-Mosqueda and Santoyo 2021). The sequence analysis of the genome gives insights into the functional and structural components of the gene (Koonin and Galperin 2002). The computational tools succor in the gene annotation using various algorithms which play vital roles in understanding the gene. *In silico* PCR assists in identifying the species from the available primer (Bellemain et al. 2010). *In silico* pi-values elucidation of translational factors revealed that the environment factors assist in translating factors such as acidic, basic and neutral (Jana and Datta 2021). The *in silico* genome-genome hybridization confers the classification of the species based on the nucleotide's resemblance provides new insights into the taxonomy, and this approach assists in the classification of species in the other genomes such as *P. psychrotolerans, P. oryzihabitans*, *P. kilonensis- P. brassicacearum* and *P. synxantha, P. libanensis* (Tran et al. 2017).

2.1.2. Computational Tools for the Understanding of RNA

RNA is a single-strand nucleic acid that assists in the various fundamental process of a living organism, such as regulation, expression, coding, and decoding of the gene. A different form of RNA permutes the information from the DNA to Protein. Wu et al. (2020) revealed that small molecule binds with RNA modulates the activity of RNA via interfering with the binding of RNA binding proteins. Transcriptional analysis revealed that endophytes metabolites assist in regulating the host through gene regulation and expression (Polonio et al. 2016). The three-dimensional structure of ITS2 revealed that the folding pattern assists in the deciphering of a structural motif using an *in silico* tool (Talukdar et al. 2021). The comparison of RNA secondary structure revealed a difference between the conserved region and variable regions of RNA secondary structure, which helped to differentiate the endophytes from the necrophytic fungi (Padhi et al. 2016).

3. Computational Biology as Endophytes Identification

Different tools of computational biology assist in identifying the endophytes species based on the derivable experimental data. The genetic data are conserved regions that play a vital role in species identification. Rhoden et al. (2013) studied ITS1-5,8S-ITS2 sequences used for the species identification of bacterial endophytes using computational tools. The 16 rRNA has been used to identify bacterial endophytes from the agronomic crop using an *in silico* tool (Bredow et al. 2015). *In silico* PCR help in the identification of different genus of the fungal endophyte by the previously available DNA sequence (Toju et al. 2012). The metabarcoding of 16S rRNA using different bioinformatics tools such as operational taxonomic units illustrates the diversity of endophytic bacterial communities that assist in understanding the organ level distribution (Alibrandi et al. 2020).

3.1. Computational Tools for Understanding Protein

Proteins are made of chains of amino acid residues. These bio-macromolecules are the fundamental part of a living organism and play vital structural and functional roles such as catalyzing metabolic reactions, DNA replications, responding to stimuli, transporting molecules, and providing structural support to cells and organisms. The study of the structure, function, and evolution of protein is imperative and obligatory. Further, the structural and functional components of proteins are greatly impacted by endophytes. The proteins can be studied *in vitro*, *in vivo,* and *in silico*. To learn, how a protein carries out its function, *in vitro* studies are performed. In contrast, to get the information about the physiological role of a protein in a cell or even in a whole organism, *in vivo* experiments are performed. However, the role of computational tools (*in silico studies*) comes into play, when one needs to study similar proteins extensively. Moreover, it is simply impossible to study all proteins experimentally and it is well-known fact that only a few of the proteins are subjected to laboratory tools. Thus, *in silico* studies employs computational methods and tools to explore all aspects of proteins. These tools have been propelled by huge genomic and proteomic data.

There are various computational tools widely used. Beginning from the general protein analysis tools such as ProtParam Tool (ExPASy) (Gasteiger et al. 2005), Translate Tool (ExPASy) (Gasteiger et al. 2003), HeliQuest Protein Helix Analysis Tool (Gautier et al. 2008). ProtParam Tool (ExPASy) is a modest yet outstanding web-based tool that allows users to arrive at a protein sequence of interest and have basic physical properties of protein such as molecular weight, theoretical isoelectric point (pI), amino acid composition, atomic composition, extinction coefficient for A280 in water, estimated half-life, instability index, aliphatic index and grand average of hydropathicity (GRAVY). For performing, three forward and three reverse reading frames of a targeted RNA or DNA sequence, the Translate Tool can be used. The physicochemical properties and amino acid composition can be readily calculated by the HeliQuest. To picturise the results, the user can opt for Helical Wheel Projection diagram.

For studying the protein structure, some most extensively used databases are Research Collaboratory for Structural Bioinformatics Protein Database (RCSB PDB) (Berman et al. 2000; Berley et al. 2021), NCBI Protein Structure database (Anjay et al. 2012), Molecular Modeling Database (MMDB) (Madej et al. 2012). The RCSB PDB, simply called "The PDB" perhaps has the most essential and substantial collection of high-resolution 3D protein structures.

The database is user-friendly as the structure files of protein can be readily downloaded in various formats, which can be used with three-dimensional structure viewing software. It has also provided the comfortability of viewing directly within java compliant browsers on site. Moreover, it has also integrated DrugBank allowing users to readily hunt for desired protein targets for drugs of concern. The other similar database is the NCBI Protein Structure Database (aka the Molecular Modeling Database (MMDB)). It includes macromolecular structure files derived from RCSB PDB. However, the data from these files is tailored in a different way and is subsequently stored in a different file format (i.e., *.cn3 for NCBI versus the *.pdb file format for RSCB). Macromolecule gateway is also derived from PDB and provides experimentally resolved structures of proteins, RNA, and DNA. NCBI has improved features for example explicit chemical graphs, computationally identified 3D domains (compact substructures), and used to characterize similar 3D structures. Moreover, it comprises links to literature, similar sequences, information about chemicals bound to the structures, and many more features embedded.

There are also several 3D structure viewers available for users, which have several benefits over the available browser-based viewers, for example, they eliminate the requirement of having internet connectivity provide the capacity to execute personalized and publication-quality images. Some of the well-known tools are Swiss-PDB Viewer (Guex et al. 1997), PyMOL, CHIMERA (Pettersen et al. 2004), and CN3D (https://www.ncbi.nlm.nih.gov/Structure/CN3D/cn3d.shtml).

Protein modeling is one of the key areas that have been addressed by computational tools. Due to its importance and vital role in protein and peptide sciences, there have been significant improvements in modeling methods. The advances in computer technology have aided the ready availability of a huge amount of biological data. These modeling tools are often used to predict the structure and to have some insights into its function and its underlying mechanism. They can also help in envisioning the design for experiments and searching for possible leads for drug design. Xu et al. (2000) has summarized the computational tools of protein modeling, including the major methods and computer programs in sequence comparison and annotation, as well as structural analysis and prediction. Further, various webservers and software used for different purposes are categorized (Tables 2, 3, and 4).

Table 2. Computational tools employed for sequence comparison

Sequence Comparison Tools				
Pairwise Sequence Alignment	Multiple Sequence Alignment	Sequence Family	Phylogenetic Classification	Search-Based on Multiple Sequence Alignment
ALIGN	BCM Search Launcher	BLOCKS	MOLPHY	HMMER
BLAST	BlockMaker	COG	PAML	PSI-BLAST
FASTA	CLUSTAL	DOMO	PASSML	SAM-T98
GCG/BESTFIT	CypData	MEGACLASS	PHYLIP	
KESTREL	GCG/PILEUP	Pfam	PUZZLE	
SSEARCH	MEME	PRINTS	TAAR	
	Multalin	ProClass	TOPAL	
	PAUP*	ProDom		
		PROSITE		
		SBASE		

These computational tools have been thoroughly used in the study of enzymes from bacterial or fungal endophytes. Also, to predict the novel bioactivities of the endophyte-based natural products. Recently, *in silico* approaches are used to predict the novel bioactivities of the fungal endophyte alkaloid, mycoleptodiscin B for human targets (Deshapriya et al. 2021). Further, the molecular characterization of β-1,3-glucanase producing *Bacillus amyloliquefaciens*-an endophyte of *Hevea brasiliensis* antagonistic to *Phytophthora meadii* was also carried out by computational tools (Abraham et al. 2013). Zadeh et al. (2020) also opted *in silico* approach to identify and characterize the anti-neoplastic asparaginase enzyme from bacterial endophytes. Hence, the usage of computational tools for studying endophytic enzymes and natural products is increasing.

Studies of interactions between a host and guest (Protein and ligand) are always a very tedious complex phenomenon. Studying such interactions are also very tedious and complex computationally as the problem lies in the prediction of interactions. However, to address these, tools such as molecular docking, and molecular mechanics, in particular, molecular dynamics were designed. The more advanced quantum dynamics methods allow the simulation of proteins in atomistic detail with an accurate description of quantum mechanical effects. For example, the multi-layer, multi-configuration, time-dependent Hartree (MCTDH) method, and the hierarchical equations of motion (HEOM) approach. It is well-known fact that both the quantum and classical mechanical simulations of biological-scale systems are extremely computationally demanding. However, if one wants to achieve a more reliable prediction, it has to be compensated by the cost of the computational exhaustiveness.

Table 3. Computational tools in use for selected sequence annotation

Selected Sequence Annotation Tools				
Pairwise Sequence Alignment	Transmembrane Segment Prediction	Motifs	Signaling Site	Secondary Structure Prediction
Johns Hopkins's Server	MEMSAT	I-sites	DictyOGlyc	PSA
Weizmann's Server	SOSUI	MOTIF	NetOGlyc	BTPRED
	TMAP		PSORT Server	NNPRED
	TMpred		SignalP	PHD
	TMHMM			IBCP Server

Table 4. Computational tools in use for selected structure annotation

Selected Structure Annotation Tools				
Visualization	Geometry Analysis	Domain Partition	Alignment / Family	Molecular Dynamics
CHIME	HBPLUS	3Dee	SCOP	AMBER
gOpenMol	NACCESS	Domain Server	CATH	CHARMM
GRASP	WAHTIF	DOMO	CE	GROMOS
LOOK		MEGACLASS	Dali Domain Dictionary	NAMD
RasMol		Pfam	FSSP	TINKER
VMD		PRINTS	HOMSTRAD	X-PLOR
		ProClass	HSSP	
		ProDom	LPFC	
		PROSITE	VAST	
		SBASE		

The field of protein design and engineering has been dramatically transformed by computational techniques and algorithms. Tailoring of protein for various biotechnological applications is done by computational approaches. To meet the escalating demands and difficulties of protein engineering, various tools and methods have been developed. A critical evaluation of recent additions to the toolbox regarding their application for semi-rational and rational protein engineering has been done to help protein engineers and bioinformaticians navigate this emerging wave of dedicated software (Sequeiros et al. 2021).

Thus, it can be thoroughly understood that protein science is frequently complemented by both experimental approaches and computational methods. The modern era relies more on computing and its potential. There are various computational tools that assist in experimental measurements and data analysis. However, this cannot be ruled out that the computational tools used to study the proteins are predictive in nature and for validation needs experimental backing. It must be taken into account that before coming to any conclusion, a general confidence level has to be attained in predictions also. This can be achieved by trying different tools available and then coming out with a consensus. Whether a conclusion drawn based on predictions is reliable or not can be found out by the various prediction approaches. Whenever possible, carrying out experiments will increase the confidence and verify the output. There is no doubt that significant advances have made protein and relatable studies easy. It is also evident that not only computational biologists but also experimentalists have tremendously benefited from these computational tools as they have provided significant information about the structure and function of a protein. However, one has to use these judiciously and cannot use these tools blindly. Using them blindly may generate inaccurate and even wrong predictions. Further, validations will always be helpful. There are still many challenges in accuracy and complete reliability, but further technical improvement will certainly enhance the reliability of the data produced by these tools.

3.2. Computational Biology in the Understanding Evolution of Species

Evolution assists in deciphering species diversity in the forms of losing/ changing the parental phenotype and transposing it into the recent species. The evolution does depend on the manifoldness of the genome of species. The decrypting of the genome comports the evolution at the species level. Computational algorithms, along with the biological aspects such as genetic drift, complex genotypic to phenotypic mapping, co-evolution, multi-objectivity, and transition in the organizational structure, play a vital role in the deciphering of genetic evolution (Miikkulainen and Forrest 2021).

Enhancement of the computational resource elevates the understanding of the biological process in a more approachable manner.

3.2.1. Genetic Drift

Genetic drift refers to the commute in the frequency of essential genes in a population by the random sampling of the organism. The essential genome set of the organism remains the same, which is essential for the survival and reproduction of different populations. Genetic drift influences the symbiotic character of the endophytic population by reducing the genome (Pinto-Carbó et al. 2018). Computer simulation may assist in interpreting genetic drift phenomena using physics principles (Katsnelson et al. 2018).

3.2.2. Complex Genotypic to Phenotypic Mapping

A genetic-phenotypic map deciphers the evolutionary tendencies among the population. Computational and experimental studies play a significant role in a better understanding of the evolutionary pattern. Nichol et al. (2019) reviewed that computational inspired algorithms, and mathematical modeling leads to inferring phenotypic and genotypic relations in the organism. Liao et al. (2019) studied that endophyte organisms deplete the gene expression of genes involved in the plant defenses response such as hormone signaling that assists in the plants' growth and development. Computational methods assist in understanding the phenotypic relationships with the genotypic by the comparative analysis of phenotype and genotype.

3.2.3. Co-Evolution

Co-evolution is a process where two or more populations take evolution by the process of natural selection. Most of the time, co-evolution happens among the closely associated populations such as prey-predator, host-endophytes, plant-pollinators, etc. Endophytes assist in the gene flow along with the host species (Sullivan and Faeth 2004). Several computational methods have been designed for the studies of co-evolution based on DNA, RNA, and protein. Co-evolution analysis using protein sequences (CAPS) deciphering the co-evolution depends on the amino acid positions in the protein sequences that result in the deviation in the 3-dimensional structure of the protein (Fares and McNally 2006). Co-evolution of Presence–Absence Patterns (CoPAP) is a computer-based program that assists in the identification of co-evolution by the presence and absence at the gene level such as patterns of methylation, and introns-exon boundaries based on the gene clustering (Cohen et al. 2013).

3.2.4. Multi-Objectivity

Multi-objectivity is a method that has multiple functions for co-related problems. Computational and bioinformatics methods help in the coarse solution of multiple problems. Multi-objectivity methods help in the optimization of the culture media whereas several factors assist in the optimal growth of the desired organism (Jamshidi et al. 2020). A multi-objectivity algorithm-based optimization assists in the species distribution in a specific geographical region (Mo et al. 2014). Multi-objectivity-inspired computational program assists in the better correlation between the endophyte's impacts upon the host and vice-versa.

3.2.5. Transition in the Organizational Structure

The interspecies and intra-species interactions lead to the transitions of organs in the interacting species. The organization complexity depends on species distribution in the host system. Organizational transitions of the host depend on the endophyte's diversity in the organ (Zuo et al. 2021). The computational-based image analysis methods assist in the explanation of organ structure (Bassel 2019). Endophytes confined inside the host assists in the organ structure changes of the host may be explained by using a computational method (Sequeiros et al. 2021).

Conclusion and Future Prospects of the Research

A variety of bacterial endophytes has a well-established capacity for promoting plant growth and increasing the resistance of the plants against several diseases along with the capability to overcome abiotic stresses. However, field trials have failed to yield positive effects in several cases which may be the outcome of ineffective colonization of the endosphere and rhizosphere by endophytes. The inadequate colonization by rhizospheric and endophytic species of the host plant arises due to the lack of various characteristics which can be considered essential for proper colonization in the soil around the plants. An in-depth analysis of the process and various factors that are involved in the colonization of different plant niches would be helpful in increasing the available information about plant-microbe interactions and assist in the development of more resilient and potent endophytic inoculants. A number of points need to be explored in detail. An example of this is the fact that many rhizosphere-limited endophytes have been found to live on and/or within plant tissues in form of viable but non-culturable cells but the reason for this switching of the state is not known (Hurek et al. 1994; Gamalero et al. 2011). The conversion of endophytic species to viable but non-culturable cells (VBNC) state during colonization might be an outcome of the stresses that endophytes experience but the exact mechanism is still not known. Due to conversion into VBNC state or lack of exact culturing conditions, it is difficult to cultivate endophytic species. There is a lack of knowledge about the functioning and effects of non-cultivable endophytes on plant growth. Therefore, computational biology along with metagenomic approaches that do not require the cultivation of microorganisms can prove to be very helpful for acquiring information about non-cultivatable micro-organisms.

It has been thoroughly understood that protein science is frequently complemented by both experimental approaches and computational methods. The modern era is more and more relying on computing and its potential. There are various computational tools that can assist in experimental measurements and data analysis. However, this cannot be ruled out that the computational tools used to study the proteins are predictive in nature and for validation needs experimental backing. It must be taken into account that before coming to any conclusion, a general confidence level has to be attained in predictions also. This can be achieved by trying different tools available and then coming out with a consensus. Whether a conclusion drawn based on predictions is reliable or not can be found out by the various differences. Whenever possible, carrying out experiments will provide confidence and verify the output results. There is no doubt that significant advances have made protein and related studies reliable and easy. It is also evident that not only computational biologists but also experimentalists have tremendously benefited from these computational tools as they have provided significant information about the structure and function of a protein. However, one has to use these judiciously and cannot use these tools blindly. Using them blindly may generate inaccurate and even wrong predictions. Further, validations will always be helpful. There are still many challenges in accuracy and complete reliability, but further technical improvement will certainly enhance the reliability of the data produced by these tools.

Most of the research work performed till now in relation to endophytes has been targeted plant growth promotion which is generally followed by colonization. We still have not developed a full understanding of the effect that endophytes have on aerial parts of the plant although several novel root colonizers are known (Andreote et al. 2009). The role of endophytes can be better understood by comparing the genomic information from plant tissue colonizing endophytes and correlating it with other factors such as species viability etc. The focus should be on endophyte species that can colonize fruits as well as seeds that are present within. They can be potential targets based on criteria of specific functions, importance to plant health and growth, and vertical transmission.

One of the complications regarding the usage of endophytes can be the fact that plants can act as hosts for various human pathogens (Berg et al. 2005; Allerberger and Sessitsch 2009). Some of these harmful pathogens also act as plant growth promotors, increase plant vitality, and colonize the rhizosphere along with other internal tissues (Berg et al. 2005). However, the latest developments can help us to differentiate between plant colonizers and clinical isolates. A study on *Stenotrophomonas maltophilia* that include species that are responsible for both clinical infection and endophytes revealed major interspecies differences even though they were isolated from the same environmental niche (Ryan et al. 2008). The absence of atypical genomic islands in endophyte species (which are usually found in clinical species) was noticeable. Simultaneously, there exists a high level of homogeneity between some endophytes and clinical isolates. A suitable explanation for this could be genome plasticity which helps organisms to adapt to different environmental niches (Ryan et al. 2008). However, few genes that are responsible for pathogenicity were found in both strains independent of source i.e., plant environment or clinical environment (Ryan et al. 2008). It is proposed that the mechanism implored by *S. maltophilia* for colonizing plants and overcoming their defense mechanism might be identical to one involved in the colonization of human tissues and antagonistic activity involved in the same. These kinds of information are important for the identification of inoculant strains that can be applied and introduced into the environment (Berg et al. 2005). Computational biology can go a long way in providing the necessary details. Some examples of similar microorganisms can be *Burkholderiales* i.e., *Burkholderia* spp. (Compant et al. 2012) and Enterobacteriaceae that have the potential to be pathogenic and use internal tissues of plants as reservoirs (Holden et al. 2009). The pathogenicity potentials of microorganisms can be explored by genetically distinguishing the colonizers of plants, animals, and humans.

The colonization process and interactions of beneficial microbes with the plants can be well understood by studying the colonization process based on techniques such as genome sequencing, identification of genes involved in colonization, and characterization of genes that are not known as of now by using various tools of computational biology. The technique of microarrays was used to identify the factors which play an integral role in the colonization of plant roots. The factors like stress adaptation and availability of specific nutrients for *Pseudomonas putida* strains in the rhizosphere of maize plants were deemed to be important (Matilla et al. 2007). The traits which can enable bacterial strains to survive in the ecological niche have been identified and can assist in understanding various aspects of plant-beneficial bacteria association such as signaling and regulatory processes (Matilla et al. 2007). There exists a high probability of the discovery of new genes, related factors, and functions that are essential for the maintenance of rhizospheric and endophytic lifestyles in times to come. The scientists need to segregate essential factors that might be required for colonization and specific conditions associated with it. It is needed because various types of adaptations and recognition strategies might be involved in several pre-existing plant-bacteria associations. In order for a better prediction of bacterial interactions with plant species, we need to develop a more detailed understanding of the colonization process. We also need to understand the potential of microbes for adaptation to field conditions during their application as biofertilizers or biological control agents.

Acknowledgments

The authors extend their gratitude to their respective research organizations.

Conflict of Interests

No conflict of interest was declared.

References

Abraham, Amith, Sunilkumar Puthenpurackal Narayanan, Shaji Philip, Divya Gopalakrishnan Nair, Aparna Chandrasekharan, and Jayachandran Kochupurackal. "*In silico* characterization of a novel β-1, 3-glucanase gene from *Bacillus amyloliquefaciens*-a bacterial endophyte of *Hevea brasiliensis* antagonistic to *Phytophthora meadii*." *Journal of molecular modeling* 19, no. 3 (2013): 999-1007.

Alibrandi, Pasquale, Sylvia Schnell, Silvia Perotto, and Massimiliano Cardinale. "Diversity and structure of the endophytic bacterial communities associated with three terrestrial orchid species as revealed by 16S rRNA gene metabarcoding." *Frontiers in Microbiology* 11 (2020): 3207.

Allerberger, F., and A. Sessitsch. "Incidence and microbiology of salad-borne disease." *CAB Rev* 4 (2009): 1-13.

Anjay, A. "*National center for Biotechnology information (NCBI).*" Bethesda, Maryland, USA (2012).

Bacon, C. W., and D. M. Hinton. "Symptomless endophytic colonization of maize by *Fusarium moniliforme*." *Canadian Journal of Botany* 74, no. 8 (1996): 1195-1202.

Barrow, J. R. "Atypical morphology of dark septate fungal root endophytes of *Bouteloua* in arid southwestern USA rangelands." *Mycorrhiza* 13, no. 5 (2003): 239-247.

Bashyal, Bharat. "*Seimatoantlerium nepalense*, an endophytic taxol producing coelomycete from Himalayan yew (*Taxus wallachiana*)." *Mycotaxon* 72 (1999): 33-42.

Bassel, George W. "Multicellular systems biology: quantifying cellular patterning and function in plant organs using network science." *Molecular plant* 12, no. 6 (2019): 731-742.

Behie, S. W., P. M. Zelisko, and M. J. Bidochka. "Endophytic insect-parasitic fungi translocate nitrogen directly from insects to plants." *Science* 336, no. 6088 (2012): 1576-1577.

Bellemain, Eva, Tor Carlsen, Christian Brochmann, Eric Coissac, Pierre Taberlet, and Håvard Kauserud. "ITS as an environmental DNA barcode for fungi: an *in silico* approach reveals potential PCR biases." *BMC microbiology* 10, no. 1 (2010): 1-9.

Berg, Gabriele, Annette Krechel, Michaela Ditz, Richard A. Sikora, Andreas Ulrich, and Johannes Hallmann. "Endophytic and ectophytic potato-associated bacterial communities differ in structure and antagonistic function against plant pathogenic fungi." *FEMS Microbiology Ecology* 51, no. 2 (2005): 215-229.

Burley, Stephen K., Charmi Bhikadiya, Chunxiao Bi, Sebastian Bittrich, Li Chen, Gregg V. Crichlow, Cole H. Christie et al. "RCSB Protein Data Bank: powerful new tools for exploring 3D structures of biological macromolecules for basic and applied research and education in fundamental biology, biomedicine, biotechnology, bioengineering, and energy sciences." *Nucleic acids research* 49, no. D1 (2021): D437-D451.

Berman, Helen M., John Westbrook, Zukang Feng, Gary Gilliland, Talapady N. Bhat, Helge Weissig, Ilya N. Shindyalov, and Philip E. Bourne. "The protein data bank." *Nucleic acids research* 28, no. 1 (2000): 235-242.

Bredow, C., J. L. Azevedo, J. A. Pamphile, C. A. Mangolin, and S. A. Rhoden. "*In silico* analysis of the 16S rRNA gene of endophytic bacteria, isolated from the aerial parts and seeds of important agricultural crops." *Genet Mol Res* 14, no. 3 (2015): 9703-9721.

Carroll, George, and Orlando Petrini. "Patterns of substrate utilization by some fungal endophytes from coniferous foliage." *Mycologia* 75, no. 1 (1983): 53-63.

Cavaglieri, L. R., A. Passone, and M. G. Etcheverry. "Correlation between screening procedures to select root endophytes for biological control of *Fusarium verticillioides* in *Zea mays* L." *Biological Control* 31, no. 3 (2004): 259-267.

Chen, Ling, Qiao-Yan Zhang, Min Jia, Qian-Liang Ming, Wei Yue, Khalid Rahman, Lu-Ping Qin, and Ting Han. "Endophytic fungi with antitumor activities: Their occurrence and anticancer compounds." *Critical reviews in microbiology* 42, no. 3 (2016): 454-473.

Chen, X. M., S. X. Guo, and C. L. Wang. "Effects of four endophytic fungi on the growth and polysaccharide content of *Anoectochilus roxburghii (Wall.) Lindl.*" *Chin. Pharm.* J 40, no. 1 (2005): 13-16.

Clement, S. L., L. R. Elberson, N. A. Bosque-Perez, and D. J. Schotzko. "Detrimental and neutral effects of wild barley–*Neotyphodium* fungal endophyte associations on insect survival." *Entomologia Experimentalis et Applicata* 114, no. 2 (2005): 119-125.

Cohen, Ofir, Haim Ashkenazy, Eli Levy Karin, David Burstein, and Tal Pupko. "CoPAP: coevolution of presence-absence patterns." *Nucleic acids research* 41, no. W1 (2013): W232-W237.

Compant, Stéphane, Angela Sessitsch, and Florence Mathieu. "The 125th anniversary of the first postulation of the soil origin of endophytic bacteria–a tribute to MLV Galippe." *Plant and Soil* 356, no. 1 (2012): 299-301.

del Carmen Orozco-Mosqueda, Ma, and Gustavo Santoyo. "Plant-microbial endophytes interactions: scrutinizing their beneficial mechanisms from genomic explorations." *Current Plant Biology* 25 (2021): 100189.

Deshapriya, Uthpala S., DL Senal Dinuka, Pamoda B. Ratnaweera, and Chinthaka N. Ratnaweera. "In silico study for prediction of novel bioactivities of the endophytic fungal alkaloid, mycoleptodiscin B for human targets." *Journal of Molecular Graphics and Modelling* 102 (2021): 107767.

Eyberger, Amy L., Rajeswari Dondapati, and John R. Porter. "Endophyte fungal isolates from *Podophyllum peltatum* produce podophyllotoxin." *Journal of natural products* 69, no. 8 (2006): 1121-1124.

Fares, Mario A., and David McNally. "CAPS: coevolution analysis using protein sequences." *Bioinformatics* 22, no. 22 (2006): 2821-2822.

Feller, Ilka C. "Effects of nutrient enrichment on growth and herbivory of dwarf red mangrove (*Rhizophora mangle*)." *Ecological Monographs* (1995): 477-505.

Firáková, Silvia, Mária Šturdíková, and Marta Múčková. "Bioactive secondary metabolites produced by microorganisms associated with plants." *Biologia* 62, no. 3 (2007): 251-257.

Freeman, Stanley, and Rusty J. Rodriguez. "Genetic conversion of a fungal plant pathogen to a nonpathogenic, endophytic mutualist." *Science* 260, no. 5104 (1993): 75-78.

Gamalero, Elisa, and Bernard R. Glick. "Mechanisms used by plant growth-promoting bacteria." In *Bacteria in agrobiology: plant nutrient management*, pp. 17-46. Springer, Berlin, Heidelberg, 2011.

Gangadevi, Venkatraman, and Johnpaul Muthumary. "Taxol production by *Pestalotiopsis terminaliae*, an endophytic fungus of *Terminalia arjuna* (arjun tree)." *Biotechnology and applied biochemistry* 52, no. 1 (2009): 9-15.

Gasteiger, Elisabeth, Alexandre Gattiker, Christine Hoogland, Ivan Ivanyi, Ron D. Appel, and Amos Bairoch. "ExPASy: the proteomics server for in-depth protein knowledge and analysis." *Nucleic acids research* 31, no. 13 (2003): 3784-3788.

Gasteiger, Elisabeth, Christine Hoogland, Alexandre Gattiker, Marc R. Wilkins, Ron D. Appel, and Amos Bairoch. "Protein identification and analysis tools on the ExPASy server." *The proteomics protocols handbook* (2005): 571-607.

Gautier, Romain, Dominique Douguet, Bruno Antonny, and Guillaume Drin. "HELIQUEST: a web server to screen sequences with specific α-helical properties." *Bioinformatics* 24, no. 18 (2008): 2101-2102.

Girlanda, Mariangela, Silvia Perotto, and Anna Maria Luppi. "Molecular diversity and ecological roles of mycorrhiza-associated sterile fungal endophytes in Mediterranean ecosystems." In *Microbial root endophytes*, pp. 207-226. Springer, Berlin, Heidelberg, 2006.

Glick, Bernard R. "The enhancement of plant growth by free-living bacteria." *Canadian Journal of microbiology* 41, no. 2 (1995): 109-117.

Glick, B. R. (2012). Plant growth-promoting bacteria: mechanisms and applications. *Scientifica*, 2012.

Glick, B. R., 2015. *Beneficial plant-bacterial interactions* (pp. 139-180). Heidelberg: Springer.

Gunatilaka, AA Leslie. "Natural products from plant-associated microorganisms: distribution, structural diversity, bioactivity, and implications of their occurrence." *Journal of natural products* 69, no. 3 (2006): 509-526.

Harper, James K., Atta M. Arif, Eugene J. Ford, Gary A. Strobel, John A. Porco Jr, David P. Tomer, Kim L. Oneill, Elizabeth M. Heider, and David M. Grant. "Pestacin: a 1, 3-dihydro isobenzofuran from *Pestalotiopsis microspora* possessing antioxidant and antimycotic activities." *Tetrahedron* 59, no. 14 (2003): 2471-2476.

Hartley, Sue E., and Alan C. Gange. "Impacts of plant symbiotic fungi on insect herbivores: mutualism in a multitrophic context." *Annual review of entomology* 54 (2009): 323-342.

Herrera-Carillo, Z., Mónica S. Torres, Ajay P. Singh, Nicholi Vorsa, Thomas Gianfagna, William Meyer, and James F. White Jr. "Phenolic, flavonoid and antioxidant profiling for cool-season grasses with and without endophyte." In *Proceedings of the 18th annual Rutgers Turfgrass symposium*, vol. 12. 2009.

Hinton, Dorothy M., and Charles W. Bacon. "*Enterobacter cloacae* is an endophytic symbiont of corn." *Mycopathologia* 129, no. 2 (1995): 117-125.

Holden, Nicola, Leighton Pritchard, and Ian Toth. "Colonization out with the colon: plants as an alternative environmental reservoir for human pathogenic enterobacteria." *FEMS Microbiology Reviews* 33, no. 4 (2009): 689-703.

Huang, Wu-Yang, Yi-Zhong Cai, Kevin D. Hyde, Harold Corke, and Mei Sun. "Endophytic fungi from *Nerium oleander* L (Apocynaceae): main constituents and antioxidant activity." *World Journal of Microbiology and Biotechnology* 23, no. 9 (2007): 1253-1263.

Huang, W. Y., Cai, Y. Z., Xing, J., Corke, H. and Sun, M., 2007. A potential antioxidant resource: endophytic fungi from medicinal plants. *Economic Botany*, 61(1), pp.14-30.

Hurek, Thomas, Barbara Reinhold-Hurek, Marc Van Montagu, and Edward Kellenberger. "Root colonization and systemic spreading of *Azoarcus sp. strain BH72* in grasses." *Journal of bacteriology* 176, no. 7 (1994): 1913-1923.

Jamshidi, Saeid, Abbas Yadollahi, Mohammad Mehdi Arab, Mohammad Soltani, Maliheh Eftekhari, and Jalal Shiri. "High throughput mathematical modeling and multi-objective evolutionary algorithms for plant tissue culture media formulation: A case study of pear rootstocks." *PloS one* 15, no. 12 (2020): e0243940.

Jana, Soma, and Partha P. Datta. "In silico analysis of bacterial translation factors reveal distinct translation event-specific pI values." *BMC genomics* 22, no. 1 (2021): 1-11.

Jumpponen, A. R. I., and James M. Trappe. "Dark septate endophytes: a review of facultative biotrophic root-colonizing fungi." *New Phytologist* 140, no. 2 (1998): 295-310.

Jumpponen, Ari, Kim G. Mattson, and James M. Trappe. "Mycorrhizal functioning of *Phialocephala fortinii* with *Pinus contorta* on glacier forefront soil: interactions with soil nitrogen and organic matter." *Mycorrhiza* 7, no. 5 (1998): 261-265.

Katsnelson, Mikhail I., Yuri I. Wolf, and Eugene V. Koonin. "Towards physical principles of biological evolution." *Physica Scripta* 93, no. 4 (2018): 043001.

Kobayashi, D. Y., and J. D. Palumbo. "Bacterial Endophytes and their applications." *Microbial endophytes* 2000 (2000): 99-233.

Koonin, Eugene, and Michael Y. Galperin. "*Sequence-evolution-function: computational approaches in comparative genomics*." (2002).

Kozyrovska, N., G. Kovtunovych, E. Gromosova, P. Kuharchuk, and V. Kordyum. "Novel inoculants for an environmentally-friendly crop production." *Resources, conservation and recycling* 18, no. 1-4 (1996): 79-85.

Kusari, S., M. Lamshöft, and M. Spiteller. "*Aspergillus fumigatus Fresenius*, an endophytic fungus from *Juniperus communis L. Horstmann* as a novel source of the anticancer pro-drug deoxypodophyllotoxin." *Journal of Applied Microbiology* 107, no. 3 (2009): 1019-1030.

Kusari, Souvik, Marc Lamshöft, Sebastian Zühlke, and Michael Spiteller. "An endophytic fungus from *Hypericum perforatum* that produces hypericin." *Journal of Natural Products* 71, no. 2 (2008): 159-162.

Kusari, Souvik, Vijay C. Verma, Marc Lamshoeft, and Michael Spiteller. "An endophytic fungus from *Azadirachta indica A. Juss.* that produces azadirachtin." *World Journal of Microbiology and Biotechnology* 28, no. 3 (2012): 1287-1294.

Kusari, Souvik, Sebastian Zühlke, and Michael Spiteller. "An endophytic fungus from *Camptotheca acuminata* that produces camptothecin and analogs." *Journal of Natural Products* 72, no. 1 (2009): 2-7.

Kusari, Souvik, Sebastian Zühlke, and Michael Spiteller. "Effect of artificial reconstitution of the interaction between the plant *Camptotheca acuminata* and the fungal endophyte *Fusarium solani* on camptothecin biosynthesis." *Journal of natural products* 74, no. 4 (2011): 764-775.

Kusari, Souvik, Sebastian Zühlke, Jan Kosuth, Eva Cellarova, and Michael Spiteller. "Light-independent metabolomics of endophytic *Thielavia subthermophila* provides insight into microbial hypericin biosynthesis." *Journal of natural products* 72, no. 10 (2009): 1825-1835.

Liao, H. L., Bonito, G., Rojas, J. A., Hameed, K., Wu, S., Schadt, C. W., Labbé, J., Tuskan, G. A., Martin, F., Grigoriev, I. V., & Vilgalys, R. (2019). Fungal Endophytes of Populus trichocarpa Alter Host Phenotype, Gene Expression, and Rhizobiome Composition. Molecular Plant-Microbe Interactions®, 32(7), 853–864.

Liu, Kaihui, Xiaowei Ding, Baiwan Deng, and Wenqiang Chen. "Isolation and characterization of endophytic taxol-producing fungi from *Taxus chinensis*." *Journal of Industrial Microbiology and Biotechnology* 36, no. 9 (2009): 1171.

Madej, T., Addess, K. J., Fong, J. H., Geer, L. Y., Geer, R. C., Lanczycki, C. J., Liu, C., Lu, S., Marchler-Bauer, A., Panchenko, A. R., Chen, J., Thiessen, P. A., Wang, Y., Zhang, D., & Bryant, S. H. "MMDB: 3D structures and macromolecular interactions." *Nucleic acids research* 40, no. D1 (2012): D461-D464.

Marler, Marilyn J., Catherine A. Zabinski, and Ragan M. Callaway. "Mycorrhizae indirectly enhance competitive effects of an invasive forb on a native bunchgrass." *Ecology* 80, no. 4 (1999): 1180-1186.

Matilla, Miguel A., Manuel Espinosa-Urgel, José J. Rodríguez-Herva, Juan L. Ramos, and María Isabel Ramos-González. "Genomic analysis reveals the major driving forces of bacterial life in the rhizosphere." *Genome biology* 8, no. 9 (2007): 1-13.

Miikkulainen, Risto, and Stephanie Forrest. "A biological perspective on evolutionary computation." *Nature Machine Intelligence* 3, no. 1 (2021): 9-15.

Mo, Hongwei, Zhidan Xu, Lifang Xu, Zhou Wu, and Haiping Ma. "Constrained multi-objective biogeography optimization algorithm." *The Scientific World Journal* 2014 (2014).

Moricca, S., and A. Ragazzi. "Fungal endophytes in Mediterranean oak forests: a lesson from *Discula quercina*." *Phytopathology* 98, no. 4 (2008): 380-386.

Narisawa, K., F. Usuki, and T. Hashiba. "Control of Verticillium yellows in Chinese cabbage by the dark septate endophytic fungus LtVB3." *Phytopathology* 94, no. 5 (2004): 412-418.

Nejad, Pajand, and Paul A. Johnson. "Endophytic bacteria induce growth promotion and wilt disease suppression in oilseed rape and tomato." *Biological control* 18, no. 3 (2000): 208-215.

Nichol, Daniel, Mark Robertson-Tessi, Alexander RA Anderson, and Peter Jeavons. "Model genotype-phenotype mappings and the algorithmic structure of evolution." *Journal of the Royal Society Interface* 16, no. 160 (2019): 20190332.

Padhi, Srichandan, Mrunmaya Kumar Panda, Devaranjan Das, and Kumananda Tayung. "ITS2 RNA secondary structure analysis reveals close affinity between endophytic and pathogenic fungi: A case study in Fusarium species." *Annals of Microbiology* 66, no. 2 (2016): 625-633.

Peters, Akira F., and Britta Schaffelke. "Streblonema (Ectocarpales, Phaeophyceae) infection in the kelp *Laminaria saccharina* (Laminariales, Phaeophyceae) in the western Baltic." *Hydrobiologia* 326, no. 1 (1996): 111-116.

Petrini, Orlando. "Fungal endophytes of tree leaves." In *Microbial ecology of leaves*, pp. 179-197. Springer, New York, NY, 1991.

Pettersen, Eric F., Thomas D. Goddard, Conrad C. Huang, Gregory S. Couch, Daniel M. Greenblatt, Elaine C. Meng, and Thomas E. Ferrin. "UCSF Chimera visualization system for exploratory research and analysis." *Journal of computational chemistry* 25, no. 13 (2004): 1605-1612.

Pinto-Carbó, Marta, Karl Gademann, Leo Eberl, and Aurelien Carlier. "Leaf nodule symbiosis: function and transmission of obligate bacterial endophytes." *Current opinion in plant biology* 44 (2018): 23-31.

Polonio, J. C., A. D. Polli, J. L. Azevedo, and J. A. Pamphile. "RNA applications for endophytic research." *Genetics and Molecular Research: GMR* 15, no. 3 (2016).

Puri, S. C., Nazir, A., Chawla, R., Arora, R., Riyaz-ul-Hasan, S., Amna, T., Ahmed, B., Verma, V., Singh, S., Sagar, R., Sharma, A., Kumar, R., Sharma, R. K., & Qazi, G. N. "The endophytic fungus *Trametes hirsuta* as a novel alternative source of podophyllotoxin and related aryl tetralin lignans." *Journal of biotechnology* 122, no. 4 (2006): 494-510.

Puri, Satish Chander, Vijeshwar Verma, Touseef Amna, Ghulam Nabi Qazi, and Michael Spiteller. "An endophytic fungus from *Nothapodytes foetida* that produces Camptothecin." *Journal of natural products* 68, no. 12 (2005): 1717-1719.

Quadt-Hallmann, A. N. D. R. E. A., J. W. Kloepper, and N. Benhamou. "Bacterial endophytes in cotton: mechanisms of entering the plant." *Canadian Journal of microbiology* 43, no. 6 (1997): 577-582.

Rashid, Shimaila, Trevor C. Charles, and Bernard R. Glick. "Isolation and characterization of new plant growth-promoting bacterial endophytes." *Applied soil ecology* 61 (2012): 217-224.

Redman, Regina S., David D. Dunigan, and Rusty J. Rodriguez. "Fungal symbiosis from mutualism to parasitism: who controls the outcome, host or invader?" *New Phytologist* 151, no. 3 (2001): 705-716.

Rhoden, S. A., A. Garcia, J. L. Azevedo, and J. A. Pamphile. "*In silico* analysis of diverse endophytic fungi by using ITS1-5, 8S-ITS2 sequences with isolates from various plant families in Brazil." *Genet. Mol. Res* 12 (2013): 935-950.

Richardson, M. D., G. W. Chapman Jr, C. S. Hoveland, and C. W. Bacon. "Sugar alcohols in endophyte-infected tall fescue under drought." *Crop Science* 32, no. 4 (1992): 1060-1061.

Rodriguez, R. J., White Jr, J. F., Arnold, A. E. and Redman, A. R. A., 2009. Fungal endophytes: diversity and functional roles. *New phytologist*, 182(2), pp. 314-330.

Rodriguez, Rusty, and Regina Redman. "More than 400 million years of evolution and some plants still can't make it on their own: plant stress tolerance via fungal symbiosis." *Journal of experimental botany* 59, no. 5 (2008): 1109-1114.

Ryan, Robert P., Kieran Germaine, Ashley Franks, David J. Ryan, and David N. Dowling. "Bacterial endophytes: recent developments and applications." *FEMS microbiology letters* 278, no. 1 (2008): 1-9.

Saikkonen, Kari, Piippa Wäli, Marjo Helander, and Stanley H. Faeth. "Evolution of endophyte–plant symbioses." *Trends in plant science* 9, no. 6 (2004): 275-280.

Schulz, Barbara, and Christine Boyle. "The endophytic continuum." *Mycological research* 109, no. 6 (2005): 661-686.

Sequeiros-Borja, Carlos Eduardo, Bartłomiej Surpeta, and Jan Brezovsky. "Recent advances in user-friendly computational tools to engineer protein function." *Briefings in bioinformatics* 22, no. 3 (2021): bbaa150.

Sherameti, Irena, Bationa Shahollari, Yvonne Venus, Lothar Altschmied, Ajit Varma, and Ralf Oelmüller. "The endophytic fungus *Piriformospora indica* stimulates the expression of nitrate reductase and the starch-degrading enzyme glucan-water dikinase in tobacco and *Arabidopsis* roots through a homeodomain transcription factor that binds to a conserved motif in their promoters." *Journal of Biological Chemistry* 280, no. 28 (2005): 26241-26247.

Shidore, Teja, Theresa Dinse, Johannes Öhrlein, Anke Becker, and Barbara Reinhold-Hurek. "Transcriptomic analysis of responses to exudates reveals genes required for rhizosphere competence of the endophyte *Azoarcus* sp. strain BH72." *Environmental Microbiology* 14, no. 10 (2012): 2775-2787.

Shunxing, Guo, and Wang Qiuying. "Character and action of good strain on stimulating seed germination of *Gastrodia elata*." *Jun wu xi Tong= Mycosystema* 20, no. 3 (2001): 408-412.

Shweta, S., S. Zuehlke, B. T. Ramesha, V. Priti, P. Mohana Kumar, G. Ravikanth, M. Spiteller, R. Vasudeva, and R. Uma Shaanker. "Endophytic fungal strains of *Fusarium solani*, from *Apodytes dimidiata E. Mey. Ex. Arn* (Icacinaceae) produce camptothecin, 10-hydroxycamptothecin, and 9-methoxycamptothecin." *Phytochemistry* 71, no. 1 (2010): 117-122.

Sieber, Thomas N. "Endophytic fungi in forest trees: are they mutualists?." *Fungal biology reviews* 21, no. 2-3 (2007): 75-89.

Sinclair, J. B., and R. F. Cerkauskas. "Latent infection vs. endophytic colonization by fungi." *Endophytic fungi in grasses and woody plants* (1996): 3-29.

Stierle, Andrea, Gary Strobel, and Donald Stierle. "Taxol and taxane production by *Taxomyces andreanae*, an endophytic fungus of Pacific yew." *Science* 260, no. 5105 (1993): 214-216.

Stone, Julie M., Jacqueline E. Heard, Tsuneaki Asai, and Frederick M. Ausubel. "Simulation of fungal-mediated cell death by fumonisin B1 and selection of fumonisin B1–resistant (fbr) *Arabidopsis* mutants." *The Plant Cell* 12, no. 10 (2000): 1811-1822.

Sullivan, T. J., and Stanley H. Faeth. "Gene flow in the endophyte *Neotyphodium* and implications for coevolution with *Festuca arizonica*." *Molecular Ecology* 13, no. 3 (2004): 649-656.

Talukdar, Rajreepa, Srichandan Padhi, Amit K. Rai, Marco Masi, Antonio Evidente, Dhruva Kumar Jha, Alessio Cimmino, and Kumananda Tayung. "Isolation and Characterization of an Endophytic fungus *Colletotrichum coccodes* producing Tyrosol from *Houttuynia cordata Thunb.* using ITS2 RNA secondary structure and molecular docking study." *Frontiers in Bioengineering and Biotechnology* 9 (2021): 484.

Tiwari, Pragya, Hanhong Bae. "Horizontal gene transfer and endophytes: an implication for the acquisition of novel traits." *Plants* 9 (2020), 305.

Tiwari, Pragya, Yashdeep Srivastava, and Hanhong Bae. "Trends of pharmaceutical design of Endophytes as anti-infective." *Current Topics in Medicinal Chemistry*, 21, no. 17 (2021): 1572-1586.

Tiwari, Pragya, Hanhong Bae. 2022. "Endophytic Fungi: Key insights, emerging prospects, and challenges in natural product drug discovery." *Microorganisms* 10, 360 (2022). https://doi.org/10.3390/microorganisms10020360.

Tofern, Britta, Macki Kaloga, Ludger Witte, Thomas Hartmann, and Eckart Eich. "Occurrence of loline alkaloids in *Argyreia mollis* (Convolvulaceae)." *Phytochemistry* 51, no. 8 (1999): 1177-1180.

Toju, Hirokazu, Akifumi S. Tanabe, Satoshi Yamamoto, and Hirotoshi Sato. "High-coverage ITS primers for the DNA-based identification of ascomycetes and basidiomycetes in environmental samples." *PloS one* 7, no. 7 (2012): e40863.

Torres, Mónica S., Ajay P. Singh, Sneha Shah, Zazil Herrera-Carrillo, Thomas Gianfagna, James F. White Jr, and Nicholi Vorsa. "LC–MS–MS identification and quantification of phenolics in symbiotic tall fescue." In *Proceedings of the 18th annual Rutgers Turfgrass symposium*, vol. 12. New Brunswick, New Jersey: Rutgers University, Center for Interdisciplinary Studies in Turfgrass Science, 2009.

Tran, Phuong N., Michael A. Savka, and Han Ming Gan. "In-silico taxonomic classification of 373 genomes reveals species misidentification and new genospecies within the genus Pseudomonas." *Frontiers in Microbiology* 8 (2017): 1296.

Unterseher, Martin, and Martin Schnittler. "Species richness analysis and ITS rDNA phylogeny revealed the majority of cultivable foliar endophytes from beech (*Fagus sylvatica*)." *Fungal Ecology* 3, no. 4 (2010): 366-378.

Ruiz, Luis Villarreal, Ian C. Anderson, and Ian J. Alexander. "Interaction between an isolate from the *Hymenoscyphus ericae* aggregate and roots of *Pinus* and *Vaccinium*." *New Phytologist* (2004): 183-192.

Wang, J. W., Li-Ping Zheng, and Ren-Xiang Tan. "The preparation of an elicitor from a fungal endophyte to enhance artemisinin production in hairy root cultures of *Artemisia annua* L." *Sheng wu Gong Cheng xue bao. Chinese Journal of Biotechnology* 22, no. 5 (2006): 829-834.

Wang, Jianfeng, Guiling Li, Huaying Lu, Zhonghui Zheng, Yaojian Huang, and Wenjin Su. "Taxol from *Tubercularia* sp. strain TF5, an endophytic fungus of *Taxus mairei*." *FEMS Microbiology Letters* 193, no. 2 (2000): 249-253.

Wang, L., L. Liu, and S. Z. Han. "Screening and identification of antimicrobe activity of endophytic fungus in *Glycyrrhiza uralensis*." *Biotechnol. Bull* 6 (2009): 034.

Wang, Yu, Chuan-Chao Dai, and Yan Chen. "Antimicrobial activity of volatile oil from *Atractylodes lancea* against three species of endophytic fungi and seven species of exogenous fungi." *Yingyong Shengtai Xuebao* 20, no. 11 (2009).

Wang, Yanhong, Lei Xu, Weiming Ren, Dan Zhao, Yanping Zhu, and Xiaomin Wu. "Bioactive metabolites from *Chaetomium globosum L18*, an endophytic fungus in the medicinal plant *Curcuma wenyujin*." *Phytomedicine* 19, no. 3-4 (2012): 364-368.

Wang, Ya, Qing Gui Zeng, Zhi Bin Zhang, Ri Ming Yan, Ling Yun Wang, and Du Zhu. "Isolation and characterization of endophytic huperzine A-producing fungi from *Huperzia serrata*." *Journal of Industrial Microbiology and Biotechnology* 38, no. 9 (2011): 1267-1278.

Waqas, Muhammad, Abdul Latif Khan, Muhammad Kamran, Muhammad Hamayun, Sang-Mo Kang, Yoon-Ha Kim, and In-Jung Lee. "Endophytic fungi produce gibberellins and indoleacetic acid and promote host-plant growth during stress." *Molecules* 17, no. 9 (2012): 10754-10773.

Weyens, Nele, Daniel van der Lelie, Safiyh Taghavi, and Jaco Vangronsveld. "Phytoremediation: plant–endophyte partnerships take the challenge." *Current opinion in biotechnology* 20, no. 2 (2009): 248-254.

Wu, Peng. "Inhibition of RNA-binding proteins with small molecules." *Nature Reviews Chemistry* 4, no. 9 (2020): 441-458.

Xu, Dong, Ying Xu, and C. E. Uberbacher. "Computational tools for protein modeling." *Current Protein and Peptide Science* 1, no. 1 (2000): 1-21.

Xu, Xinran, Runye Huang, and Wen-Bing Yin. "An Optimized and Efficient CRISPR/Cas9 System for the endophytic fungus *Pestalotiopsis fici*." *Journal of Fungi* 7, no. 10 (2021): 809.

Zadeh Hosseingholi, Elaheh, Neda Neghabi, Ghader Molavi, Seyed Mohammad Gheibi Hayat, and Hamed Shahriarpour. "In silico identification and characterization of antineoplastic asparaginase enzyme from endophytic bacteria." *IUBMB life* 72, no. 5 (2020): 991-1000.

Zhang, Chunying, Lijuan Yin, and Silan Dai. "Diversity of root-associated fungal endophytes in *Rhododendron fortunei* in subtropical forests of China." *Mycorrhiza* 19, no. 6 (2009): 417-423.

Zhang, Hua Wei, Yong Chun Song, and Ren Xiang Tan. "Biology and chemistry of endophytes." *Natural product reports* 23, no. 5 (2006): 753-771.

Wenxin, Zou, and Tan Renxiang. "Recent advances on endophyte research." *Acta Botanica Sinica* 43, no. 9 (2001): 881-892.

Zuo, Yiling, Xia Li, Jingya Yang, Jiaqiang Liu, Lili Zhao, and Xueli He. "Fungal endophytic community and diversity associated with desert shrubs driven by plant identity and organ differentiation in the extremely arid desert ecosystem." *Journal of Fungi* 7, no. 7 (2021): 578.

Chapter 10

Multi-Omics Approaches for Identification of Endophytes in Agriculture

Bhawana Mishra[1] **and Muktesh Chandra**[2,*]

[1]Department of Environmental Sciences, Central University of Himachal Pradesh, Dharamshala, Himachal Pradesh, India

[2]SL Education Institute, Moradabad, MJP Rohilkhand University, Bareilly, UP, India

Abstract

Plant diseases, insect pests, and different abiotic stressors cause significant agricultural production losses. Agriculture is the world's largest economic sector; thus, it is critical to identify and implement the best plan for sustainable agriculture and crop growth. Endophytes reside inside the host microenvironment, where they are protected from external stressors, compete for less with other microbes, and have more access to nutrients. Endophytes are microorganisms that thrive asymptomatically within plant tissues without harming the host. The plant microbiome is described as a community of microbial symbionts (microbiota) plus their collective genetic material, which controls the characteristics of interactions between the microorganisms and their host. Plant health and agricultural productivity are heavily influenced by the plant microbiome. Understanding and managing the plant microbiome has the potential to reduce plant diseases while increasing agricultural output. This can reduce chemical fertilizers and pesticides in fields, boosting food production to feed the world's population while also safeguarding the environment and human health. As a result, numerous scientific research has concentrated on uncovering the mysteries of the plant microbiome,

*Corresponding Author's Email: mchandra33@gmail.com.

In: Endophytes
Editor: Pragya Tiwari
ISBN: 979-8-88697-045-6

and the advent of many omics methods has tremendously aided in this endeavor. An omics-based approach could be used to develop predictive models for how microbes and plants will respond to stress as a result of environmental change. This chapter will go through the methodology for using high-throughput sequencing tools to undertake metagenomics research on the microbiota as a component of the plant microbiome. We will offer an overview of the most recent research on this issue and a comprehensive discussion of various omics techniques critical to decoding and understanding the plant microbiome, outlining some of the essential goals tackled to date.

Keywords: endophytes, microbiome, multi-omics, high-throughput, sustainable agriculture

1. Introduction

In nature, plants do not thrive as axenic creatures but rather as hosts to a varied community of microbes known as the plant microbiota. According to growing evidence, the plant microbiome functions in plant development and can protect plants from invading diseases. Humankind has been reshaping its environment for food production for ten millennia. During the green revolution, farmers intentionally altered their microbial habitats by using chemical inputs such as pesticides. Crop-associated microbial populations have been radically imbalanced as agricultural techniques have become more intensive. First appearing in the 1980s, plant growth-promoting rhizobacteria (PGPR) were designed to address microbial population dynamics and their potential benefits for plants in general (Kloepper et al. 1986). A number of helpful microbes have been discovered since then. Most PGPRs and biocontrol are attributed to rhizosphere and root endosphere bacteria, which are likely to derive from bulk soil microbiota (Tiwari et al. 2021a). Understanding the destiny of such microbiota is critical for building smart agricultural techniques; however, determining how to do so will take a significant time. Abiotic and biotic variables influence the microbial populations in the rhizosphere (Tiwari et al. 2021b). The rhizosphere effect, mediated by bipartite interactions, is one of several underlying processes that explain belowground bacteria's composition, organization, and destiny (Hardoim et al. 2012). Some bacteria can be passed vertically from one generation to the next and survive in the early phases of root growth. Plants and their epiphytic microbiota, which can be acquired from seeds, soil, or air, also interact during the formation of aboveground microbial communities, implying a close relationship (Vorholt 2012; Hardoim et al. 2015). The development of field-effective bio-based plant protection products has not occurred yet despite significant advancements in the research of crop-beneficial microorganisms. In order to evaluate the possibility of efficient crop-associated microbiomes, it is necessary to analyze their fate in the agroecosystems into which they have been introduced and the biological functions they may perform in agroecosystems. In addition to changing the tilling regime and growing cover crops, grow cash crops to minimize soil erosion and boost soil fertility (Abdalla et al. 2019). Researchers studied the effects of sowing cash crops between mixed legume and non-legume cover crops to reduce nitrogen leaching and boost soil organic carbon sequestration and grain yields during legume–non-legume cover crops.

The DNA of microbial populations occupying discrete ecological niches and interacting through different and specific roles was initially termed microbiome (Berg et al. 2020). Berg et al. (2020) expanded on this description by including all microorganisms, such as prokaryotes and eukaryotes, their habitats, and the "theatre of action" mediated by microbial structures, metabolites, and genetic elements. Meta 'omics is a term that encompasses metagenomic, transcriptomic, proteomic, and metabolomic research that are used to analyze microbial populations without the need for *invivo* experiments (Figure 1). Genome/transcriptome assembly, as well as annotation to extract the full meaning of data, are examples of next-generation technology (Shukla et al. 2021; Mishra et al. 2021). According to these findings, the endophytic, root-associated, and soil fungal populations are highly influenced by related plant species. The following chapter focuses on the endophytes of the agriculture system to gain deeper insights into bacterial and fungal ecology, which are taxonomically diverse but functionally redundant. The use of next-generation sequencing technologies has significantly impacted the area of genetic research as they offer a low-cost, high-throughput alternative to classic capillary sequencers. A diverse range of utility-based applications has been developed using these novel sequencing technologies that have enabled a more extensive assessment of microbial genome structure and content than could previously only be imagined. As the commercialization of third-generation sequencing technology approaches, we review the uses of existing next-generation sequencing methods and evaluate how they impact microbial genome research.

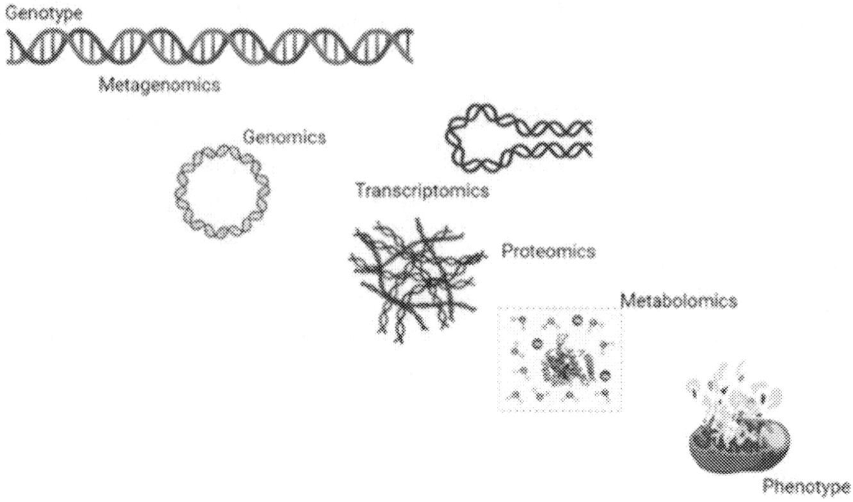

Figure 1. Multi-omics approaches utilized for endophyte studies.

2. The Plant Microbiota: An Investigation

Plant-derived nutrients benefit colonial microorganisms, which develop taxonomically coherent patterns. Two different approaches may produce such microbial architectures, which are not mutually exclusive. During a plant growth, there are vacant niches for invading microbes that can utilize the nutrients available, leading to stochastic colonization episodes. Coevolution between plants and microbes, however, may serve as the basis for an organic selection process that results in passive recruitment of microbes or, a minimum, keystone species that serve the plant host (Tiwari and Bae 2020). As a result, this might contribute to shaping the eventual community during plant growth. They distinguish between coevolved relationships and stochastic possibilities in a system as complex as the host microbiota. Metabolomics is a branch of omics research that examines the composition of metabolites in biological systems under certain conditions over time. Metabolites include amino acids, sugars, organic acids, steroids, nucleic acid bases, fatty acids, sugar alcohols, and many others. From the beginning of the 2000s, metabolomics has been widely used to identify and examine the responses of microbes to external stressors and environmental conditions, such as severe temperature and Ph, and the activity of organic chemicals (Table 1).

Table 1. Abiotic stress is mediated by plants and the associated microbes.

Plant	Stress/toxicity	Strategy for tolerance	Micro-organism associated	References
Arabidopsis thaliana	Salt	Sodium transporter regulation	*Bacillus subtilis*	Zhang et al. 2008
Glycine max	Salt	Soyabean seed germination	*Pseudomonas simiae*	Vaishnav et al. 2018
Arabidopsis thaliana	Salt	pathways of salicylic acid	*Pseudomonas syringae*	Barriuso et al. 2008
Oryza sativa	Salt	Plant gene expression salt stress	PGPR	Jha et al. 2014
Oryza sativa, Triticum aestivum, Zea mays, Gossypium hirsutum	Salt	Elicitor molecules -phytohormones	*Cyanobacteri*a	Singh 2014
Glycine max	Salt	Root increase	*Glomus etunicatum*	Sharifi et al. 2007
Phragmites australis	Salt	Carbohydrates accumulation	*Glomus fasciculatum*	Al-Garni 2006
Zea mays	Osmotic stress	increased root expression - ZmPIP	*Bacillus megaterium*	Marulanda et al. 2010
Phaseolus vulgaris	Osmotic stress	Root conductance enhanced	*Glomus intraradices*	Aroca et al. 2007
Phaseolus vulgaris	Drought	Stress tolerance gene upregulation	*Rhizobium tropici*	Figueiredo et al. 2008
Capsicum annuum	Salinity	Photosynthesis	*Azospirillum brasilense*	Del Amor and Cuadra-Crespo 2012
Vigna radiata	Salinity	Increased K^+ in roots	*Glomus clarum*	Rabie 2005
Arabidopsis	Salinity	decreasing root Na^+ import	*Bacillus subtilis*	Zhang et al. 2008
Lactuca sativa	Salinity	Abscisic acid conc. Reduces	*Glomus intraradices*	Aroca et al. 2008
Gossypium hirsutum	Salinity	Salinity prevented	*Pseudomonas putida*	Yao et al. 2008
Lactuca sativa	Salinity	stimulation of shoot biomass	*Bacillus subtilis*	Arkhipova et al. 2007
Brassica juncea	Arsenic toxicity	available phosphorus	*Staphylococcus arlettae*	Srivastava et al. 2013
Sedum plumbizincicola	Pb/Zn toxicity	Resistance to metals	*Phyllobacterium myrsinacearum*	Ma et al. 2013
Triticum aestivum	Zn toxicity	Improved biomass	*Pseudomonas aeruginosa*	Islam et al. 2014

Table 1. (Continued)

Plant	Stress/toxicity	Strategy for tolerance	Micro-organism associated	References
Sinapis alba	Zn toxicity	P solubilization	*Enterobacter intermedius*	Plociniczak et al. 2013
Miscanthus sinensis	Cd, AS, Cu, Pb and Zn toxicity	IAA production	*Pseudomonas koreensis*	Babu et al. 2015
Phragmites australis	Hg toxicity	Mercury reductase activity	*Photobacterium spp.*	Mathew et al. 2015

A metabolomics method has also been demonstrated to be useful for studying bacteria metabolic problems that are treated with antibiotics and various bioactive substances. Nevertheless, metabolomics has yet to fully investigate the mechanisms of action of bioactive chemicals produced from plants and endophytic microbes. It is often possible to compare metabolic profiles of the examined bioactive substances to those generated with antibiotics, whose mechanisms of action are well understood. There is a possibility that the relative abundance of beneficial rhizosphere and root endospheric bacteria groups varies between resistant and susceptible cultivars (Wei et al. 2019).

Plant genotype, on the other hand, is not the most essential determinant in the early establishment of a rhizosphere microbiome. The composition of the rhizosphere microbiome is influenced by soil type rather than cultivar, according to various studies (Hardoim et al. 2015). This characteristic has been shown in fungi (Nallanchakravarthula et al. 2014), bacteria (Schlemper et al. 2017), and arbuscular mycorrhizal fungi (Santos-González et al. 2011).

Genotype becomes a strong predictor of root-associated microbiomes as plants mature (Nceolu et al. 2010; Schlemper et al. 2017). For a particular growth stage, different cultivars may have varied exudate release patterns (Micallef, Shiaris, and Colón-Carmona 2009; Mönchgesang et al. 2016; Sasse, Martinoia, and Northen 2018), affecting rhizosphere microbial populations in different ways. In the endosphere, the effect of the host genotype on microbial populations is substantially more profound (Urbina et al. 2021). Given that co-evolutionary processes have favored populations well-suited to this niche, this is not surprising. The microbial population of the spermosphere in the early stages of a plant life, around germination, is made up of microorganisms from inside and outside the seed, as well as germs imbibed from the soil. The seed genotype is currently the most critical feature defining the colonizer populations of the spermosphere (Wassermann et al. 2019). Plant growth, nitrogen fixation, and disease control may all be impacted by microbial communities associated with germinating seeds (Walitang et al. 2017). When rootlets in the spermosphere are turned into roots in the rhizosphere, the plant genotype becomes less significant in community growth than soil type. The phytohormone profiles of two cultivars with distinct endophytic populations were notably different in a recent investigation on tomato root endospheric fungus (Manzotti et al. 2020). More study is required to discover whether the hormonal profile effects the composition of endophytic communities (as it does with root exudates) or if it is a consequence of mycobiome composition. In terms of bacteria, most bacterial endophytes have been found to interact with plant hormonal systems (Jasim et al. 2015).

In mixed cropping, plant genotypes are considered to be crucial. For root-associated bacterial and fungal populations, the most effective were the two pea varieties. Rhizosphere fungal diversity is enhanced by polyculture (LeBlanc, Kinkel, and Kistler 2015), which might be influenced by plant species identification and environmental factors. (Schlatter et al. 2015). The differences in bacterial and fungal populations may be due to the microbial interactions encompassing changes in soil characteristics, root exudates, and even intervarietal competition, plant communication, and mineralization (Hinsinger et al. 2011; Reiss and Drinkwater 2018).

3. Bacterial Community Catalogues Using Next-Generation Sequencing

Nearly 20 years after Sanger developed the chain termination sequencing strategy, J. Craig Venter Institute (http://www.jcvi. Org) sequenced the first cellular organism genomes, *Haemophilus influenzae* (Fleischmann et al. 1995) and *Mycoplasma genitalium* (Fraser et al. 1995) bacterial species. The publication of these genomes not only provided a glimpse of a living organism's full genomes, but it also revolutionized genomics by advanced sequencing methodologies such as paired-end sequencing (Edwards et al. 1990; Roach et al. 1995) and the adoption of the whole genome shotgun methodology (Weber and Myers 1997). Larger genomes sequences following these were *Bacillus subtilis* (Kunst et al. 1997) and *Escherichia coli* (Blattner et al. 1997), as well as the genomes of the eukaryotes *Saccharomyces cerevisiae* (Goffeau 1998), *Caenorhabditis elegans* (Schumacher et al. 2001), *Arabidopsis thaliana* (The Arabidopsis Genome Initiative 2000), *Drosophila melanogaster* (Myers et al. 2000), and eventually the genome of *Homo sapiens* (Lander et al. 2001; Venter et al. 2001). Despite breakthroughs in technologies, the cost of sequencing has remained high and prohibitive for most research organizations. The capillary electrophoresis (CE) sequencing platforms' high cost per base and limited throughput encouraged the development of next-generation sequencing (NGS). These NGS technologies had provided high throughput at low cost bringing a paradigm shift.

More than, 1566 bacterial genomes have been sequenced, and are present in public repositories. Additionally, out of 5156 in process' genome projects, 4226 are bacterial (http://www. Genomesonline.org). The genome sequencing business prior to NGS was controlled by approaches based on the Sanger sequencing methods and chemicals. For regular sequencing of bacterial genomes, automated Sanger capillary-based sequencing methods that rely on clone libraries were costly, time-consuming, and labor-intensive. For the *de novo* sequencing and re-sequencing of bacterial genomes, NGS technologies have shown to be viable tools with high throughput and low cost.

On the other hand, cultivable bacteria make up a modest portion of the world's overall microbial diversity (Pace 2009). To comprehend better and analyze microbial diversity scientists trust is inclining towards the discipline of metagenomics. Metagenomics studies population dynamics, genetic diversity, and microbial communities interactions in their habitats using culture-independent approaches. Using standard sequencing technology, early metagenomics investigations focused on 16S Rrna gene amplicons to examine microbial diversity by targeted sequencing (Delgado, Suárez, and Mayo 2006) or whole community shotgun metagenomics (Tyson et al. 2004).

Despite being conserved across all bacteria, the 16S Rrna gene contains enough interspecies variability to be employed as a molecular tool for bacterial identification (Janda and Abbott 2007). With long enough reads acquired on CE platforms, bacterial amplicons may be reliably identified to genus and, in some instances, species level. Sanger sequencing, on the other hand, is time-consuming and labor-intensive, and the need for a cloning step may lead to a bias against cloned sequences that are not stable in the heterologous host. As a consequence, there has been a slow move to metagenomics for microbial identification. For microbiological identification, however, just a small portion of the 16S Rrna gene is required. The hypervariable regions of the 16S Rrna gene range in length from 50 to 200 nucleotides. Microbial identification can be done quickly, cheaply, and with minimal effort by using high-throughput sequencing of a subset of these areas (Luna et al. 2007; Petrosino et al. 2009; Ventura et al. 2009).

Despite its use in identifying bacterial species in populations, 16S Rrna gene sequencing has limits. While 16S Rrna gene sequencing may offer a plethora of information on microbial diversity in a niche and how community composition influences health and sickness, it can only provide a limited amount of information about each species' contribution to the ecosystem. To fully comprehend a microbiome's genetic potential, whole shotgun sequencing of the community *i.e.,* metagenomics is required. Thanks to the vast quantity of data collected, WS metagenomics has permitted the annotation of a diverse range of microbial genes as well as the discovery of many novel genes encoding new functionalities. Large-scale metagenome projects like the Global Ocean Survey (http://www.jcvi.org/cms/research/projects/gos/), HMP (http://www.hmpdacc.org), MetaHIT (http://www.meta hit.eu), now have a much better understanding of microbial gene diversity.

The root bacterial microbiome frequently (but not always) has the greatest diversity of microorganisms when compared to other plant organs (Berg et al. 2020). *Verrucomicrobia*, *Acidobacteria*, *Proteobacteria*, *Bacteroidetes*, *Actinobacteria*, and *Planctomycetes*, are the most prevalent taxa, and the majority of them may be located in the rhizosphere (Hardoim et al. 2015). Only a small percentage of rhizosphere bacteria may penetrate roots. The most prevalent phyla found in grapevine roots were *Chloroflexi*, *Proteobacteria*, *Planctomycetes*, *Acidobacteria*, *Firmicutes*, *Bacteroidetes*, *Actinobacteria*, *Verrucomicrobia* and *Gemmatimonatedes* (Samad et al. 2017). *Firmicutes*, *Proteobacteria*, and *Bacteroidetes* are discovered as prominent phyla inside maize roots (Correa-Galeote, Bedmar, and Arone 2018). Distinct plant species have different microbiota compositions, but community structure and diversity are also influenced by plant genotype and developmental stage. Furthermore, various organisms' changes in the variety and abundance are influenced by biotic or abiotic pressures, the soil environment, and its history (Samad et al. 2017; Sasse, Martinoia, and Northen 2018; Correa-Galeote, Bedmar, and Arone 2018).

For researching microbial diversity, high-throughput sequencing of the taxonomically significant 16S Rrna gene is a useful approach. The study of bacterial communities found in environmental and biological specimens has become more popular. PCR amplification of a portion of the taxonomically informative 16S Rrna gene and subjecting this product to next-generation DNA sequencing, which allows individual reads to be classified to specific taxa, is one common method for classifying microbial species present in a sample, which is based on previous cloning-based methods. Unlike shotgun sequencing of genomic DNA extracted from a sample, which sequences and classifies random fragments of bacterial genomes (and potentially contaminating DNA from host species or other organisms present), 16S Rrna amplicon sequencing (Janda and Abbott 2007) can be targeted specifically against bacteria, requires no reference genome sequences, and can be used in cases where only trace amounts or poor-quality bacterial DNA templates are available. Therefore, sequencing employed in a variety of metagenomic profiling studies is 16S Rrna sequencing, especially in cases where the input material is limited.

The makeup of root microbiota may be altered because of fungi, and bacterial microbial interactions, the chapter will further focus on these aspects.

4. Analyses of Fungal Communities by Next-Generation Sequencing

Fungal endophytes reside in the internal tissues of living plants and may be found in almost every plant on the globe, from the arctic to the tropics. The endophyte–host relationship is a well-balanced symbiotic continuum that ranges from mutualism through commensalism to parasitism. In most terrestrial ecosystems, fungi play an essential role in nutrient cycling, mostly via symbiotic interactions with plants such as mycorrhiza and the breakdown of wood and plant waste (Stajich et al. 2009). Because mushrooms spend most of their lives underground or within substrates, scientists only have a rudimentary grasp of their variety, and only a few million existing species of fungus have been documented (Wasser 2011). The lack of association between the availability of fungal fruiting bodies or other macroscopic structures and the complete diversity of fungal tumors at specific sample sites shifts the focus of fungal ecology from fruiting bodies to molecular data. Sequence data broadly describes all recent attempts to characterize fungal communities (Taylor et al. 2010). The high cost and effort associated with traditional Sanger sequencing of large numbers of samples have traditionally limited the depth of sequencing of such studies. However, recent methodological advances in the form of next-generation sequence (NGS) technology provide solutions to these problems (Shendure and Ji 2008). One of these NGS approaches is massively parallel ('454') pyrosequencing, which can produce over a million sequences in a day, making it a game-changing tool for fungal environmental sequencing (Margulies et al. 2005). In the absence of universally agreed standards, all of the research opportunities provided by NGS in general, and pyrosequencing in particular, remain difficult to pursue and may even be damaging to fungal ecology and mycology in general. At various stages of the analysis, various forms of incompletely understood biases are introduced (Quince et al. 2009; Bellemain et al. 2010). These are frequently ignored throughout the interpretation of the results. Users have different opinions about delimiting species or operational taxonomic units (OTUs) using DNA data and how to treat abundance data, taxonomic standards, and ecological classifications (Hibbett et al. 2011). Modern web software, such as SCATA (http://scata.mykopat.slu.se), CLOTU (http://www.bioportal.uio.no), and PlutoF (http://unite.ut.ee), has been developed specifically for fungal pyrosequencing data sequence processing and analysis, and these resources approach sequence clustering and identification in different ways.

5. Other Omics Methods for Investigation of the Plant Microbiome

To study the functional role of endophytic communities, a variety of -omics methods are now accessible, such as (meta)transcriptomics, (meta)genomics, (meta)proteomics, culturomics and metabolomics (Kaul, Sharma, and K Dhar 2016) (Figure 2). These technologies advancement at an alarming pace, results in high-dimensional, heterogeneous datasets that must be reduced and merged utilizing contemporary bioinformatics and statistical approaches in order to extract usable information (Rohart et al. 2017). These omics technologies are being investigated in plant species to better understand their unique nature, whether for remediation, pharmaceutical, or other purposes (Mishra et al. 2021; Chandra et al. 2020). All of these methods give more information on the function of bacterial endophytes. For example, metagenomics, also known as shotgun metagenomics, pioneered the field by uncovering genes, features, and metabolic pathways required for the endophytic lifestyle (Nesme et al. 2016). The metabolic pathways and activities triggered in an endophytic bacterial strain that improves sugarcane growth when exposed to apoplast fluid are determined using transcriptomic and proteomic methods (Terra et al. 2019). To find the over-expressed metabolites in plant tissues colonized by a bacterial endophyte, researchers coupled confocal and fluorescent imaging with *in situ* metabolomics. The next challenge in plant endophyte research will be to relate research problems to conceptual frameworks, not just to learn and integrate new technologies (Saikkonen, Nissinen, and Helander 2020). Multi-omics methods are unquestionably the way of the future, and they will enable microbiology-based agricultural breakthroughs (Compant et al. 2019).

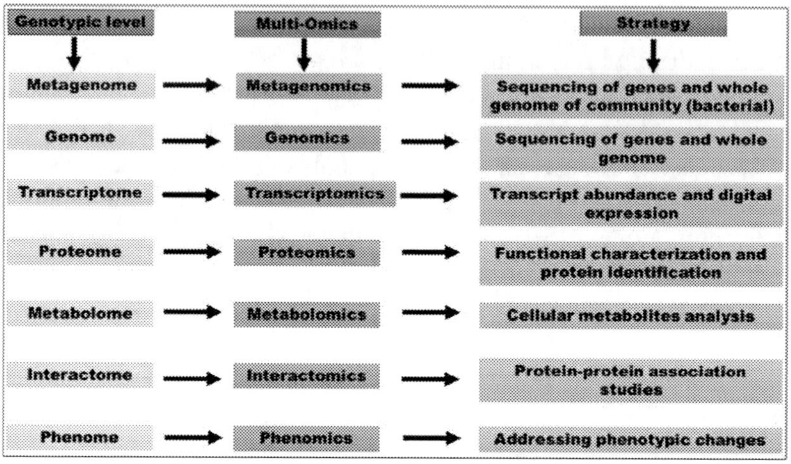

Figure 2. Components at the cellular level, multi-omics techniques to address various levels, and ways for identifying the consequence of microbial interactions.

On the other hand, endophyte research should be driven not just by technological advancements, data flows, and bioeconomy, but also by ecological and evolutionary ideas (Saikkonen, Nissinen, and Helander 2020). Further, we will need to develop conceptual frameworks that account for the effects of the plant microbial population on plant phenotypes and scale these discoveries across space and time. The host plant, the surrounding microbial flora, other organisms within the plant environment, and the environment with its structural and functional properties and polynutrient interactions must all be considered within such a framework. Integrating evolutionary ecology, plant physiology & pathology, plant genetics &epigenetics, and earth & ecosystem sciences into these frameworks will need interdisciplinary methodologies beyond omics methodology. They will help us better understand the function and development of plant holobionts, as well as forecast their responses and feedback effects in response to environmental changes on a global scale (Saikkonen, Nissinen, and Helander 2020).

Microbial populations and communities provide the majority of ecosystem services and products on which human populations depend. Microbes are responsible for a wide range of supplying, regulating, and supporting functions, including food, enzymes for industrial processes, water quality, bioactive metabolites, pollution reduction, and biological processes including plant-microbial symbioses (Tiwari et al. 2021c; Tiwari et al. 2022). To conserve and safeguard these ecosystems, which are now confronting climate change and human activity-related difficulties, and to harness some of the naturally occurring microbial activities, a detailed understanding of the microbiology underlying ecosystem functioning is essential. To this goal, metagenomics and other meta-omics technologies have been used in experimental procedures. The increased understanding of results leads to the creation of novel applications in many sectors spanning all microbial ecosystem services and commodities.

Conclusion

Understanding microbial populations and their activity in the plant endosphere is challenging due to methodological restrictions. Advances in transcriptomes and genomes make it easier to examine endophytes from the plant host or bacteria in the phyllosphere and rhizosphere in-planta. These limits, together with the huge amounts of data produced by -omics research, have created a major need for procedure standardization, such as surface sterilization or sample processing. Hypothesis-driven concepts and research approaches coupled with multi-disciplinary analyses will enhance endophyte activity and adaptation to their host plant, plant–endophyte interactions, and the evolution of the plant holobiont. This information will be crucial in developing microbial consortia with synergistic activities that will lead to more sustainable microbe-based agricultural production.

Acknowledgments

The authors extend their gratitude to their respective research organizations.

Conflict of Interests

No conflict of interest was declared.

References

Abdalla, Mohamed, Astley Hastings, Kun Cheng, Qian Yue, Dave Chadwick, Mikk Espenberg, Jaak Truu, Robert M Rees, and Pete Smith. "A critical review of the impacts of cover crops on Nitrogen leaching, net greenhouse gas balance and crop productivity." *Global Change Biology* 25, no. 8 (2019): 2530–43.

Amor, Francisco M. del, and Paula Cuadra-Crespo. "Plant growth-promoting bacteria as a tool to improve salinity tolerance in sweet pepper." *Functional Plant Biology* 39, no. 1 (2012): 82. https://doi.org/10.1071/FP11173.

Arkhipova, T. N., E. Prinsen, S. U. Veselov, E. V. Martinenko, A. I. Melentiev, and G. R. Kudoyarova. "Cytokinin producing bacteria enhance plant growth in drying soil." *Plant and Soil* 292, no. 1-2 (2007): 305–15. https://doi.org/10.1007/s11104-007-9233-5.

Aroca, Ricardo, Rosa Porcel, and Juan Manuel Ruiz-Lozano. "How does Arbuscular mycorrhizal symbiosis regulate root hydraulic properties and plasma membrane Aquaporins in *Phaseolus vulgaris* under drought, cold, or salinity stresses?" *New Phytologist* 173, no. 4 (2007): 808–16. https://doi.org/10.1111/j.1469-8137.2006.019 61.x.

Barriuso, Jorge, Beatriz Ramos Solano, Rupert G. Fray, Miguel Cámara, Anton Hartmann, and F. Javier Gutiérrez Mañero. "Transgenic tomato plants alter quorum sensing in plant growth-promoting Rhizobacteria." *Plant Biotechnology Journal* 6, no. 5 (2008): 442–52. https://doi.org/10.1111/j.1467-7652.2008.00331.x.

Bellemain, Eva, Tor Carlsen, Christian Brochmann, Eric Coissac, Pierre Taberlet, and Håvard Kauserud. "ITS as an environmental DNA barcode for Fungi: an in silico approach reveals potential PCR biases." *BMC Microbiology* 10, no. 1 (2010): 1–9.

Berg, Gabriele, Daria Rybakova, Doreen Fischer, Tomislav Cernava, Marie-Christine Champomier Vergès, Trevor Charles, Xiaoyulong Chen, Luca Cocolin, Kellye Eversole, and Gema Herrero Corral. "Microbiome definition re-visited: old concepts and new challenges." *Microbiome* 8, no. 1 (2020): 1–22.

Blattner, Frederick R., Guy Plunkett, Craig A. Bloch, Nicole T. Perna, Valerie Burland, Monica Riley, Julio Collado-Vides, Jeremy D. Glasner, Christopher K. Rode, and George F. Mayhew. "The complete genome sequence of *Escherichia coli K-12*." *Science* 277, no. 5331 (1997): 1453–62.

Chandra, Muktesh, Kushwaha, Shiwani, and Sangwan, Neelam Singh, «Comparative transcriptome analysis to identify putative genes related to trichome development in Ocimum species.» *Molecular Biology Reports* 47 (2020), 6587-6598 https://doi.org/ 10.1007/s11033-020-05710-1.

Compant, Stéphane, Abdul Samad, Hanna Faist, and Angela Sessitsch. "A review on the plant microbiome: ecology, functions, and emerging trends in the microbial application." *Journal of Advanced Research* 19 (2019): 29–37.

Correa-Galeote, David, Eulogio J. Bedmar, and Gregorio J. Arone. "Maize endophytic bacterial diversity as affected by soil cultivation history." *Frontiers in Microbiology* 9 (2018): 484.

Delgado, Susana, Adolfo Suárez, and Baltasar Mayo. "Identification of dominant bacteria in feces and colonic mucosa from healthy Spanish adults by culturing and by 16S rDNA sequence analysis." *Digestive Diseases and Sciences* 51, no. 4 (2006): 744–51.

Goffeau, A. "The Yeast genome." *Pathologie-Biologie* 46, no. 2 (1998): 96–97.

Hardoim, Pablo R., Cristiane C. P. Hardoim, Leonard S. Van Overbeek, and Jan Dirk Van Elsas. "Dynamics of seed-borne Rice endophytes on early plant growth stages." *PloS One* 7, no. 2 (2012): e30438.

Hardoim, Pablo R., Leonard S. Van Overbeek, Gabriele Berg, Anna Maria Pirttilä, Stéphane Compant, Andrea Campisano, Matthias Döring, and Angela Sessitsch. "The hidden world within plants: Ecological and evolutionary considerations for defining functioning of microbial endophytes." *Microbiology and Molecular Biology Reviews* 79, no. 3 (2015): 293–320.

Hibbett, David S., Anders Ohman, Dylan Glotzer, Mitchell Nuhn, Paul Kirk, and R. Henrik Nilsson. "Progress in molecular and morphological taxon discovery in Fungi and options for formal classification of environmental sequences." *Fungal Biology Reviews* 25, no. 1 (2011): 38–47.

Janda, J. Michael, and Sharon L. Abbott. "16S rRNA gene sequencing for bacterial identification in the diagnostic laboratory: pluses, perils, and pitfalls." *Journal of Clinical Microbiology* 45, no. 9 (2007): 2761–64.

Jha, Yachana, Gaurav Sablok, Naidu Subbarao, Raja Sudhakar, M. H. U. Turabe Fazil, R. B. Subramanian, Andrea Squartini, and Sunil Kumar. "Bacterial-induced expression of RAB18 protein in *Orzya sativa* salinity stress and insights into molecular interaction with GTP Ligand." *Journal of Molecular Recognition* 27, no. 9 (2014): 521–27. https://doi.org/10.1002/jmr.2371.

Kaul, Sanjana, Tanwi Sharma, and Manoj K. Dhar. "'Omics' tools for better understanding the plant–endophyte interactions." *Frontiers in Plant Science* 7 (2016): 955.

Kloepper, J. W., F. M. Scher, M. Laliberte, and B. Tipping. "Emergence-promoting Rhizobacteria: description and implications for Agriculture." In *Iron, Siderophores, and Plant Diseases* (1986):155–64. Springer.

Kunst, F., N. Ogasawara, I. Moszer, A. M. Albertini, G. O. Alloni, Vasco Azevedo, M G. Bertero, Philippe Bessières, A. Bolotin, and Sea Borchert. "The complete genome sequence of the Gram-positive bacterium *Bacillus subtilis*." *Nature* 390, no. 6657 (1997): 249–56.

Lander, Eric S., Lauren M. Linton, Bruce Birren, Chad Nusbaum, Michael C. Zody, Jennifer Baldwin, Keri Devon, Ken Dewar, Michael Doyle, and William FitzHugh. "Initial sequencing and analysis of the human genome." 2001.

Luna, Ruth Ann, Lea R. Fasciano, Shaunte C. Jones, Bobby L. Boyanton Jr., Trang T. Ton, and James Versalovic. "DNA pyrosequencing-based bacterial pathogen identification in a pediatric hospital setting." *Journal of Clinical Microbiology* 45, no. 9 (2007): 2985–92.

Ma, Ying, Mani Rajkumar, Yongming Luo, and Helena Freitas. "Phytoextraction of heavy metal polluted soils using *Sedum plumbizincicola* inoculated with metal mobilizing *Phyllobacterium myrsinacearum RC6b*." *Chemosphere* 93, no. 7 (2013): 1386–92. https://doi.org/10.1016/j.chemosphere.2013.06.077.

Margulies, Marcel, Michael Egholm, William E. Altman, Said Attiya, Joel S. Bader, Lisa A. Bemben, Jan Berka, Michael S. Braverman, Yi-Ju Chen, and Zhoutao Chen. "Genome sequencing in microfabricated high-density picolitre reactors." *Nature* 437, no. 7057 (2005): 376–80.

Marulanda, Adriana, Rosario Azcón, François Chaumont, Juan Manuel Ruiz-Lozano, and Ricardo Aroca. "Regulation of plasma membrane Aquaporins by inoculation with a *Bacillus megatherium* strain in Maize (*Zea mays* L.) plants under unstressed and salt-stressed conditions." *Planta* 232, no. 2 (2010): 533–43. https://doi.org/10.1007/s00425-010-1196-8.

Mathew, Dony Chacko, Ying-Ning Ho, Ronnie Gicaraya Gicana, Gincy Marina Mathew, Mei-Chieh Chien, and Chieh-Chen Huang. "A rhizosphere-associated symbiont, photobacterium spp. strain MELD1, and its targeted synergistic activity for phytoprotection against mercury." *PLOS ONE* 10, no. 3 (2015): e0121178. https://doi.org/10.1371/journal.pone.0121178.

Meena, Kamlesh K., Ajay M. Sorty, Utkarsh M. Bitla, Khushboo Choudhary, Priyanka Gupta, Ashwani Pareek, Dhananjaya P. Singh, et al. "Abiotic stress responses and microbe-mediated mitigation in plants: The omics strategies." *Frontiers in Plant Science* 8 (2017). https://doi.org/10.3389/fpls.2017.00172.

Mishra, Bhawana, Chandra, Muktesh, and Pant, Deepak. « Genome-mining for stress-responsive genes, profiling of antioxidants and radical scavenging metabolism in hyperaccumulator medicinal and aromatic plants «*Industrial Crops and Products* 173 (2021), 114107 https://doi.org/10.1016/j.indcrop.2021.114107.

Mishra, Pallavi, Maurya, Ranjeet, Avashthi, Himanshu, Mittal, Shikha, Chandra, Muktesh, and Pramod Wasudeo Ramteke «Genome assembly and annotation « In *Bioinformatics Methods and Applications* 1 (2022) 49-66, https://doi.org/10.1016/B978-0-323-89775-4.00013-4

Myers, Eugene W., Granger G. Sutton, Art L. Delcher, Ian M. Dew, Dan P. Fasulo, Michael J. Flanigan, Saul A. Kravitz, Clark M. Mobarry, Knut H. J. Reinert, and Karin A Remington. "A whole-genome assembly of *Drosophila*." *Science* 287, no. 5461 (2000): 2196–2204.

Nesme, Joseph, Wafa Achouak, Spiros N. Agathos, Mark Bailey, Petr Baldrian, Dominique Brunel, Åsa Frostegård, Thierry Heulin, Janet K. Jansson, and Edouard Jurkevitch. "Back to the future of soil metagenomics." *Frontiers in Microbiology* 7 (2016): 73.

Pace, Norman R. "Mapping the Tree of Life: Progress and prospects." *Microbiology and Molecular Biology Reviews* 73, no. 4 (2009): 565–76.

Petrosino, Joseph F., Sarah Highlander, Ruth Ann Luna, Richard A. Gibbs, and James Versalovic. "Metagenomic pyrosequencing and microbial identification." *Clinical Chemistry* 55, no. 5 (2009): 856–66.

Płociniczak, T., A. Sinkkonen, M. Romantschuk, and Z. Piotrowska-Seget. "Characterization of *Enterobacter intermedius MH8b* and Its use for the enhancement of heavy metals uptake by *Sinapis alba* L." *Applied Soil Ecology* 63 (2013): 1–7. https://doi.org/10.1016/j.apsoil.2012.09.009.

Quince, Christopher, Anders Lanzén, Thomas P. Curtis, Russell J. Davenport, Neil Hall, Ian M. Head, L. Fiona Read, and William T. Sloan. "Accurate determination of microbial diversity from 454 pyrosequencing data." *Nature Methods* 6, no. 9 (2009): 639–41.

Rabie, G. H. "Influence of Arbuscular mycorrhizal fungi and kinetin on the response of Mungbean plants to irrigation with seawater." *Mycorrhiza* 15, no. 3 (2005): 225–30. https://doi.org/10.1007/s00572-004-0345-y.

Rohart, Florian, Benoit Gautier, Amrit Singh, and Kim-Anh Lê Cao. "MixOmics: An R Package for 'omics Feature Selection and multiple data integration." *PLoS Computational Biology* 13, no. 11 (2017): e1005752.

Saikkonen, Kari, Riitta Nissinen, and Marjo Helander. "Toward comprehensive plant microbiome research." *Frontiers in Ecology and Evolution* 8 (2020): 61.

Samad, Abdul, Friederike Trognitz, Stéphane Compant, Livio Antonielli, and Angela Sessitsch. "Shared and host-specific microbiome diversity and functioning of grapevine and accompanying weed plants." *Environmental Microbiology* 19, no. 4 (2017): 1407–24.

Sasse, Joelle, Enrico Martinoia, and Trent Northen. "Feed your friends: Do plant exudates shape the root microbiome?" *Trends in Plant Science* 23, no. 1 (2018): 25–41.

Schumacher, Björn, Kay Hofmann, Simon Boulton, and Anton Gartner. "The *C. elegans* homolog of the P53 tumor suppressor is required for DNA damage-induced apoptosis." *Current Biology* 11, no. 21 (2001): 1722–27.

Sharifi, Mozafar, Mahlagha Ghorbanli, and Hassan Ebrahimzadeh. "Improved growth of salinity-stressed soybean after inoculation with salt pre-treated Mycorrhizal fungi." *Journal of Plant Physiology* 164, no. 9 (2007): 1144–51. https://doi.org/10.1016/j.jplph.2006.06.016.

Shendure, Jay, and Hanlee Ji. "Next-generation DNA sequencing." *Nature Biotechnology* 26, no. 10 (2008): 1135–45. https://doi.org/10.1038/nbt1486.

Shukla, Samvedna, Mishra, Bhawana, Avashthi, Himanshu, and Chandra, Muktesh. Biological sequence analysis, In *Bioinformatics Methods and Applications* 1 (2022) 33-47. https://doi.org/10.1016/B978-0-323-89775-4.00003-1.

Singh, S. "A review on possible elicitor molecules of cyanobacteria: their role in improving plant growth and providing tolerance against biotic or abiotic stress." *Journal of Applied Microbiology* 117, no. 5 (2014): 1221–44. https://doi.org/10.1111/jam.12612.

Stajich, Jason E., Mary L. Berbee, Meredith Blackwell, David S Hibbett, Timothy Y. James, Joseph W. Spatafora, and John W. Taylor. "Primer-the fungi." *Current Biology: CB* 19, no. 18 (2009): R840.

Taylor, D. Lee, Ian C. Herriott, Kelsie E. Stone, Jack W. McFarland, Michael G. Booth, and Mary Beth Leigh. "Structure and resilience of fungal communities in Alaskan boreal forest soils." *Canadian Journal of Forest Research* 40, no. 7 (2010): 1288–1301.

Terra, Leonardo A., Cleiton P. de Soares, Carlos H. S. G. Meneses, Michelle Z. Tadra Sfeir, Emanuel M. de Souza, Vanildo Silveira, Márcia S. Vidal, José I. Baldani, and Stefan Schwab. "Transcriptome and proteome profiles of the diazotroph *Nitrospirillum amazonense strain CBAmC* in response to the sugarcane apoplast fluid." *Plant and Soil* (2019): 1–24.

The *Arabidopsis* Genome Initiative. "Analysis of the genome sequence of the flowering plant *Arabidopsis Thaliana*." *Nature* 408, no. 6814 (2000): 796–815. https://doi.org/10.1038/35048692.

Tiwari, Pragya, Subir Kumar Bose, and Hanhong Bae. "Plant growth-promoting soil microbiomes: beneficial attributes and potential applications." In *Soil Microbiomes for Sustainable Agriculture - Volume 2: Functional Annotation* edited by Dr. Ajar Nath Yadav, (2021a), Springer Publication.

Tiwari, Pragya, Mangalam Bajpai, Lalit Kumar Singh, Ajar Nath Yadav, Hanhong Bae. "Portraying Fungal mechanisms in stress tolerance: perspective for sustainable agriculture." In Springer-*Nature book on Recent Trends in Mycological Research*, 2021b.

Tiwari, Pragya, Yashdeep Srivastava, and Hanhong Bae. "Trends of pharmaceutical design of Endophytes as anti-infective." *Current Topics in Medicinal Chemistry*, 21, no. 17 (2021c): 1572-1586.

Tiwari, Pragya, Hanhong Bae. 2022. "Endophytic Fungi: Key insights, emerging prospects, and challenges in natural product drug discovery." *Microorganisms* 10, 360 (2022). https://doi.org/10.3390/microorganisms10020360.

Tyson, Gene W., Jarrod Chapman, Philip Hugenholtz, Eric E. Allen, Rachna J. Ram, Paul M. Richardson, Victor V. Solovyev, Edward M. Rubin, Daniel S. Rokhsar, and Jillian F. Banfield. "Community structure and metabolism through reconstruction of microbial genomes from the environment." *Nature* 428, no. 6978 (2004): 37–43.

Vaishnav, Anukool, Sarita Kumari, Shekhar Jain, Ajit Varma, Narendra Tuteja, and Devendra Kumar Choudhary. "PGPR-mediated expression of salt tolerance gene in soybean through volatiles under Sodium nitroprusside." *Journal of Basic Microbiology* 56, no. 11 (2016): 1274–88. https://doi.org/10.1002/jobm.201600188.

Venter, J. Craig, Mark D. Adams, Eugene W. Myers, Peter W. Li, Richard J. Mural, Granger G. Sutton, Hamilton O. Smith, Mark Yandell, Cheryl A. Evans, and Robert A. Holt. "The sequence of the Human genome." *Science* 291, no. 5507 (2001): 1304–51.

Ventura, Marco, Sarah O'flaherty, Marcus J. Claesson, Francesca Turroni, Todd R. Klaenhammer, Douwe Van Sinderen, and Paul W. O'toole. "Genome-scale analyses of health-promoting bacteria: probiogenomics." *Nature Reviews Microbiology* 7, no. 1 (2009): 61–71.

Vorholt, Julia A. "Microbial life in the phyllosphere." *Nature Reviews Microbiology* 10, no. 12 (2012): 828–40.

Wasser, Solomon P. "Current findings, future trends, and Unsolved problems in studies of medicinal mushrooms." *Applied Microbiology and Biotechnology* 89, no. 5 (2011): 1323–32.

Wei, Xiaomeng, Zhenke Zhu, Liang Wei, Jinshui Wu, and Tida Ge. "Biogeochemical cycles of key elements in the paddy-rice rhizosphere: microbial mechanisms and coupling processes." *Rhizosphere* 10 (2019): 100145.

Yao, Lixia, Zhansheng Wu, Yuanyuan Zheng, Imdad Kaleem, and Chun Li. "Growth promotion and protection against salt stress by *Pseudomonas putida Rs-198* on Cotton." *European Journal of Soil Biology* 46, no. 1 (2010): 49–54. https://doi.org/10.1016/j.ejsobi.2009.11.002.

Zhang, Huiming, Mi-Seong Kim, Yan Sun, Scot E. Dowd, Huazhong Shi, and Paul W. Paré. "Soil bacteria confer plant salt tolerance by tissue-specific regulation of the sodium transporter HKT1." *Molecular Plant-Microbe Interactions* 21, no. 6 (2008): 737–44. https://doi.org/10.1094/MPMI-21-6-0737.

Chapter 11

Plant-Associated Endophytes: Molecular Mechanisms and Significance in Promoting Sustainable Agriculture

Pragya Tiwari[1,*], Muhammad Adil[2], Nagaraj Basavegowda[1] and Jen-Tsung Chen[3,†]

[1]Department of Biotechnology, Yeungnam University, Gyeongsan, Gyeongbuk, Republic of Korea

[2]Pharmacology and Toxicology Section, University of Veterinary and Animal Sciences, Lahore, Jhang Campus, Pakistan

[3]Department of Life Sciences, National University of Kaohsiung, Kaohsiung, Taiwan

Abstract

Plant-associated endophytes comprise microbes residing within plant tissues and demonstrating multi-faceted roles in the environment, agriculture, and healthcare. Their co-evolution and the integral association with the plant partners contributed to multiple plant growth-promoting attributes including biomass accumulation and enhanced plant growth, nutrient acquisition, and recycling to confer abiotic/biotic stress tolerance to plant species. In the recent perspective, endophytes highlight significant use as eco-friendly and cost-effective bio-stimulants for plant growth promotion and as efficient biocontrol agents against phytopathogens. To reduce the use of agro-chemicals, bio-formulation of endophytes for promoting crop productivity defines a prospective approach towards sustainable agriculture. With the extensive exploration

---

*Corresponding Author's Email: pragyatiwari@ynu.ac.kr.

†Corresponding Author's Email: jentsung@nuk.edu.tw.

In: Endophytes
Editor: Pragya Tiwari
ISBN: 979-8-88697-045-6

of endophytes in promoting plant growth and stress tolerance, these are poised as emerging eco-friendly candidates for improving crop productivity highlighting prospects in sustainable agriculture. Though the dynamics of plant-endophyte interactions still need to be completely understood, employing endophytes is envisaged as a prospective approach to reduce economic burdens and the use of agrochemicals in agriculture. With an overview of endophyte biology and their classification, the chapter provides a detailed account of plant-associated endophytes and their emerging importance in arrays of plant growth benefits for plant partners. The potential mechanisms of endophytes in plant growth and development necessitate bio-prospection of endophytes and the use of upcoming scientific technologies to address the associated bottlenecks in full utilization of endophyte potential and establishing a sustainable approach to crop improvement.

Keywords: agrochemicals, bio-stimulants, fungal endophytes, phytohormones, plant growth promotion, sustainable agriculture, stress tolerance

1. Plant-Endophyte Associations in Nature

With declining natural resources, there is a need to protect and conserve threatened plant species and depleting novel metabolic entities in nature and search for alternative therapeutics from natural sources (Tiwari et al. 2021a; Tiwari et al. 2021b; Tiwari et al. 2022a). On a global scale, approximately 300,000 plant species exhibit endophyte associations (Strobel and Daisy 2003), accounting for major functional roles in ecosystems (Sridhar et al. 2019; Tiwari and Bae 2020a). Endophyte communities worldwide exhibit enormous biodiversity; for example, rainforests in Brazil are a rich source of diverse flora and fauna but are relatively less explored for their biotechnological potential. Due to their multi-faceted attributes and benefits, endophytes are gaining importance as natural resources of bioactive metabolites, in addition to prospects in agriculture and the environment (Tiwari et al. 2021c; Tiwari et al. 2022b).

Endophytes constitute bacterial or fungal microorganisms, inhabiting inter/intracellular spaces within plant tissues, and showing ubiquitous presence. Since the plants restrict the growth of endophytes, endophytes evolve mechanisms for adaptation to the environment (Das and Varma 2009; Dudeja et al. 2012) including the production of metabolites for plant growth and survival (Das and Varma 2009). The fungal endophytes isolated from plants and algae belong to Ascomycota, while very few reports of Basidiomycetes are available (Rungjindamai et al. 2008).

Some of the plant-associated microbes, namely *Penicillium glandicola*, *Phoma tropica*, and *Acremonium terricola*, are classified as fungal endophytes (Bezerra et al. 2012). The colonization of endophytes within a plant tissue may either be specific (known as non-systemic) or systemic having variable effects on plant-endophyte associations (Boyle et al. 2001). Moreover, the plant tissue also influences the colonization pattern of endophytes, for example, endophytes of the tribe Balansieae (family Clavicipitaceae) colonize plants in Juncaceae, Poaceae, and Cyperaceae family (Clay 1990). The *Epichloe* species demonstrates systemic colonization of all the above-ground tissues of grasses (Pooideae sub-family) and is highly synchronized with the colonizing tissue and growth (Tan and Zhou 2001; Christensen and Voisey 2007; Christensen et al. 2008). Moreover, these fungi grow within the intercellular spaces of grasses and show less hyphal branching (Christensen and Voisey 2007; Christensen et al. 2008). Some endophytes exhibit non-systemic colonization patterns, for example, some endophytes show limited inter/intracellular growth within plant tissue (Stergiopoulos and Gordon 2014). In some studies, researchers have tried to address whether the endophyte colonization pattern governs its biocontrol activity considering many endophytes do not come in direct contact with a pathogen. Yan et al. (2015) suggested that systemic colonization was not essential for endophyte-pathogen interaction and "antibiosis" as a mechanism of action is essentially predominant in most endophytes as biocontrol agents (BCAs). Therefore, it would be more relevant to study the movement and concentration of these secondary metabolites, which in turn would determine the biocontrol efficacy of an endophyte against a pathogen attack. Additionally, in the endophytes which act by inducing systemic resistance (e.g., *P. indica*), the distribution and plant tissue colonization may be less relevant in terms of biocontrol activity (Stergiopoulos and Gordon 2014).

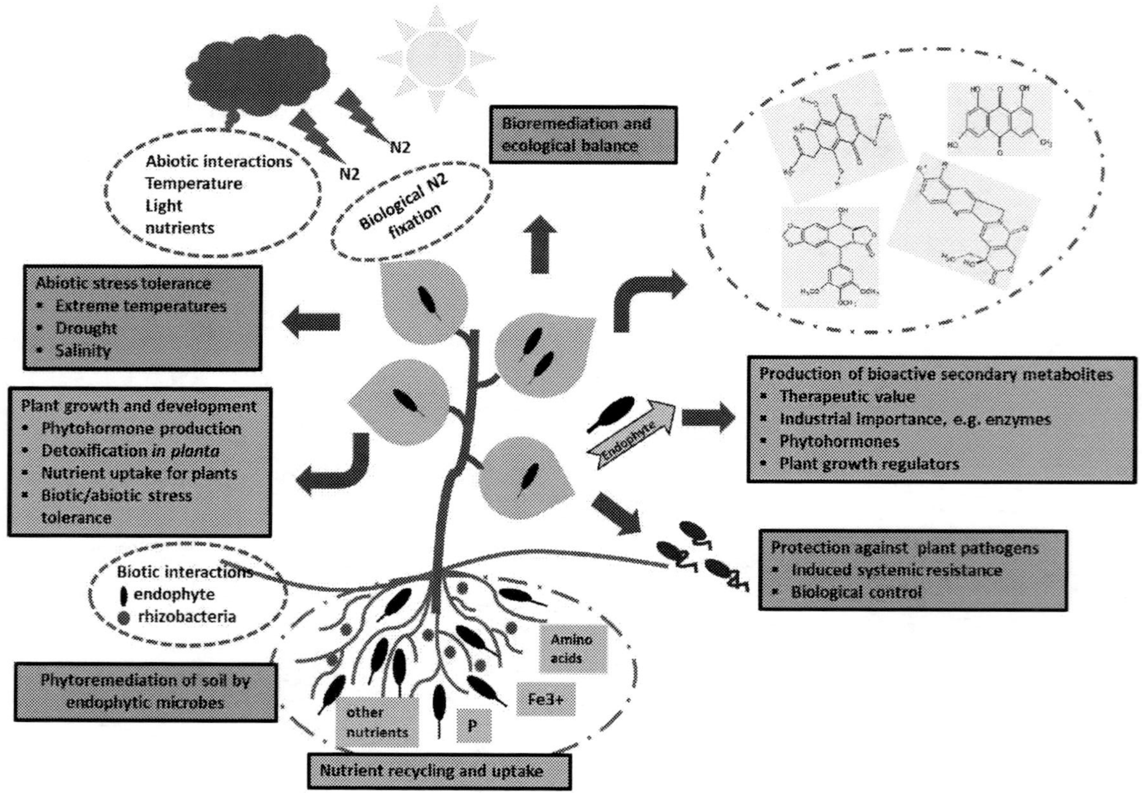

Figure 1. The multi-faceted role of endophytes in plant growth promotion and sustainable agriculture.

In research studies, the majority of the described endophytes were found to demonstrate non-systemic plant colonization (Canals et al. 2014). However, for some endophytes, there is little evidence about the mechanism of plant colonization, in *B. subtilis*, the endophyte exhibits systemic colonization as revealed by the strain isolation from plant vascular tissues (Hall et al. 1986; Zhang et al. 2009). In this area, techniques like isolation of axenic cultures of plant tissues and fluorescence microscopy were employed to delineate the colonization pattern of endophyte of interest, not reported for most of the plant-microbial associations (Reinhold-Hurek and Hurek 2011). Another strategy for understanding endophyte colonization patterns in plants was put forth by Rodriguez and Redman (1997) a classification based on endophyte dynamics in plant tissues, endophyte which colonize a small region of host tissue, endophyte which remains dormant within-host the, fungi that show extensive colonization and endophytes which are metabolically active inside host. An understanding of the distribution and colonization pattern of endophytes would aim at the development of techniques for endophyte detection *in situ* and the creation of unique associations.

Plant-endophyte interactions are found to have a substantial impact on the environment and agriculture. Endophyte promotes plant health and provides tolerance against biotic/abiotic stress conditions (Gkouletsos T, 2018). Considering the fluctuating climatic conditions and rising demands for agricultural production, endophytes are emerging as key players in plant protection against adverse environmental conditions and enhancing crop productivity in agriculture (Figure 1). However, information about the molecular mechanism of endophytes *in planta* is limited, the beneficial plant-endophyte associations have not been completed exploited for their socio-economic benefits. Considering the present trends and the emerging significance of endophytes, the chapter discusses the role of endophytes in the augmentation and improvement of plant health and their beneficial role in sustainable agriculture. The molecular mechanisms of plant-associated endophytes leading to plant growth promotion, enhanced crop productivity, sustainable agriculture, associated challenges in endophyte research, and prospects of future success are discussed.

2. Ecological Implications of Plant-Endophyte Associations

Endophytes exhibit diverse biological functions, namely plant adaptation to environmental stresses, plant growth promotion, and bioremediation, highlighting significant ecological applications. Additionally, endophytes may produce diverse metabolites including growth hormones, which may have a profound effect on plant physiology. Plants are not single entities but are regarded as complex communities demonstrating multiple associations (Rodriguez et al. 2009; Rey and Schornack 2013). Studies have shown that the plant-microbe association dates back to more than 400 million years ago, its existence as a “natural phenomenon” is considered essential for associated mutualistic organisms (Strobel et al. 2004; Partida-Martínez and Heil 2011).

Plants inhabiting different geographical locations are likely to face various environmental challenges, which influence the endophyte association and their variation within plants (Arnold AE, 2007; Vega et al. 2010). The latitude changes may affect the occurrence and diversity of endophytes; for instance, studies have shown the presence of a few fungal species at high altitudes (Arnold and Lutzoni 2007). Moreover, plants growing in similar environmental conditions also do not harbor the same endophyte species; an example suggests that isolated endophytes from willow and cottonwood trees (inhabiting the same climatic zones) were different from each other (Doty et al. 2009). Such studies and several others indicated that the genetic organization of any plant species influences its colonization by a particular endophyte species (Ulrich et al. 2008). In plants, the occurrence and composition of microbial populations vary with genotype, soil type, geographical conditions, etc. which may have shown co-evolution with the associated plant species (Saikkonen et al. 2004). Studies have suggested that the co-evolution of endophytes with plants may have been affected by the interaction of both the genotypes, as well as the inter-genomes of the interacting partners (Wade MJ, 2007). However, an understanding of how the genetic organization governs the differential association of microbial communities needs to be understood.

The interaction between microbial communities and plants has positive outcomes in terms of plant nutrition and tolerance to stress conditions (Tiwari et al. 2020b; Tiwari et al, 2021d; Tiwari et al, 2021e), and such beneficial traits have led to genetic manipulations of endophytes. The use of endophytes for environmental remediation, stress tolerance (Acuña-Rodríguez et al. 2019), and nutrient recycling in sustainable agriculture (Yan et al. 2019) would be economical as well as eco-friendly, reducing the use of fertilizers and pesticides. Additionally, an interesting application of endophytes as BCAs for pest management and insect control has multiple advantages (Park et al. 2018). They are directly applied to the seeds and large quantities are not required. Some key examples of fungal endophytes are *Trichoderma harzianum*,

Fusarium oxysporum, *Paecilomyces lilicinus*, etc., which have been commercially produced and used as BCAs on large scale (Mendoza and Sikora 2009; Sikora 2010; Silva et al. 2017). As there has been extensive use of synthetic chemicals and fertilizers in agriculture, it has put forth a growing concern for the employment of safe and effective alternatives, and the use of endophytes for crop production provides a distinct solution.

In recent years, considerable progress has been made in employing endophytes for environmental bioremediation either by increasing plant growth and phytoremediation (Domka et al. 2019) or through biodegradation of environmental pollutants/hazards (Stepniewska and Kuzniar 2013). Some other significant roles of endophytes in the atmosphere account for nitrogen acquisition by direct uptake from the atmosphere (Sherameti et al. 2005) or through insects (Behie et al. 2012) and cycling, thereby enhancing nutrient uptake, e.g., phosphorous utilization and uptake (Yadav et al. 2010) and phytohormone-mediated growth in plants (Waqas et al. 2012). Additionally, volatile compounds, which affect the dynamics of the ecosystem by generating information, are produced by microbial endophytes (Wheatley RE, 2002). An example includes the endophyte *Muscodor* spp. which kills plant pathogens and is employed in agricultural practices (Strobel 2006; Bitas et al. 2013). Studies have shown that microorganisms producing volatile compounds confer tolerance against pathogens, and are likely to form symbiotic associations with plants. Therefore, the ecological implications of endophytes are significant and hold potential in sustainable agriculture and environmental bioremediation.

2.1. Bacterial Endophytes

Bacterial endophytes constitute an important category of plant-bacteria association and perform key roles in plant growth and development. The diverse bacterial endophytes associated with plants include *Proteobacteria*, *Actinobacteria*, and *Bacteroidetes*, while the most common colonizing species include *Pseudomonas*, *Bacillus*, *Micrococcus,* and *Burkholderia* sp. (Sturz and Nowak 2000; Sun et al. 2008; Marquez-Santacruz et al. 2010; Romero et al. 2014). Studies have suggested differences in endophyte genomes compared to rhizospheric bacteria and the expression of gene families during colonization and establishment in plants. Moreover, certain genes namely nifA (nitrogen assimilation), wspE-wspR (polysaccharide production), barAuvrY (carbon storage), tenA (thiamine metabolism), ampR (beta-lactamase resistance), and ATP-Transporters showed higher levels of expression in endophytes (Hardoim et al. 2015). The mechanism of entry and colonization has been documented suggesting that the bacteria enter the plant through primary roots via plant injury (Sorensen and Sessitsch 2015). Initially, the bacteria enter the plant through the root and proceed to the stem and leaf, exhibiting a varying degree of densities in different parts of the plant. Another suggested mechanism is entry through stomatal openings on leaves, roots, and seeds, and lenticels on stems (Compant et al. 2010), triggered by the effect of various chemicals. Several mechanisms of endophytes have been suggested from facultative to obligate and passive for plant colonization. For example, *Burkholderia phytofirmans* infect the plant by the production of endoglucanase and endopolygalacturonase (the cell wall degradation), moving through the endodermis and xylem tissue, as well as establishing itself in the plant roots (Compant 2005; Compant 2008).

Several species of bacterial endophytes have been found to colonize internal plant tissues, namely *Azospirillum* sp. B510 in rice (Kaneko et al. 2010), *Burkholderia* spp. KJ006 in rice (Kwak et al. 2012), and *Enterobacter* sp. 638 in *Poplar* (Taghavi et al. 2010). Studies have suggested root colonization by bacterial endophytes is more frequent compared to other plant tissues such as seeds, roots, leaves, fruits, etc. (Rosenblueth and Martínez-Romero 2006). The dynamic nature of endophytes is due to multiple factors ranging from environmental conditions, physiological state, and growth stages of the respective plant (Ardanov et al. 2012; Mercado-Blanco and Lugtenberg 2014). Moreover, the presence of a bacterial endophyte confers several advantages to the plant that it colonizes, specifically plant "growth" and "promotion" and tolerance to plant pathogens (Eljounaidi et al. 2016), demonstrating an effective interaction with the plant, under diverse environmental conditions (Ali et al. 2012; Coutinho et al. 2015). Moreover, a host plant benefits from the bacterial association through various mechanisms such as growth promotion, induced systemic resistance, production of antibiotics, etc. (Collins and Jacobsen 2003; Jorjani et al. 2012; Mansoori et al. 2013).

The significance of bacterial endophytes as biocontrol agents (BCAs) was studied for their multiple advantages to host plants as well as the environment; however, several challenges need to be addressed. During exogenous application to plant roots, BCAs may struggle for survival with soil microflora, requiring early colonization for adaptation (Ting 2014). Additional bottleneck includes poor storage viability, low yield in cultures, and toxicity concerns (Yu et al. 2010; Ting 2014). Bacteria that are pathogenic to humans may also be present in endophytes. A comprehensive insight into the antagonistic mechanism of bacterial endophytes projects effective control of diseases in plants. Moreover, scientific technologies such as recombination-based in vivo expression technology (RIVET) will demonstrate bacterial colonization, plant growth, and resistance against pathogens (Mercado-Blanco and Lugtenberg 2014). Additionally, employing fluorescence methods to study the colonization mechanism of bacterial endophytes would be significant to understand their potential as BCAs (Malfanova 2013).

2.2. Fungal Endophytes and Their Classification

The earliest report highlighting the existence of fungal endophytes is available from a 400-million-year-old fossil from the early Devonian deposits (Krings et al. 2007) suggesting the co-evolution of higher plants with fungal endophytes (Sridhar KR, 2019). Since then, bioactive metabolites with therapeutic properties were increasingly isolated and studied from plant sources, although over-harvesting concerns account for threatened or endangered plant species. Therefore, plant-associated fungal species have drawn special attention as production platforms for novel secondary metabolites after the remarkable discovery of taxol from *Taxomyces andreanae* (Stierle et al. 1993). The breakthrough in the production of the anticancer drug from endophytes has opened new avenues in the bio-prospecting of fungal endophytes as sustainable alternative "production factories" for novel secondary metabolites and their commercial exploitation.

Plant roots are colonized by different fungi which exhibit symbiotic associations and are non-pathogenic to the host plant, namely dark septate endophytes, ectomycorrhizae, and arbuscular mycorrhizae (Deng and Cao 2016), dark septate endophytes are sub-divided into septate and conidial endophytes, forming intra- and intercellular hyphae, during root colonization. These represent ubiquitous distribution, related to different mycorrhizal fungi (Upson et al. 2009). The ectomycorrhizae are mainly members of Basidiomycetes, with few exceptions (Zygomycetes and Ascomycetes), and comprise around 6,000 species. The arbuscular mycorrhizae include members of Zygomycetes and are found in association with woody plants and herbs (Mandyam and Jumpponen 2005). The fungal endophytes are distributed worldwide and define a broad range of host categories, such as lichens (Li et al. 2007), mosses (Daveyand Currah 2006), ferns (Swatzell et al. 1996), grasses (Müller and Krauss 2005) and coniferous plants (Albrectsen et al. 2010).

The plant-fungal communities are generally host-specific; however, several endophytes of the genus *Colletotrichum* and *Phoma* show colonization of a wide range of host plants. Some significant biological roles of fungi include the cycling of elements, mineral transformations, and phytoremediation (Gadd 2007; Yan et al. 2019). Although several species of fungal endophytes enhance bioremediation, however, the knowledge regarding the functional roles and dynamics of fungal endophytes is still limited. The plant-fungal endophytes perform multiple functions, namely tolerance to herbivores (Tikhonovich and Provorov 2009), heat tolerance (Ali et al. 2019), plant biodiversity (Arnold and Lutzoni 2007), uptake of nutrients (Malinowski et al. 2000), drought tolerance (Dey et al. 2019; Gange et al. 2019), salinity tolerance (Acuña-Rodríguez et al. 2019), insect control (Gonda et al. 2016; Rabiey et al. 2019), metal toxicity (Domka et al. 2019) and biotransformation (Azevedo and Araujo 2007). Studies have highlighted extensive research on plant-associated fungi and their significant role in the production of metabolites of socio-economic importance. Endophytes influence the physiology of the host plant, thereby leading to plant resistance to pathogen attack and insect control (Mantzoukas and Lagogiannis 2019), growth and development, and agricultural yields, as well as confer tolerance to abiotic stress (Gadd 2007; Acuña-Rodríguez et al. 2019).

2.3. Fungal Endophytes and Secondary Metabolite Production

Fungal endophytes are gaining momentum as the biological reservoir of diverse bioactive metabolites with applications in agriculture, industries, and healthcare. The secondary metabolites biosynthesized by endophytes are classified into different categories, consisting of phenols, terpenoids, xanthones, isocoumarins, cytochalasins, and xanthones, and display multiple biological activities (Tiwari et al. 2022b). Fungal endophytes are found to be associated with plants in diverse ecological niches and found to be a potential reservoir of therapeutically important metabolites demonstrating antifungal, antiviral, antioxidant, and antineoplastic properties (Deshmukh et al. 2018). *Pestalotiopsis* sp. is regarded as "the *E. coli* of the rain forests" with the biosynthesis of the plethora of bioactive molecules comprising Torreyanic acid, Taxol, Ambuic acid, Subglutinol A and B, among others (Deshmukh et al. 2018). *Pestalotiopsis* sp. defines a key example of genus coelomycetes and produces antifungal metabolites including (3R,4R,6R,7S)-7-hydroxyl-3,7-dimethyl-oxabicyclo[3.3.1]nonan-2-one (monoterpene lactone) and (3R,4R)-3-(7-methylcyclohexenyl)-propanoic acid which demonstrated antifungal activity against *Phytophthora nicotianae* and *Botrytis cinerea* (Xu et al. 2016). Different species of the genus *Pestalotiopsis* produce different bioactive metabolites. Another fungal endophyte, *Phomopsis* is an important genus that produces Cytochalasin H with potent antifungal activity (Chapla et al. 2014b).

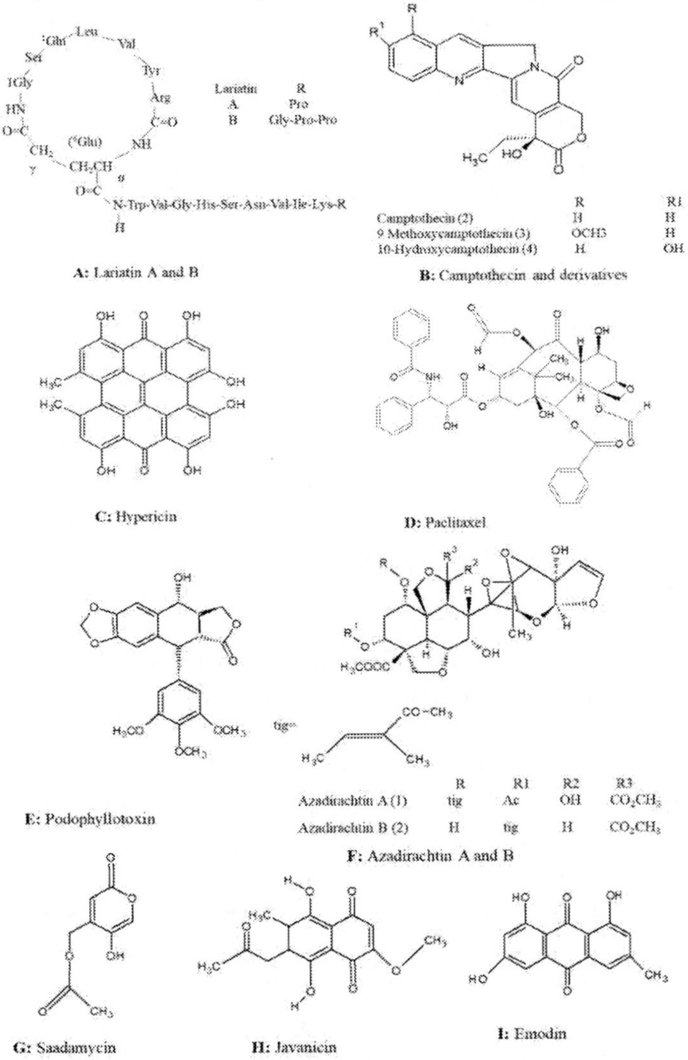

Figure 2. Some therapeutically important metabolites are produced by endophytes.

Guignardia sp., a fungal endophyte from *Euphorbia sieboldiana* produced guignardic acid and guignardone N with potent fractional inhibitory concentration (FIC) index of 0.19 and 0.23 against *Candida albicans* (Li et al. 2015). Taxol production by *Taxomyces andreanae* (Stierle et al. 1993) was a landmark discovery in the production of a therapeutic metabolite from fungal endophyte and gained huge success as a therapeutic for cancer treatment. In addition, different fungal endophytes were found to produce the anticancer drug, Paclitaxel, and comprised of *Aspergillus fumigatus* (Sun et al. 2008), *Alternaria* sp. (Kim S, 1999), *Fusarium solani* (Chakravarthi et al. 2008), *Pestalotiopsis guepinii* (Strobel et al. 1997), *Phyllosticta citricarpa* (Senthil Kumaran et al. 2008), *Taxomyces* sp. (Qiu et al. 1994) and others. Rohitukine (anticancer alkaloid) was isolated from *Amoora rohituka* (Harmon et al. 1979) and *Dysoxylum binectariferum*. Camptothecin, an antineoplastic agent is used for the treatment of colon, ovarian, and uterine cancer and was isolated from *Fusarium solani* and *Entrophospora infrequens* (Puri et al. 2005; Shweta et al. 2010). Hypericin was produced by *Chaetomium globosum* (Kusari et al. 2013), Berberine (antibiotic) from *Alternaria* sp. (Duan et al. 2009), azadirachtin (insecticide) from *Eupenicillium parvum* (Kusari et al. 2012). Besides, as an anti-inflammatory and antioxidant agent, Borneol was produced by *Cochliobolus nisikadoi* (Chen et al. 2011). These and several other studies showed the importance of fungal endophytes in the production of diverse metabolites of pharmacological attributes. Figure 2 provides some key examples of endophyte-mediated production of bioactive metabolites.

3. Genetic and Molecular Mechanisms of Endophytes in Alleviating Stress and Plant Growth Promotion

Plants constitute a complex community associated with diverse microbes and their interaction, known as the "plant microbiome", affects plant physiology (Conn et al. 2008). A number of microorganisms may live inside plant tissues for a long time and co-integrate with the genetic organization of the respective host plant considered "true endophytes" (Moricca and Ragazzi 2008). Researchers have tried to delineate the molecular mechanism of plant-endophyte associations, but it remains relatively less understood (Gundel et al. 2012). The challenge in the elucidation of the molecular aspect of endophytes include difficulties in studying their mechanisms *in planta* and the complex interaction between plant and microbes. Studies have suggested that the association between a microbe and its dynamics within the host plant is governed by the microbial species and genetic organization of the plant (Qawasmeh et al. 2012; Bacon and White 2016). Moreover, the endophyte microbes, the functional component of the plant microbiome, is comprised of 1) Microbes involved in the host plant nutrition 2) Microbes that alleviate abiotic stress 3) Microbes that confers biotic stress tolerance (White and Torres 2010). These functions demonstrated by endophytes define potential in ecological perspective and sustainable agriculture. Several mechanisms for endophyte responses in the host plant were discussed as follows:

3.1. Antioxidant Production and Oxidative Stress

Studies have shown the functional role of endophytes in enhancing the expression of enzymes and phenolic compounds showing antioxidant activity. The increased antioxidant levels would counter-balance reactive oxygen species, produced during stress conditions. *Epichloe*-endophyte mediated stress tolerance in grasses was attributed to the antioxidant compounds production and mitigation of oxidative stress was discussed (Schulz et al. 2002) hypothesis was further supported by studies on antioxidant compound production by endophytes (Huang et al. 2007; Hamilton and Bauerle 2012). Hamilton and Bauerle (2012) showed the increase in antioxidant activity during abiotic stress conditions as compared to endophyte absence in the host plant suggesting that protection from abiotic stress was a measure of oxidative stress protection mechanism.

Endophytes produce novel compounds with diverse chemical structures, such as phytohormones, volatile compounds, antibiotics, and reactive oxygen species (ROS), among a few more (Griffin et al. 2010; Khan et al. 2012; Tiwari et al. 2021c) for carrying out various biological functions. Studies have demonstrated the effectiveness of the production of ROS on the interactions of endophytes with the host plant, which induce antioxidant production and thereby protect the host from oxidative stress (Schulz et al. 2002). Moreover, multiple biological activities of the host plant are also influenced by endophytes, for example, *Neotyphodium lolii*, symbiotically associated with *Lolium perenne*, affects the antioxidant activity of the plant (Bacon and White 2016). Some endophytes affect the regulation of nutrition and its homeostasis (Singh et al. 2013).

3.2. Endophytes and Altered Gene Expression

The molecular mechanism of endophyte dynamics within the host plant is due to altered gene expression of defense genes and their metabolic pathways (Sherameti et al. 2008; Mathys et al. 2012). It has been demonstrated by researchers that systemic and local changes in the host may be induced due to association with an endophyte (Glazebrook J, 2005; Ownley et al. 2010). The plants perceive the presence of metabolites and a signal transduction pathway is triggered, leading to defense responses of the plant to pathogens (Bacon and White 2016). The rhizobacterial endophytes present in the living tissues, stimulate defense pathways and confer increased resistance to diseases (Moricca and Ragazzi 2008). Research has suggested that pathogens induce systemic resistance in *Arabidopsis* in response to infection and disease symptoms (Glazebrook J, 2005). However, sometimes, endophytes are considered "mini pathogens" and plant defense genes are activated in the absence of a pathogen (Moricca and Ragazzi 2008). In recent times, the co-integration of "omics" technologies, comprising computational biology and metabolomics coupled with approaches in systems biology, would open new avenues in understanding the physiological mechanisms of endophytes in plants.

3.3. Regulation of Ethylene and Enhanced Stress Tolerance

Another mechanism for increased tolerance to abiotic stress by the host plant was by regulation of increased production of ethylene during stress conditions. In plants, ethylene is necessary for normal growth and development, and an increased level results in growth inhibition (Saravanakumar and Samiyappan 2007). Moreover, certain endophytes promote stress tolerance by preventing ethylene formation (degradation of 1-aminocyclopropane-1-carboxylic acid (ACC) by ACC deaminase production). ACC deaminase is also produced by some fungal symbionts and non-pathogenic bacteria and cleaves ACC, enriching nutrient uptake by the plants. Studies have suggested the growth promotion function of the endophyte is by inhibiting ethylene production and promoting auxin-mediated plant growth. However, the knowledge about the mechanism of microbe-mediated stress tolerance and plant growth is not yet understood and needs to be further studied in detail.

3.4. Protection against Plant Pathogens

Plant-associated endophytes produce diverse metabolites for protection against pathogen attack. Bacterial endophytes (genus *Bacillus*) produce antibacterial and antifungal lipopeptides (Gond et al. 2014) which inhibit the pathogens and provide protection to the host plant. The lipopeptides inhibit viruses by surfactin, which dissolves the lipid layer and protein coat of the viruses (Ongena and Jacques 2008). Moreover, studies have suggested that many wild plant species show a lipoprotein-mediated disease protection mechanism (Gond et al. 2014). Plant infection by endophytes leads to the expression of defense genes as shown in the corn plant (Gond et al. 2014). A key study by Moy et al. (2000) suggested an exclusive mechanism without pathogen infection for endophyte conferring fungal resistance to the *Epichloe* host plant. The defense mechanism was based on a network of mycelium on leaf blades of grasses which acted like 'defensive nets', limiting infection by a fungal pathogen (Tadych and White 2007). Moreover, studies have demonstrated the endophyte-mediated defense against fungal diseases in grasses, *Festuca arundinaceae, F. rubra, Elymus cylindricus* (Gwinn and Gavin 1992; Clarke et al. 2006) highlighting that endophyte mediated defense is an important mechanism in plant-microbe interaction.

3.5. Detoxification of Ammonia by Endophyte

The plant-associated microbes (bacteria and fungi) have ammonia transporters that may absorb any excess ammonia present in the cell. The photorespiration process results in ammonia generation leading to increased reactive oxygen species and oxidative stress (Kiraly et al. 2013). Some endophyte shows a higher affinity for ammonia (Kiraly et al. 2013) and may aid in the detoxification of ammonia. Thereby reduction in ammonia levels would lead to a decrease in oxidative stress in plants, a mechanism by which the microbe extracts organic nitrogen present in cells. The removal of excess ammonia by endophyte defines a detoxification mechanism whereby any excess metabolites was removed and plant growth is maintained in mutualistic association (White and Torres 2010).

3.6. Anti-Herbivory and Endophyte

In-plant endophyte interaction, the existence of anti-herbivory has been documented in the literature (Gwinn and Gavin 1992). Endophyte colonizing grasses deterred various herbivorous organisms namely nematodes, arachnids, mammals, and insects (Siegel et al. 1991; Rowen et al. 1986; Belesky and Bacon 2009; Bacetty et al. 2009a; Bacetty et al. 2009b). Moreover, the endophyte microbe produces compounds in response to pathogen attack, which deters grazing animals. The anti-herbivory effect highlights a mechanism of ecological importance and may have a role in the speciation of the host. Some prime studies in this area showed that ergot alkaloids in endophyte-colonized grasses play a defense role in deterring all kinds of herbivores (Lyons et al. 1986). The plant-associated microbes produce a diverse range of biologically active compounds that repel pathogens. The microbial strains producing defense compounds were mostly employed as biocontrol agents to reduce crop infection and loss. In low concentrations, surfactants serve as biocontrol agents and induce systemic resistance in plants (Ongena and Jacques 2008). In addition to the above mechanisms, endophytes are also associated with nutrient cycling and uptake by the associated host plant (Pankievicz et al. 2015) enhanced drought or salt stress mitigation by osmotic adjustment (Gond et al. 2014), phosphate solubilization by endophytes (Puente et al. 2009) besides other. These studies show that plant-associated endophytes have ecological and agricultural importance in the ecosystem.

4. Bioremediation by Endophytes: An Eco-Friendly Approach for Maintaining the Ecological Balance

Endophyte-mediated bioremediation of organic pollutants, heavy metals, and toxic substances define a key application in the environment and agriculture. The fungal endophytes display good potential in removing environmental contaminants and maintaining ecological balance. The industrial and agricultural wastes cause an imbalance in ecosystem functioning and endophyte mediated bioremediation highlights an environment-friendly approach to address this concern (Tiwari and Bae 2020). For example, *Nicotiana tabacum* plants inoculated with the endophytes showed improved tolerance to cadmium (Cd) stress and increased plant biomass (Mastretta et al.

2009). Moreover, endophytes were effective in the biodegradation of plastics, in recent times. Several endophytes were screened for degradation of polyurethane, a synthetic polymer; however, the genus *Pestalotiopsis* displayed promising activity and secreted serine hydrolase enzyme for bio-degradation of polyurethane (Russell et al. 2011). Siciliano et al. (2001) suggested that endophytes having phytoremediation-associated genes occur naturally in plants present in the contaminated areas. The endophyte-mediated bioremediation in an environment contaminated with hazardous plastics offers a significant alternative to chemical methods in biotechnological applications. Moreover, the role of endophytes in the bioremediation of oil-contaminated saline soil suggests its significance in the mitigation of stress conditions and improvement of plant tolerance (Wu et al. 2019). The presence of heavy metals in soil (namely lead, cadmium and zinc) leads to poor growth and productivity of the plant. Plant-associated endophytes not only regulate the heavy metal contamination by enhancing plant tolerance but also promotes plant growth, thus offering a highly beneficial alternative in the bioremediation of soil. In this area, several studies demonstrated the prospect of endophyte-mediated bioremediation, some key examples comprise phytoremediation of arsenic-contaminated soil (Wu et al. 2018), biodegradation of petroleum hydrocarbons in oil-contaminated soils (Wu et al. 2019), degradation of Chlorpyrifos and Bifenthrin in polluted water (Chen et al. 2012) among other significant ones. These and several other studies demonstrate the prospects of employing fungal endophytes for bioremediation in the environment and agriculture.

5. Contributions of Computational Biology in Endophyte Research

With the sequencing of whole genomes and the availability of "big data," computational biology has made significant contributions in key areas of biological sciences. Endophytes from diverse associations with plants (mostly symbiotic), and a knowledge of plant microbiomes and endophyte dynamics, would help understand these complex interactions. With the development of computational tools and web resources, identification and annotation of genes/gene families involved in different metabolic functions have been achieved. Modern "omics" tools and methods, such as whole-genome sequencing, comparative genomics, microarray, and transcriptomics were employed to understand plant-microbe associations (Kaul et al. 2016).

In a study, a comparative genome analysis was carried out to identify genes responsible for endophyte associations from the genus *Burkholderia* and those of different genera of endophytes (Ali et al. 2014). Genes from nine species of endophytes, namely *Burkholderia phytofirmans* PsJN, *Klebsiella pneumoniae* 342, *Azoarcus* sp. BH72, *Serratia proteamaculans* 568, *Klebsiella pneumoniae* 342, *Enterobacter cloacae* ENHKU01, *Azospirillum lipoferum* 4B, and *Burkholderia* spp. strain JK006, were screened and studied for genes involved in endophyte associations with plants, and a set of bacterial gene orthologs were identified. Furthermore, the comparative genome analysis suggested that genes of different bacteria might perform similar functions (Ali et al. 2014). Therefore, different genome comparisons may lead to the identification of genes involved in endophyte response. A similar study on transcriptome profiling and TaqMan quantitative PCR studies of the endophyte *Burkholderia phytofirmans* PsJN revealed its gene expression in endophyte and the physiological mechanism of plant-microbe interaction. The whole-genome sequencing of *B. phytofirmans* PsJN associated with *Symphytum tuberosum* L. revealed that genes involved in diverse functions related to drought stress response, namely cellular homeostasis, detoxification of ROS, and transcriptional regulation, were upregulated, showing a stress response mechanism (Sheibani-Tezerji et al. 2015). Several other *in silico* studies on plant-associated endophytes have provided crucial insights into the identification of genes and their expression in endophyte interactions, and their physiological mechanisms in host plants. Furthermore, whole-genome studies on endophytes screened several key genes governing various processes, such as plant growth and development, resistance to antibiotics, surface attachment of endophytes, and metabolic processes (Kaul et al. 2016). In recent times, significant environmental benefits of fungi (order Sebacinales) in plant stress tolerance and increasing plant growth, have prompted researchers to study them (Weiss et al. 2011). An example includes genome analysis of *P. indica* (order Sebacinales) which revealed several key genes involved in plant-microbe interactions.

Table 1. Genome sequencing of plant-associated endophytes and their functional significance in plant growth promotion

Endophyte species	**Genome size (Mb)**	**Sequencing method**	**Functional importance**	**Reference**
Bacillus sp. strain WR11	4.20	Single-Molecule Real-Time (SMRT)	Plant growth promotion (PGP) Abiotic stress tolerance	Chen et al. 2020
Pseudarthrobacter phenanthrenivorans strain MHSD1	4.45	Illumina MiSeq platform	Phytohormone production Bioremediation	Tshishonga and Serepa-Dlamini 2020
Alternaria alternata JS-1623	33.67	SMRT	Secondary metabolite production (SMP)	Park et al. 2020
Sphingomonas sp. LK11	3.78	SMRT	PGP Bioremediation	Asaf et al. 2018
Penicillium aurantiogriseum NRRL 62431	32.7	Illumina sequencing	Paclitaxel production	Yang et al. 2014
Enterobacter sp. 638	4.51	Whole-genome shotgun sequencing	PGP Nutrient uptake Tolerance of heavy metals	Taghavi et al. 2010
Sarocladium brachiariae	31.86	SMRT	SMP	Yang et al. 2019
Burkholderia sp. strain KJ006	6.60	Illumina and Solexa platforms	PGP Bioremediation of aromatic compounds	Kwak et al. 2012
Cellulosimicrobium sp. JZ28	4.37	PacBio *RSII* (Pacific Biosciences) sequencing platform	Production of osmoprotectants Abiotic and biotic stress tolerance	Eida et al. 2020
Enterobacter sp. J49	4.96	Illumina MiSeq System	Biofertilizer PGP	Luduena et al. 2018
Cadophora sp. *Periconia macrospinosa*	70.46 54.99	Illumina sequencing platform	Carbohydrate active enzymes Melanin biosynthesis Aquaporin production	Knapp et al. 2018
Rhodanobacter glycinis T01E-68	4.17	SMRT	PGP SMP Biocontrol function	Lee et al. 2019
Fusarium solani strain JS-169	45.81	HiSeq 2000 platform and PacBio RSII	Antifungal activity	Kim et al. 2017
Burkholderia sp. strain SSG	8.60	MinION device (Oxford Nanopore Technologies)	PGP Biocontrol function	Kong and Hon 2020
Variovorax paradoxus S110	6.75	Sanger sequencing	PGP	Han et al. 2011
Pseudomonas putida TJI51	5.80	Illumina next-generation sequencing	SMP Bioremediation	Asif et al. 2016

Table 1. (Continued)

Endophyte species	Genome size (Mb)	Sequencing method	Functional importance	Reference
Gluconacetobacter diazotrophicus Pal5	3.90	Whole-genome shotgun sequencing	PGP Nitrogen fixation	dos Santos et al. 2009
Bacillus thuringiensis KB1	861.64	Illumina HiSeq 2500 platform	PGP Biocontrol function	Jeong et al. 2016

Moreover, other significant *in silico* studies on endophytes comprise comparative genome analysis of *B. phytofirmans* PsJN and other bacterial endophytes (Mitter et al. 2013), genome-wide studies on the identification of lovastatin biosynthesizing genes (Bhargavi et al. 2014), transcriptome and gene-expression studies on *Azospirillum brasilense* forming endophyte association with *Triticum aestivum* roots (Camilios-Neto et al. 2014), comparative genomics of three strains of *P. ananatis* (*Zea mays* endophytes) (Sheibani-Tezerji et al. 2015), bioinformatics analysis of bacterial endophytes and berberine production in the Chinese medicinal plant *Coptis teeta* (Liu et al. 2019) and genome and transcriptome analyses of *P. indica* colonizing roots of *Hordeum vulgare* (Zuccaro et al. 2011). *In silico* tools and resources in computational biology have been employed to unravel the complex physiological mechanisms of plant-endophyte associations, thus, providing a better understanding of the role of microbial communities in the plant microbiome (Table 1).

6. Strategies and Perspectives for Successful Development/Applications of Endophytes in Agriculture

With scientific breakthroughs in high throughput technologies and the availability of complete genomes, research on plant endophytes is gaining considerable momentum. Worldwide, the utilization of alternative strategies to manage depleting resources and ecological balance is a key concern. In agriculture, crops are constantly exposed to climatic changes and global warming, which adversely affects crop yield and production. "Plant microbiomes" were genetically engineered for enhancing plant growth and production, by monitoring pathogen infections (Rowen et al. 1986; Mousa et al. 2016). The need to restore these symbiotic relationships for ecological balance highlights a key concern and sequencing of "plant microbiomes" was undertaken for many plant species (Doty SL, 2017). For instance, metagenome analysis of an endophyte from *Oryza sativa* roots provided information about the genes associated with the endophyte's lifestyle and functional dynamics, genes involved in phytohormone production, cell signaling, plant colonization, nitrogen fixation, and other functions, highlighting key features of plant-associated endophytes (Sessitsch et al. 2012; Burton et al. 2014). For sustainability of the environment and agriculture, and in turn, socio-economic returns, a precise and clear understanding of plant microbiomes is essential. Steps have been taken in this direction; in agriculture, the application of "bio-stimulants" for plant growth and its global marketing has seen a tremendous upsurge, with a projected income of USD 3.6 billion by 2022 (Timmusk et al. 2017). Moreover, considerable progress has been witnessed in the reduction of greenhouse emissions and the use of biological inoculants for crop production, in place of chemical pesticides. Some significant studies have extensively reviewed the significance of endophytes in sustainable agriculture and their environmental benefits (Lugtenberg et al. 2016; Le Cocq et al. 2017; Timmusk et al. 2017). The use of plant-associated endophytes projects certain advantages, such as the microbes are easier to grow and maintain, and provide several benefits in stress tolerance, improvement in plant health, and nutrient uptake by the host plant, thereby reducing the cost of agriculture and improving crop production.

High throughput technologies have made significant contributions to understanding plant-microbe interactions, adding to our existing knowledge. These scientific tools have demonstrated novel functions of microbes, as well as their role in maintaining ecological balance, suggesting that "endophytes" are key players in sustainable agriculture. Understanding how plant-associated microbes confer stress tolerance would define a platform to employ scientific technologies aiding in plant biomass production. One striking example of a biotechnological approach is the production of cellulosic bioethanol and other chemicals from *Populus* sp., an important bioenergy crop (Weyens et al. 2009; Hacquard and Schadt 2017). The application of an endophyte on a hybrid poplar tree showed better plant traits, namely, accumulation of root mass, nitrogen fixation, and tolerance to drought conditions (Knoth et al. 2014). These traits increased drought tolerance in the poplar tree, suggesting enhanced biomass production for sustainable agriculture.

7. Role of Endophytes in Promoting Crop Productivity and Sustainable Agriculture

The rising population across the globe and increasing nutritional demands are major concerns attributed to agricultural output. So far, the agriculture systems have employed the use of chemical fertilizers and agrochemicals to boost agricultural output, however, research has shown the beneficial effect of nitrogen-fixing bacteria and mycorrhizal fungi in plant growth and development (Santoyo et al. 2016).

Presently, a more sustainable approach has been undertaken, employing environment-friendly practices and scientific methods comprising biological organisms. Plant-microbe associations have been extensively studied due to their multi-faceted environmental significance such as enhancement of plant growth and stress tolerance, and BCAs have been highlighted as prospective biotechnological tools for a more sustainable approach to agriculture. Plants are prone to a number of environmental stresses, particularly salinity and drought conditions, which severely hamper their survival and productivity (Ruiz-Lozano and Aroca 2010; Liu et al. 2019). The agronomic factors, plant growth-promoting effects, and increased agricultural production (Khan et al. 2015) are the key reasons for employing endophytes in agriculture. It has been suggested that mycorrhizal fungi may enhance water uptake and storage in drought-stressed crops, and an example is corn plants associated with mycorrhizal fungi, in which water retention was doubled as compared to that in the control plants, and a higher transpiration rate was witnessed (Kothari et al. 1990). Furthermore, it was shown that stomatal efficiency was increased due to root colonization by soil fungi, and water uptake from the soil by associated plant roots was enhanced, leading to better access to water and its uptake by the plant (Ruiz-Lozano and Aroca 2010). In plants, exposure to soil salinity causes the accumulation of ROS thereby damaging the plants, and the production of antioxidants by endophytes leads to tolerance to salinity stress (Yan et al. 2019). In another study, an endophyte, *Metarhizium* formed a symbiotic association with *Glycine max* roots and conferred salinity tolerance to the plant (Khan et al. 2011). Studies have suggested that associations of Antarctic extremophiles could mitigate the adverse conditions of soil salinity, thereby acting as a key biotechnological approach to soil remediation (Yan et al. 2019).

Genetic engineering of plant-associated endophytes would provide a better understanding of the beneficial traits conferred to the host plant, thereby improving crop production. The genetically modified crops with improved traits would comprise one or more of the following: increased tolerance to abiotic and biotic stress and tolerance to plant pathogens employing endophytes as "biocontrol agents", besides other traits aimed at increasing the crop yield and quality (Behie and Bidochka 2014). Additionally, the best way to employ endophytes in agriculture is through direct application on crops, by adding endophyte inoculant to the seed or soil, and this method improved the yield of sugarcane (Silva et al. 2012). However, this method has not been successful in the field on a large scale, when the crops are required to be inoculated each time with the microbial inoculants and challenges governing the establishment of endophytes (O'Callaghan M, 2016). Through seed dressings, microbes associated with seed will colonize the plant roots, improving the chances of endophyte associations. Another method is to promote plant growth by gaining insights into the "soil microbiome" leading to better crop yields.

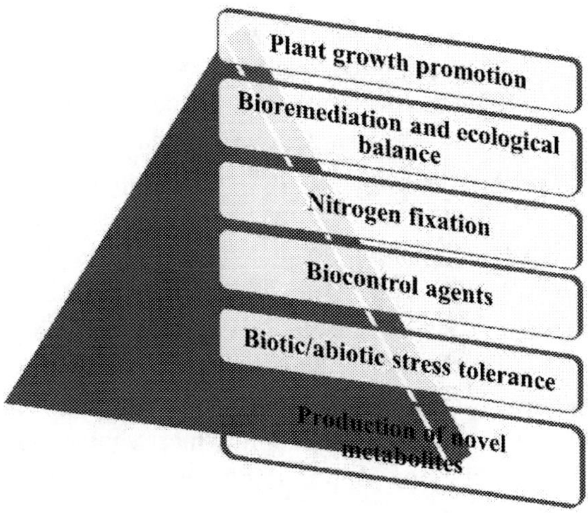

Figure 3. Diagrammatic representation of multiple beneficial functions of endophytes in enhancing plant growth and survival.

Several challenges exist that limit the efficient utilization of endophytes for sustainable agriculture. Bacterial or fungal endophytes, which confer resistance against herbivores, should be checked for toxicity to human health. The behavior of endophytes and their potentially harmful effects, if any, should be carefully monitored at all stages of their entire life cycle (Redman et al. 2001). Additionally, bacterial and fungal endophytes may be regarded together as part of the "soil microbiome," and a better understanding of the interaction between microbial endophytes, and their respective host plants, would lead to better exploitation of beneficial endophytes in agriculture.

Conclusion and Perspectives

Plant-endophyte associations define an evolutionary favored association, significantly impacting the environment and agriculture. Endophytes demonstrate multi-faceted benefits in bioremediation, nutrient uptake, plant growth promotion, and enhanced crop productivity. Displaying the potential biocontrol of phytopathogens and promoting plant growth, endophytes are key players with a broad range of socio-economic applications. With the direct application on plants or through seed dressing, endophytes are emerging as the potential candidates for promoting plant growth toward more sustainable agriculture. Understanding how plant-associated microbes confer stress tolerance would define a platform to employ scientific technologies aiding in plant biomass production. The rising global population and an ever-increasing demand for food supply project a requirement for increased food production. To date, chemical fertilizers including pesticides and agro-chemicals were employed for crop protection but nowadays, fungal endophytes are emerging as natural biofertilizers/bio inoculants for plant growth promotion in an eco-friendly manner. Additionally, the fluctuating climatic conditions and global warming adversely affect crop yield and production and it necessitates a need to restore symbiotic plant-endophyte associations (Khan et al. 2015). Ever-increasing research on deciphering the multi-faceted benefits and dynamics of plant-endophyte associations is providing new insights and opening new avenues in increasing crop production. Given the rising importance of endophytes, the article discusses the molecular mechanism of endophytes in plant growth promotion towards a bio-based economy, opening new avenues in sustainable agriculture.

Acknowledgments

The authors are thankful to their organizations for their encouragement and support.

Conflict of Interest

None to declare.

References

Acuña-Rodríguez, Ian S., Hermann Hansen, Jorge Gallardo-Cerda, Cristian Atala and Marco A. Molina-Montenegro. "Antarctic extremophiles: biotechnological alternative to crop productivity in saline soils." *Front. Bioeng. Biotechnol.* 7 (2019): 22. Doi: 10.3389/fbioe.2019.00022.

Albrectsen, Benedicte R., Lars Björkén, Akkamahadevi Varad, Åsa Hagner, Mats Wedin, Jan Karlsson, Stefan Jansson. "Endophytic fungi in European aspen (*Populus tremula*) leaves: diversity, detection, and suggested correlation with herbivory resistance." *Fungal Diversity* 41 (2010): 17–28.

Ali, Abeer H., Usama Radwan, Soad El-Zayat, and Magdi A. El-Sayed, 2019. "The role of the endophytic fungus, *Thermomyces lanuginosus*, on mitigation of heat stress to its host desert plant *Cullen plicata*." Biol. Fut. 70 (2019): 1–7.

Ali, S., T.C. Charles, B.R. Glick. "Delay of flower senescence by bacterial endophytes expressing 1-aminocyclopropane-1-carboxylate deaminase." *J. Appl. Microbiol.* 113 (2012): 1139–1144.

Ali, Shimaila, Jin Duan, Trevor C. Charles, Bernard R. Glick. "A bioinformatics approach to the determination of genes involved in endophytic behavior in *Burkholderia* spp." *J. Theor. Biol.* 343 (2014):193–198.

Ardanov, Pavlo, Angela Sessitsch, Hely Häggman, Natalia Kozyrovska, Anna Maria Pirttilä. "*Methylobacterium*-induced endophyte community changes correspond with protection of plants against pathogen attack." *PLoS One* 7, (2012):e46802. https://doi.org/10.1371/journal.pone.0046802.

Arnold, A. Elizabeth, F. Lutzoni. "Diversity and host range of foliar fungal endophytes: are tropical leaves biodiversity hotspots?" *Ecology* 88 (2007): 541–549.

Arnold, A. Elizabeth. "Understanding the diversity of foliar endophytic fungi: progress, challenges, and frontiers." *Fungal. Biol. Rev.* 21 (2007):51–66.

Asaf, Sajjad, Abdul Latif Khan, Muhammad Aaqil Khan, Ahmed Al-Harrasi and In-Jung Lee. "Genome sequencing and analysis of endophytic *Sphingomonas sp. LK11* and its potential in plant growth." *3 Biotech* 8, no. 9 (2018):389. doi:10.1007/s13205-018-1403-z.

Asif, Huma, David J. Studholme, Asifullah Khan, M. Aurongzeb, Ishtiaq A. Khan, M. Kamran Azim. "Comparative genomics of an endophytic Pseudomonas putida isolated from mango orchard." *Genetics and Molecular Biology*, 39, no. 3 (2016):465-473.

Azevedo, Joao Lucio, Welington Luiz Araújo. "Diversity and applications of endophytic fungi isolated from tropical plants." in: Ganguli, B.N., Deshmukh, S.K. (Eds.), *Fungi Multifaceted Microbes*; Anamaya: New Delhi, India, (2006): pp. 189–207.

Bacetty, A.A., M.E. Snook, A.E. Glenn, J.P. Noe, N. Hill, A. Culbreath, P. Timper, P. Nagabhyru, C.W. Bacon. "Toxicity of endophyte-infected tall fescue alkaloids and grass metabolites on *Pratylenchus scribneri*." *Phytopathology* 99 (2009a):1336–1345.

Bacetty, Ada A., Maurice E. Snook, Anthony E. Glenn, James P. Noe, Padmaja Nagabhyru, Charles W. Bacon. "Chemotaxis disruption in *Pratylenchus scribneri* by tall fescue root extracts and alkaloids." *J. Chem. Ecol.* 35 (2009b):844–850.

Bacon, Charles W., and James F. White Jr. "Functions, mechanisms, and regulation of endophytic and epiphytic microbial communities of plants." *Symbiosis* 68 (2016): 87–98.

Behie, S.W., P.M. Zelisko, M.J. Bidochka. "Endophytic insect-parasitic fungi translocate nitrogen directly from insects to plants." *Science* 336 (2012): 1576–1577.

Behie, Scott W., and Michael J. Bidochka. "Nutrient transfer in plant-fungal symbioses." *Trends in Plant Science* 19 (2014):734–740.

Belesky, David P., and Charles W. Bacon. "Tall fescue and associated mutualistic toxic fungal endophytes in agroecosystems." *Tox. Rev.* 28 (2009):102–117.

Bezerra, J.D.P., M.G.S. Santos, V.M. Svedese, D.M. Lima, M.J. Fernandes, L.M. Paiva, C.M. Souza-Motta. "Richness of endophytic fungi isolated from *Opuntia ficus-indica* Mill. (Cactaceae) and preliminary screening for enzyme production." *World J. Microbiol. Biotechnol.* 28 (2012): 1989–1995.

Bhargavi, S.D., Praveen, V.K., Savitha, J. "Bioinformatic comparative analysis of Lovastatin gene cluster in endophytic fungi and a soil fungus, *Aspergillus terreus*." *MOJ Proteomics Bioinformatics* 1 (2014): 00026. Doi: 10.15406/mojpb.2014.01.00026.

Bitas, Vasileios, Hye-Seon Kim, Joan W. Bennett, Seogchan Kang. "Sniffing on microbes: diverse roles of microbial volatile organic compounds in plant health." *Mol. Plant Microbe Interaction* 26 (2013): 835–843.

Boyle, C., M. Gotz, T.U. Dammann, B. Schulz. "Endophyte-host interactions III. Local vs. systemic colonization." *Symbiosis* 31 (2001):259–281.

Burton, Joshua N., Ivan Liachko, Maitreya J. Dunham, Jay Shendure. "Species-level deconvolution of metagenome assemblies with Hi-C–Based contact probability maps." *G3: Genes, Genomes, Genetics* vol. 4, no. 7 (2014):1339-1346.

Camilios-Neto, Doumit, Paloma Bonato, Roseli Wassem, Michelle Z. Tadra-Sfeir, Liziane C.C. Brusamarello-Santos, Glaucio Valdameri, Lucélia Donatti, Helisson Faoro, Vinicius A. Weiss, Leda S Chubatsu, Fábio O. Pedrosa, Emanuel M. Souza. "Dual RNA-seq transcriptional analysis of wheat roots colonized by *Azospirillum brasilense* reveals up-regulation of nutrient acquisition and cell cycle genes." *BMC Genomics* 15 (2014):378. Doi: 10.1186/1471-2164-15-378.

Canals, Rosa M., Leticia San-Emeterio, Salud Sanchez-Marquez, Igor Ruiz de los Mozos, Pablo Pujol and Iñigo Zabalgogeazcoa. "Nonsystemic fungal endophytes in *Carex brevicollis* may influence the toxicity of the sedge to livestock." *Span. J. Agric. Res.* 12 (2014):623–632.

Chakravarthi, B.V.S.K., Prasanta Das, Kalpana Surendranath, Anjali A Karande, Chelliah Jayabaskaran. "Production of paclitaxel by *Fusarium solani* isolated from *Taxus celebica*." *Journal of Bioscience* 33, no. 2 (2008):259.

Chapla, Vanessa Mara, Maria Luiza Zeraik, Valdecir F Ximenes, Lisinéia Maria Zanardi, Márcia N Lopes, Alberto José Cavalheiro, Dulce Helena S Silva, Maria Cláudia M Young, Luiz Marcos da Fonseca, Vanderlan S Bolzani, Angela Regina Araújo. "Bioactive secondary metabolites from *Phomopsis* sp., an endophytic fungus from *Senna spectabilis*." *Molecules* 19 (2014b): 6597–6608.

Chen, Can, Zonghao Yue, Cuiwei Chu, Keshi Ma, Lili Li, and Zhongke Sun. "Complete genome sequence of *Bacillus* sp. strain WR11, an endophyte isolated from Wheat root providing genomic insights into its plant Growth-promoting effects." *Molecular Plant-Microbe Interactions* (2020). https://doi.org/10.1094/MPMI-02-20-0030-

Chen, Meilan, Li Yang, Qin Li, Ye Shen, Aijuan Shao, Shufang Lin, Luqi Huang. "Volatile metabolites analysis and molecular identification of endophytic fungi bn12 from *Cinnamomum camphora* chvar. borneol." *China J Chin Materia Medica* 36, no. 23 (2011):3217–3221.

Chen, W-M, Tang Y-Q, Mori K, Wu X-L. "Distribution of culturable endophytic bacteria in aquatic plants and their potential for bioremediation in polluted waters." *Aquatic Biology*15 (2012): 99–110.

Christensen, Michael J., Raymond J. Bennett, Helal A. Ansari, Hironori Koga, Richard D. Johnson, Gregory T. Bryan, Wayne R. Simpson, John P. Koolaard, Elizabeth M. Nickless, Christine R. Voisey. "*Epichloe* endophytes grow by intercalary hyphal extension in elongating grass leaves." *Fungal Genet. Biol.* 45 (2008):84–93.

Christensen, M.J., C.R. Voisey. "The biology of the endophyte/grass partnership." In: Popay A, Thom ER (eds). *Proceedings of the 6th International Symposium on Fungal Endophytes of Grasses.* Christchurch: New Zealand Grassland Association, (2007):123–133.

Christine B. Müller, Jochen Krauss. "Symbiosis between grasses and asexual fungal endophytes." *Curr. Opin. Plant Biol.* 8 (2005): 450–456.

Clarke, Bruce B., James F. White Jr., Richard H. Hurley, Mónica S. Torres, S. Sun, David R. Huff. "Endophyte-mediated suppression of dollar spot disease in fine fescues." Plant Disease 90 (2006):994–998.

Clay, Keith. "Fungal endophytes of grasses." *Annu. Rev. Ecol. Evol. Syst.* 21 (1990): 275–295.

Cocq, Kate Le, Sarah J, Gurr, Penny R, Hirsch, Tim H, Mauchline. "Exploitation of endophytes for sustainable agricultural intensification." *Mol. Plant Pathol.* 18 (2017): 469–473.

Collins, Douglas P., Barry J. Jacobsen. "Optimizing a *Bacillus subtilis* isolate for biological control of sugar beet *Cercospora* leaf spot." *Biol. Control J.* 26 (2003): 153–161.

Compant, Stéphane, Birgit Reiter, Angela Sessitsch, Jerzy Nowak, Christophe Clément, Essaïd Ait Barka. "Endophytic colonization of *Vitis vinifera* L. by plant growth-promoting bacterium *Burkholderia* sp. strain PsJN." *Appl. Environ. Microbiol.* 71 (2005):1685–1693.

Compant, Stéphane, Christophe Clément, Angela Sessitsch. "Plant growth-promoting bacteria in the rhizo-and endosphere of plants: their role, colonization, mechanisms involved and prospects for utilization." *Soil Biol. Biochem.* 42 (2010): 669–678.

Compant, Stéphane, Hervé Kaplan, Angela Sessitsch, Jerzy Nowak, Essaïd Ait Barka, Christophe Clément. "Endophytic colonization of *Vitis vinifera* L. by *Burkholderia phytofirmans*strain PsJN: from the rhizosphere to inflorescence tissues." *FEMS Microbiol. Ecol.* 63 (2008): 84–93.

Conn, V.M., A.R. Walker, C.M.M. Franco. "Endophytic actinobacteria induces defense pathways in *Arabidopsis thaliana*." *Mol. Plant Microbe Interact.* 21 (2008): 208–218.

Coutinho, Bruna G., Danilo Licastro, Lucia Mendonça-Previato, Miguel Cámara, Vittorio Venturi. "Plant-influenced gene expression in the rice endophyte *Burkholderia kururiensis* M130." *Mol. Plant-Microbe Interact.* 28 (2015): 10–21.

Das, Aparajita, Ajit Varma. "Symbiosis: the art of living." in Varma A., Kharkwal A.C. (Eds.), *Symbiotic Fungi Principles and Practice*, Springer, Berlin, Germany, (2009): pp. 1–28.

Davey, Marie L., and Randolph S. Currah. "Interactions between mosses (Bryophyta) and fungi." *Can. J. Bot.* 84 (2006): 1509–1519.

Deng, Zujun, Lixiang Cao. "Fungal endophytes and their interactions with plants in phytoremediation: a review." *Chemosphere* 168 (2016). Doi: 10.1016/j.chemosphere. 2016.10.097.

Deshmukh, Sunil K., Manish K. Gupt, Ved Prakash, Sanjai Saxena. "Endophytic Fungi: a source of potential antifungal compounds." *J. Fungi* 4 (2018): 77; doi:10.3390/jof403 0077.

Dey, Prajjal, Diptanu Datta, Debanjana Saha, Selukash Parida and Darshan Panda. "Plant-endophyte interaction and its application to abiotic stress management of crop plants". *Int. J. Curr. Microbiol. App. Sci.* 8 (2019): 2708–2716.

Domka, Agnieszka Małgorzata, Piotr Rozpądek and Katarzyna Turnau. "Are fungal endophytes merely mycorrhizal copycats? The role of fungal endophytes in the adaptation of plants to metal toxicity." *Front. Microbiol.* 10 (2019): 371.

dos Santos, M.B.Q., R. Albano, V. de Padua, L. Rouws, C. Rojas, A. Hemerly, K. Teixeira, S. Schwab, J. Araujo, A. Oliveira, L. Franca, V. Magalhaes, S. Alqueres, A. Cardoso, W. Almeida, M.M. Loureiro, E. Nogueira, D. Cidade, D. Oliveira, P. Ferreira, "Complete genome sequence of the sugarcane nitrogen-fixing endophyte *Gluconacetobacter diazotrophicus Pal5*." *BMC Genomics* 10 (2009): 450. https://doi. org/10.1186/1471-2164-10-450

Doty, Sharon L., Brian Oakley, Gang Xin, Jun Won Kang, Glenda Singleton, Zareen Khan, Azra Vajzovic and James T. Staley. "Diazotrophic endophytes of native black cottonwood and willow." *Symbiosis* 47 (2009): 23–33.

Doty, Sharon Lafferty. "Functional Importance of the Plant Endophytic Microbiome: Implications for Agriculture, Forestry, and Bioenergy." Springer International Publishing AG 2017, S.L. Doty (ed.), *Functional Importance of the Plant Microbiome*, (2017). DOI 10.1007/978-3-319-65897-1_1.

Duan, L., G. Liwei Y. Hong. "Isolation and identification of producing endophytic fungi of berberine from the plant *Phellodendron amurense*." *J Anhu.i Agric Sci.* 22, no. 007 (2009). https://doi.org/10.3969/j.issn.0517-6611.2009.22.007

Dudeja, S.S., Rupa Giri, Ranjana Saini, Pooja Suneja-Madan, Erika Kothe. "Interaction of endophytic microbes with legumes." *J. Basic Microbiol.* 52 (2012): 248–260.

Eida, Abdul Aziz, Salim Bougouffa, Intikhab Alam, Maged M Saad, Heribert Hirt. "Complete genome sequence of the endophytic bacterium *Cellulosimicrobium* sp. JZ28 isolated from the root endosphere of the perennial desert tussock grass *Panicum turgidum*." *Arch Microbiol* 202 (2020): 1563–1569. https://doi.org/10.1007/s00203-020-01859-2

Eljounaidi, Kaouthar, Seung Kyu Lee, Hanhong Bae. "Bacterial endophytes as potential biocontrol agents of vascular wilt diseases-review and future prospects." *Biol. Control* 103 (2016): 62–68.

Gadd, Geoffrey M. "Geomycology: biogeochemical transformations of rock, minerals, metals, and radionuclides by fungi, bioweathering, and bioremediation." *Mycol. Res.* 111 (2007): 3–49.

Gange, A.C., J. Koricheva, A.F. Currie, L.R. Jaber, S. Vidal. "Meta-analysis of the role of entomopathogenic and unspecialized fungal endophytes as plant bodyguards." *New Phytol.* 223 (2019): 2002–2010.

Gkouletsos, T. "The role of endophytes in agriculture: Abiotic and biotic factors affecting their communities." MSc dissertation thesis (2018).

Glazebrook, Jane. "Contrasting mechanisms of defense against biotrophic and necrotrophic pathogens." *Annu. Rev. Phytopathol.* 43 (2005): 205–227.

Gond, Surendra K., Marshall S. Bergen, Mónica S. Torres, James F. White Jr. "Endophytic *Bacillus* spp. produce antifungal lipopeptides and induce host defense gene expression in maize." *Microbiol. Res.* 172 (2014):79–87.

Gonda, Sándor, Zsolt Szűcs, Borbála Balla, Gábor Vasas, Attila Kiss-Szikszai. "Efficient biotransformation of non-steroid anti-inflammatory drugs by endophytic and epiphytic fungi from dried leaves of a medicinal plant, *Plantago lanceolata* L." *Int. Biodeter. Biodegr.* 108 (2016): 115–121.

Griffin, Meghan A., Daniel J. Spakowicz, Tara A. Gianoulis, Scott A. Strobel. "Volatile organic compound production by organisms in the genus *Ascocoryne* and a re-evaluation of myco-diesel production by NRRL 50072." *Microbiol.* 156 (2010): 3814–3829.

Gundel, Pedro E., María A. Martínez-Ghersa, Marina Omacini, Romina Cuyeu, Elba Pagano Raúl Ríos, and Claudio M. Ghersa. "Mutualism effectiveness and vertical transmission of symbiotic fungal endophytes in response to host genetic background." *Evol. Appl.* 5 (2012): 838–884.

Gwinn, K.D., A.M. Gavin. "Relationship between endophyte infestation level of tall fescue seed lots and *Rhizoctonia* zeae seedling disease." *Plant Dis* 76 (1992):911–914.

Hacquard, Stéphane, Christopher W. Schadt. "Towards a holistic understanding of the beneficial interactions across the *Populus* microbiome." *New Phytol.* 205 (2015): 1424–1430.

Hall, T., L. Schreiber, C. Leben. "Effects of xylem-colonizing *Bacillus* spp. on *Verticillium* wilt in maples." *Plant Dis.* 70 (1986):521–524.

Hamilton C.E., T.L. Bauerle. "A new currency for mutualism? Fungal endophytes alter antioxidant activity in hosts responding to drought." *Fungal Divers* 54 (2012):39–49.

Han, Jong-In, Hong-Kyu Choi, Seung-Won Lee, Paul M. Orwin, Jina Kim, Sarah L. Laroe, Tae-Gyu Kim, Jennifer O'Neil, Jared R. Leadbetter, Sang Yup Lee, Cheol-Goo Hur, Jim C. Spain, Galina Ovchinnikova, Lynne Goodwin, Cliff Han. "Complete genome sequence of the metabolically versatile plant growth-promoting endophyte *Variovorax paradoxus S110.*" *Journal of Bacteriology* 193, no. 5 (2011):1183-1190.

Hardoim, Pablo R., Leonard S. van Overbeek, Gabriele Berg, Anna Maria Pirttilä, Stéphane Compant, Andrea Campisano, Matthias Döring, Angela Sessitsch. "The hidden world within plants: ecological and evolutionary considerations for defining functioning of microbial endophytes." *Microbiol. Mol. Biol. Rev.* 79 (2015): 293–320.

Harmon A.D., U. Weiss, J. Silverton. "The structure of rohitukine, the main alkaloid of *Amoora rohituka* (Syn. *Aphanamixis polystachya*) (meliaceae)." *Tetrahedron Lett.* 20, no. 8 (1979): 721–724.

Huang, Wu Yang, Yi Zhong Cai, Kevin D. Hyde, Harold Corke, Mei Sun. "Endophytic fungi from *Nerium oleander* L (Apocynaceae): main constituents and antioxidant activity." *World J. Microbiol. Biotechnol.* 23 (2007): 1253–1263.

Jeong, Haeyoung, Sung Hee Jo, Chi Eun Hong, Jeong Mee Park. "Genome sequence of the endophytic bacterium *Bacillus thuringiensis* strain KB1, a potential biocontrol agent against phytopathogens." *Genome Announcments* 4, no. 2 (2016):e00279-16. doi:10.1128/genomeA.00279-16.

Jorjan, Marjan, Asghar Heydari, Hamid Reza Zamanizadeh, Saeed Rezaee, Laleh Naraghi, Parima Zamzami. "Controlling sugar beet mortality disease by application of new bioformulations." *J. Plant Prot. Res.* 52 (2012): 303–307.

Kaneko, T., K. Minamisawa, T. Isawa, H. Nakatsukasa, H. Mitsui, Y. Kawaharada, Y. Nakamura, A. Watanabe, K. Kawashima, A. Ono, Y. Shimizu, C. Takahashi, C. Minami, T. Fujishiro, M. Kohara, M. Katoh, N. Nakazaki, S. Nakayama, M. Yamada, S. Tabata, S. Sato. "Complete genomic structure of the cultivated rice endophyte *Azospirillum* sp. B510." *DNA Res.* 17, no. 1 (2010):37.

Kaul, Sanjana, Tanwi Sharma and Manoj K. Dhar. "Omics tools for better understanding the Plant-endophyte interactions." *Front. Plant Sci.* 7 (2016): 955.

Khan, Abdul Latif, Javid Hussain, Ahmed Al-Harrasi, Ahmed Al-Rawahi, In-Jung Lee. "Endophytic fungi: resource for gibberellins and crop abiotic stress resistance." *Crit Rev Biotechnol.* 35 (2015):62–74.

Khan, Abdul Latif, Muhammad Hamayun, Javid Hussain, Sang-Mo Kang, In-Jung Lee. "The newly isolated endophytic fungus *Paraconiothyrium* sp. LK1 produces ascotoxin." *Molecules* 17 (2012): 1103–1112.

Khan, Abdul Latif, Muhammad Hamayun, Nadeem Ahmad, Javid Hussain, Sang-Mo Kang, Yoon-Ha Kim, Muhammad Adnan, Dong-Sheng Tang, Muhammad Waqas, Ramalingam Radhakrishnan, Young-Hyun Hwang, In-Jung Lee. "Salinity stress resistance offered by endophytic fungal interaction between *Penicillium minioluteum LHL09* and *Glycine max*. L." *J Microbiol Biotechnol.* 21, no. 9 (2011):893-902.

Kim, S. "Screening of taxol-producing endophytic fungi from *Ginkgo biloba* and *Taxus cuspidate* in Korea." *Agric. Chem. Biotechnol.* 42 (1999):97–99.

Kim, Jung A, Jongbum Jeon, Sook-Young Park, Ki-Tae Kim, Gobong Choi,b Hyunjung Lee, Yangsun Kim, Hee-sun Yang, Joo-Hong Yeo, Yong-Hwan Lee, and Soonok Kim. "Genome sequence of an endophytic fungus, *Fusarium solani JS-169*, which has antifungal activity." *Genome Announcements* 5, no. 42 (2017) e01071-17; DOI: 10.1128/genomeA.01071-17.

Kiraly, K.A., Pilinszky,K., Bittsanszky, A., Gyulai, G., Komives, T. "Importance of ammonia detoxification by plants in phytoremediation and aquaponics." (2013): pp 99–102, In: *Proceedings 12th Alps-Adria Scientific Workshop*, suppl. doi:10.12666/ Novenyterm. 62.2013.

Knapp, Dániel G., Julianna B. Németh, Kerrie Barry, Matthieu Hainaut, Bernard Henrissat, Jenifer Johnson, Alan Kuo, Joanne Hui Ping Lim, Anna Lipzen, Matt Nolan, Robin A. Ohm, László Tamás, Igor V. Grigoriev, Joseph W. Spatafora, László G. Nagy & Gábor M. Kovács. "Comparative genomics provides insights into the lifestyle and reveals functional heterogeneity of dark septate endophytic fungi." *Sci Rep* 8 (2018): 6321. https://doi.org/10.1038/s41598-018-24686-4.

Knoth, Jenny L., Soo-Hyung Kim, Gregory J. Ettl, Sharon L. Doty. "Biological nitrogen fixation and biomass accumulation within poplar clones as a result of inoculations with diazotrophic endophyte consortia." *New Phytol.* 201, no. 2 (2014):599-609.

Kong, Ping, and Chuanxue Hong. "Complete genome sequence of a boxwood endophyte *Burkholderia sp. SSG* with broad biotechnological application potential." *Biotechnology reports* (Amsterdam, Netherlands) (2020), *DOI:* 10.1016/j.btre.2020. e00455.

Kothari, S.K., H. Marschner, E. George. "Effect of VA mycorrhizal fungi and rhizosphere microorganisms on root and shoot morphology, growth and water relations in maize." *New Phytol.* 116 (1990): 303–311.

Krings, Michael, Thomas N. Taylor, Hagen Hass, Hans Ker, Nora Dotzler, Elizabeth J. Hermsen. "Fungal endophytes in a 400-million-yr-old land plant: infection pathways, spatial distribution, and host responses." *New Phytol.* 174 (2007): 648–657.

Kusari, Souvik, Marc Lamshöft, Sebastian Zühlke, Michael Spiteller. "An endophytic fungus from *Hypericum perforatum* that produces hypericin." *J Nat. Prod.* 71, no. 2 (2008):159–162.

Kusari, Souvik, Vijay C. Verma, Marc Lamshoeft, Michael Spiteller. "An endophytic fungus from *Azadirachta indica A. Juss.* that produces azadirachtin." *World J Microbiol. Biotechnol.* 28, no. 3 (2012):1287–1294.

Kwak, Min Jung, Ju Yeon Song, Seon Young Kim, Haeyoung Jeong, Sung Gyun Kang, Byung Kwon Kim, Soon Kyeong Kwon, Choong Hoon Lee, Dong Su Yu, Seung Hwan Park, Jihyun F. Kima. "Complete genome sequence of the endophytic bacterium *Burkholderia sp. strain KJ006*." *Journal of Bacteriology* 194, no. 16 (2012): https://doi.org/10.1128/JB.00821-12.

Lee, Shin Ae, Bashistha Kumar Kanth, Hyeon Su Kim, Tae-Wan Kim, Mee Kyung Sang, Jaekyeong Song, and Hang-Yeon Weon. "Complete genome sequence of the plant growth-promoting endophytic bacterium *Rhodanobacter glycinis T01E-68* isolated from tomato (*Solanum lycopersicum* L.) plant roots." *Korean J. Microbiol.* 55 ():422-424.

Li-Wen-Chao, Guo-Shou-Yu, Guo-Liang-Dong. "Endophytic fungi associated with lichen *Physcia stellaris* using different surface sterilization methods." *J. Fungal Res.* 5 (2007): 202–206.

Li, T.X., M.H. Yang, X.B. Wang, Y. Wang, L.Y. Kong. "Synergistic antifungal meroterpenes and dioxolanone derivatives from the endophytic fungus *Guignardia* sp." *J. Nat. Prod.* 78 (2015): 2511–2520.

Liu, Chongxi, Xiaoxin Zhuang, Zhiyin Yu, Zhiyan Wang, Yongjiang Wang, Xiaowei Guo, Wensheng Xiang, and Shengxiong Huang. "Community structures and antifungal activity of root-associated endophytic actinobacteria of healthy and diseased Soybean." *Microorganisms* 7, no. 8 (2019):243. doi:10.3390/microorganisms7080243.

Luduena, L.M., M.S. Anzuay, J.G. Angelini, M. Mclntosh, A. Becker, A. Rupp, O. Goesmann, A. Blom, J. Fabra, A.T. Taurian. "Genome sequence of the endophytic strain *Enterobacter sp. J49*, a potential biofertilizer for peanut and maize." *Genomics* 111, no. 4 (2018): 913e920.

Lugtenberg, B.J.J., J.R. Caradus, and L.J. Johnson. "Fungal endophytes for sustainable crop production." *FEMS Microbiology Ecology* vol. 92 (2016).

Lyons, P.C., R.D. Plattner, C.W. Bacon. "Occurrence of peptide and clavine ergot alkaloids in tall fescue." *Science* 232 (1986):487–489.

Malfanova, N.V. "*Endophytic bacteria with plant growth promoting and biocontrol abilities.*" Doctoral thesis, Institute Biology of Leiden (IBL). Faculty of Science, Leiden University, (2013): http://hdl.handle.net/1887/20732.

Malinowski, D.P., G.A. Alloush, D.P. Belesky. "Leaf endophyte *Neotyphodium coenophialum* modifies mineral uptake in tall fescue." *Plant Soil* 227 (2000): 115–126.

Mandyam, Keerthi, Ari Jumpponen. "Seeking the elusive function of the root colonizing dark septate endophytic fungi." *Stud. Mycol.* 53 (2005): 173–189.

Mansoori, Mahdiyeh, Asghar Heydari, Nader Hassanzadeh, Saeed Rezaee, Laleh Naraghi. "Evaluation of *Pseudomonas* and *Bacillus* bacterial antagonists for biological control of cotton Verticillium Wilt Disease." *J. Plant Prot. Res.* 53 (2013): 11–14.

Mantzoukas, S., I. Lagogiannis. "Endophytic colonization of Pepper (*Capsicum annum*) controls Aphids (*Myzus persicae Sulzer*)." *Appl. Sci.* 9 (2019): 2239, doi: 10.3390/app9112239.

Marquez-Santacruz, H.A., R. Hernandez-Leon, M.C. Orozco-Mosqueda, I. Velazquez-Sepulveda, G. Santoyo. "Diversity of bacterial endophytes in roots of Mexican husk tomato plants (*Physalisixocarpa*) and their detection in the rhizosphere." *Genet. Mol. Res.* 9 (2010): 2372–2380.

Mastretta, Chiara, Safiyh Taghavi, Daniel van der Lelie, Alessio Mengoni, Francesca Galardi, Christina Gonnelli, Tanja Barac, Jana Boulet, Nele Weyens, Jaco Vangronsveld. "Endophytic bacteria from seeds of *Nicotiana tabacum* can reduce cadmium phytotoxicity." *Int. J. Phytoremediation* 11 (2009): 251–267.

Mathys, Janick, Kaat De Creme, Pieter Timmermans, Stefan Van Kerckhove, Bart Lievens, Mieke Vanhaecke, Bruno P. A. Cammue and Barbara De Coninck. "Genome-wide characterization of ISR induced in *Arabidopsis thaliana* by *Trichoderma hamatum* T382 against *Botrytis cinerea* infection." *Front. Plant Sci.* 3 (2012): 108.

Mendoza, A.R., R.A. Sikora. "Biological control of *Radopholus similis* by co-application of the mutualistic endophyte *Fusarium oxysporum* strain 162, the egg pathogen *Paecilomyces lilacinus*s train 251 and the antagonistic bacteria *Bacillus firmus*." *Bio. Control* 54 (2009):263–272.

Mercado-Blanco, Jesús and Ben J.J. Lugtenberg. "Biotechnological applications of bacterial endophytes." *Curr. Biotech.* 3 (2014): 60–75.

Michael Weiß, Zuzana Sýkorová, Sigisfredo Garnica, Kai Riess, Florent Martos, Cornelia Krause, Franz Oberwinkler, Robert Bauer, Dirk Redecker. "Sebacinales everywhere: previously overlooked ubiquitous fungal endophytes." *PLoS One* 6, no. 2 (2011):e16793. doi: 10.1371/journal.pone.0016793.

Mitter, B., A. Petric, M.W. Shin, P.S. Chain, L. Hauberg-Lotte, B. Reinhold-Hurek, J. Nowak, A. Sessitsch. "Comparative genome analysis of *Burkholderia phytofirmans* PsJN reveals a wide spectrum of endophytic lifestyles based on interaction strategies with host plants." *Front. Plant Sci.* 4 (2013):120.

Moricca, S., A. Ragazzi. "Fungal endophytes in Mediterranean oak forests: a lesson from Disculaquercina." *Phytopathology* 98 (2008): 380–386.

Moy, M., Belanger, F., Duncan, R., Freehof, A., Leary, C., Meyer, W., Sullivan, R., White, J.F. "Identification of epiphyllous mycelial nets on leaves of grasses infected by clavicipitaceous endophytes." *Symbiosis* 28 (2000):291–302.

O'Callaghan, M. "Microbial inoculation of seed for improved crop performance: issues and opportunities." *Appl. Microbiol. Biotechnol.* 100 (2016): 5729–5746.

Ongena, Marc, Philippe Jacques. "Bacillus lipopeptides: versatile weapons for plant disease biocontrol." *Trends Microbiol.* 16 (2008):115–125.

Ownley, Bonnie H., Kimberly D. Gwinn and Fernando E. "Vega Endophytic fungal entomo-pathogens with activity against plant pathogens: ecology and evolution." *BioControl* 55 (2010): 113–128.

Pankievicz, Vânia C.S., Fernanda P. do Amaral, Karina F.D.N. Santos, Beverly Agtuca, Youwen Xu, Michael J. Schueller, Ana Carolina M. Arisi, Maria B.R. Steffens, Emanuel M. de Souza, Fábio O. Pedrosa, Gary Stacey, Richard A. Ferrieri. "Robust biological nitrogen fixation in a model grass-bacterial association." *Plant Journal* 81 (2015):907–919.

Park S.-Y., J. Jeon, J.A. Kim, M.J. Jeon, M.-H. Jeong, Y. Kim, Y. Lee, H. Chung, Y.-H. Lee, S. Kim. "Draft genome sequence of *Alternaria alternata* JS-1623, a fungal fndophyte of *Abies koreana*." *Mycobiology* 48, no. 3 (2020): 240-244.

Park, Young-Hwan, Ratnesh Chandra Mishra, Sunkyung Yoon, Hoki Kim, Changho Park, Sang-Tae Seo, Hanhong Bae. "Endophytic *Trichoderma citrinoviride* isolated from mountain-cultivated ginseng (*Panax ginseng*) has great potential as a biocontrol agent against ginseng pathogens." *Journal of Ginseng Research* 43, no. 3 (2019):408-420.

Partida-Martínez, L.P., M. Heil. "The microbe-free plant: fact or artifact?" *Front. Plant Sci.* 2 (2011): 100.

Puente, M. Esther, Ching Y. Li, Yoav Bashan. "Rock-degrading endophytic bacteria in cacti." *Environ. Exp. Bot.* 66 (2009):389–401.

Puri, S.C., V. Verma, T. Amna, G.N. Qazi, M. Spiteller. "An endophytic fungus from *Nothapodytes foetida* that produces Camptothecin." *J Nat. Prod.* 68, no. 12 (2005):1717–1719.

Qawasmeh, Abdelqader, Hassan K. Obied, Anantanarayanan Raman, Warwick Wheatley, 2012. "Influence of fungal endophyte infection on phenolic content and antioxidant activity in grasses: interaction between *Lolium perenne* and different strains of *Neotyphodium lolii*." *J. Agric. Food Chem.* 60 (2012): 3381–3388.

Qiu, D., M. Huang, X. Fang, C. Zhu, Z. Zhu. "Isolation of an endophytic fungus associated with *Taxus yunnanensis*." *Cheng et LK Fu. Acta Mycol Sin* 13, no. 4 (1994):314–316.

Rabiey, M., L.E. Hailey, S.R. Roy, K. Grenz, M.A.S. Al-Zadjali, G.A. Barrett, R.W. Jackson. "Endophytes vs tree pathogens and pests: can they be used as biological control agents to improve tree health?" *Eur. J. Plant Pathol.* 155 (2019): 711–729.

Redman, Regina S., David D. Dunigan, Rusty J. Rodriguez. "Fungal symbiosis from mutualism to parasitism: who controls the outcome, host or invader?" *New Phytol.* 151 (2001): 705–716.

Reinhold-Hurek B., T. Hurek. "Living inside plants: bacterial endophytes." *Curr. Opin. Plant Biol.* 14 (2011):435–443.

Rey, Thomas, and Sebastian Schornack. "Interactions of beneficial and detrimental root-colonizing filamentous microbes with plant hosts." *Genome Biol.* 14 (2013): 121.

Rodriguez, R., R. Redman. "Fungal life-styles and ecosystem dynamics: biological aspects of plant pathogens, plant endophytes and saprophytes." *Adv. Bot. Res.* 24 (1997):169–195.

Rodriguez, R.J., J.F. White Jr., A.E. Arnold, R.S. Redman. "Fungal endophytes: diversity and functional roles." *New Phytol.* 182 (2009): 314–330.

Romero, Fernando M., María Marina, Fernando L. Pieckenstain. "The communities of tomato (*Solanum lycopersicum* L.) leaf endophytic bacteria, analyzed by 16S-ribosomal RNA gene pyrosequencing." *FEMS Microbiol. Lett.* 351 (2014): 187–194.

Rosenblueth, M., E. Martínez-Romero. "Bacterial endophytes and their interactions with hosts." *Mol. Plant Microbe Interact.* 19 (2006): 827–837.

Rowen, D.D., M.B. Hunt, D.L. Gaynor. "Peramine, a novel insect feeding deterrent from ryegrass infected with the endophyte *Acremonium loliae*." *Chem. Commun.* (1986):935–936.

Ruiz-Lozano, J.M., R. Aroca. "Host response to osmotic stresses: stomatal behaviour and water use efficiency of arbuscular mycorrhizal plants." in: Koltai, H., Kapulnik, Y. (Edn.) *Arbuscular Mycorrhizas: Physiology and Function.* Berlin: Springer Science and Business Media, (2010):pp. 239–256.

Rungjindamai, N., U. Pinruan, R. Choeyklin, T. Hattori, E.B.G. Jones. "Molecular characterization of basidiomycetous endophytes isolated from leaves, rachis and petioles of the oil palm, *Elaeis guineensis*, in Thailand." *Fungal Divers.* 33 (2008): 139–161.

Russell, J.R., J. Huang, P. Anand, K. Kucera, A.G. Sandoval, K.W. Dantzler, D. Hickman, J. Jee, F.M. Kimovec, D. Koppstein, D.H. Marks, P.A. Mittermiller, S.J. Núñez, M. Santiago, M.A. Townes, M. Vishnevetsky, N.E. Williams, M.P.N. Vargas, L.A. Boulanger, C.B. Slack, S.A. Strobel. "Biodegradation of polyester polyurethane by endophytic fungi." *Appl. Environ. Microbiol.* 77 (2011): 6076–6084.

Saikkonen, Kari, Piippa Wäli, Marjo Helander, Stanley H. Faeth. "Evolution of endophyte-plant symbioses." *Trends Plant Sci.* 9 (2004):275–280.

Santoyo, Gustavo, Gabriel Moreno-Hagelsieb, Ma del Carmen Orozco-Mosqueda, Bernard R Glick. "Plant growth-promoting bacterial endophytes." *Microbiol. Res.* 183 (2016): 92–99.

Saravanakumar, D., R. Samiyappan. "ACC deaminase from *Pseudomonas fluorescens* mediated saline resistance in groundnut (*Arachis hypogea*) plants." *J Appl. Microbiol.* 102 (2007):1283-1292.

Schulz, Barbara, Christine Boyle, Siegfried Draeger, Anne-Katrin Römmert, Karsten Krohn. "Endophytic fungi: a source of novel biologically active secondary metabolites." *Mycol. Res.* 106 (2002): 996–1004.

Senthil Kumaran, R., J. Muthumary B. Hur. "Production of taxol from *Phyllosticta spinarum*, an endophytic fungus of *Cupressus* sp." *Eng. Life Sci.* 8, no. 4 (2008):438–446.

Sessitsch, A., P. Hardoim, J. Döring, A. Weilharter, A.Krause, T. Woyke, B. Mitter, L. Hauberg-Lotte, F. Friedrich, M. Rahalkar. "Functional characteristics of an endophyte community colonizing rice roots as revealed by metagenomic analysis." *Mol. Plant-Microbe Interact.* 25 (2012); 28–36.

Sheibani-Tezerji, Raheleh, Thomas Rattei, Angela Sessitsch, Friederike Trognitz, and Birgit Mitter. "Transcriptome profiling of the endophyte *Burkholderia phytofirmans* PsJN indicates sensing of the plant, environment, and drought stress." *MBio.* 6 (2015): 1–15.

Sherameti, I., B. Shahollari, Y. Venus, L. Altschmied, A. Varma, R. Oelmuller. "The endophytic fungus *Piriformospora indica* stimulates the expression of nitrate reductase and the starch-degrading enzyme glucan-water dikinase in tobacco and *Arabidopsis* roots through a homeodomain transcription factor which binds to a conserved motif in their promoters." *J. Biol. Chem.* 280 (2005): 26241–26247.

Sherameti, I., S. Tripathi, A. Varma, R. Oelmüller. "The root colonizing endophyte *Piriformospora indica* confers drought tolerance in *Arabidopsis* by stimulating the expression of drought stress–related genes in leaves." *Mol. Plant Microbe Interact.* 21 (2008):799–800.

Shweta S., S. Zuehlke, B. Ramesha, V. Priti, P.M. Kumar, G. Ravikanth, M. Spiteller, R. Vasudeva, R.U. Shaanker. "Endophytic fungal strains of *Fusarium solani*, from *Apodytes dimidiata E. Mey. ex. Arn* (Icacinaceae) produce camptothecin, 10-hydroxycamptothecin and 9-methoxycamptothecin." *Phytochemistry* 71, no. 1 (2010):117–122.

Siegel, M.R., G.C.M. Latch, L.P. Bush, N.F. Fammin, D.D. Rowen, B.A. Tapper, C.W. Bacon, M.C. Johnson. "Alkaloids and insecticidal activity of grasses infected with fungal endophytes." *J Chem. Ecol.* 16 (1991):3301– 3315.

Sikora, R.A., A. ZumFelde, A.Mendoza, R. Menjivar, L. Pocasangre. "*In planta* suppressiveness to nematodes and long-term root health stability through biological enhancement - do we need a cocktail?" *Acta Hort* 879 (2010):553–560.

Silva, G.H., M.L. Zeraik, C.M. de Oliveira, H.L. Teles, H.C. Trevisan, L.H. Pfenning, C.P. Nicolli, M.C.M. Young, Y.P. Mascarenhas, L.M. Abreu, A.C. Saraiva, A.I. Medeiros, V. da S. Bolzani, A. R. Araujo. "Lactone derivatives produced by a *Phaeoacremonium* sp., an endophytic fungus from *Senna spectabilis*." *J. Nat. Prod.* 80 (2017): 1674–1678. \

Silva, M.F., C.S. Antônio, P.J. Oliveira, G.R. Xavier, N.G. Rumjanek, L.H.B. Soares, and V.M. Reis, "Survival of endophytic bacteria in polymer-based inoculants and efficiency of their application to sugarcane." *Plant Soil* 356 (2012): 231–243.

Singh, Ramesh K., Namrata Malik and Surendra Singh. "Improved nutrient use efficiency increases plant growth of rice with the use of IAA-overproducing strains of endophytic *Burkholderia cepacia* strain RRE25." *Microb. Ecol.* 66 (2013): 375–384.

Sørensen, J., A. Sessitsch. "Plant-associated bacteria lifestyle and molecular interactions". in: van Elsas, J.D., Trevors, J.T., Jansson, J.K., Nannipieri, P. (Eds.) *Modern soil microbiology*, 2nd edn. CRC Press, Boca Raton, (2015):pp. 211–236.

Sridhar, K.R. "Diversity, ecology, and significance of fungal endophytes". in: Jha, S., (ed.), *Endophytes and Secondary Metabolites*, Springer Nature Switzerland, (2019):pp. 61-100.

Stępniewska, Z., A. Kuźniar. "Endophytic microorganisms–promising applications in bioremediation of greenhouse gases." *Appl. Microbiol. Biotechnol.* 97 (2013):9589–9596.

Stergiopoulos, Ioannis, and Thomas R. Gordon. "Cryptic fungal infections: the hidden agenda of plant pathogens." *Front. Plant Sci.* 5 (2014):506.

Steven D. Siciliano, Nathalie Fortin, Anca Mihoc, Gesine Wisse, Suzanne Labelle, Danielle Beaumier, Danielle Ouellette, Real Roy, Lyle G. Whyte, M. Kathy Banks, Paul Schwab, Ken Lee, and Charles W. Greer. "Selection of specific endophytic bacterial genotypes by plants in response to soil contamination." *Appl. Environ. Microbiol.* 67 (2001): 2469–2475.

Stierle, A., G. Strobel, D. Stierle. "Taxol and taxane production by *Taxomyces andreanae*, an endophytic fungus of Pacific yew." *Science* 260 (1993): 214–216.

Strobel, G.A., W. Hess, J-Y. Li, E. Ford, J. Sears, R.S. Sidhu, B. Summerell. "*Pestalotiopsis guepinii*, a taxol-producing endophyte of the Wollemi pine, *Wollemia nobilis*." *Aust. J. Bot.* 45, no. 6 (1997):1073–1082.

Strobel, Gary and Bryn Daisy. "Bioprospecting for microbial endophytes and their natural products." *Microbiol. Mol. Biol. Rev.* 67 (2003): 491–502.

Strobel, Gary. "*Muscodor albus* and its biological promise." *J. Ind. Microbiol. Biotechnol.* 33 (2006): 514–522.

Strobel, Gary, Bryn Daisy, Uvidelio Castillo, James Harper. "Natural products from endophyte microorganisms." *J. Nat. Prod.* 67 (2004): 257–268.

Sturz, A.V., J. Nowak, "Endophytic communities of rhizobacteria and the strategies required to create yield enhancing associations with crops." *Appl. Soil Ecol.* (2000):183–190.

Sun, D., X. Ran, J. Wang. "Isolation and identification of a taxol-producing endophytic fungus from *Podocarpus*." *Acta Microbiol. Sin.* 48, no. 5 (2008):589–595.

Sun, Leni, Linyan He, Yanfeng Zhang, Wenhui Zhang, Qi Wang, Xiafang Sheng. "Isolation and biodiversity of copper-resistant bacteria from rhizosphere soil of *Elsholtzia splendens*." *Wei Sheng Wu Xue Bao* 49 (2008):1360–1366.

Swatzell, L.J., M.J. Powell, J.Z. Kiss. "The relationship of endophytic fungi to the gametophyte of the fern *Schizaea pusilla*." *Int. J. Plant Sci.* 157 (1996): 53–62.

Tadych, M., J.F. White. "Ecology of epiphyllous stages of endophytes and implications for horizontal dissemination." *Proc 6th Inter Symp Fungal Endophytes Grasses*, New Zealand Grass Association –Grassland Research and Practice Series 13 (2007): 157–161.

Taghavi, S., D. van der Lelie, A. Hoffman, Y.B. Zhang, M.D. Walla, J. Vangronsveld, L. Newman, S. Monchy. "Genome sequence of the plant growth promoting endophytic bacterium *Enterobacter sp. 638*." *PLoS Genet.* 6 no. 5 (2010):e1000943. doi:10.1371/journal.pgen.1000943

Tan, R.X., W.X. Zhou. "Endophytes: a rich source of functional metabolites." *Natural Product Reports* 18 (2001): 448-459.

Tikhonovich, I.A., Provorov, N.A. "From plant-microbe interactions to symbiogenesis: a universal paradigm for the interspecies genetic integration." *Ann. Appl. Biol.* 154 (2009): 341–350.

Timmusk, S., L. Behers, J. Muthoni, A. Muraya, A.C. Aronsson. "Perspectives and challenges of microbial application for crop improvement." *Frontiers in Plant Science* 8 (2017): DOI: 10.3389/fpls.2017.00049.

Ting, A.S.Y. "Biosourcing endophytes as biocontrol agents of wilt diseases." In: Verma, V.C., Gange, A.C. (Eds.), *Advances in Endophytic Research.* Springer, India, (2014): pp. 283–300.

Tiwari, Pragya, Hanhong Bae. "Horizontal gene transfer and Endophytes: an implication for acquisition of novel traits." *Plants* 9 (2020a): 305, DOI: 10.3390/plants9030305.

Tiwari, Pragya, Mangalam Bajpai, Lalit Kumar Singh, Shashank Mishra, and Ajar Nath Yadav. "Phytohormones producing fungal communities: Metabolic engineering for abiotic stress tolerance in plants." In *Agriculturally important fungi for sustainable agriculture*. edited by Dr. Vijai Kumar Gupta and Dr. Maria Tuohy, Springer publication, 2020b.

Tiwari, Pragya, Yashdeep Srivastava, Vinay Kumar. "Antimicrobial peptides as effective agents against drug-resistant pathogens." In *Antimicrobial Resistance*; Kumar, V., Shriram, V., Paul, A., Thakur, M., Eds.; Springer: Singapore, (2022a): pp. 289-322.

Tiwari, Pragya, Hanhong Bae. "Endophytic fungi: Key insights, emerging prospects, and challenges in natural product drug discovery." *Microorganisms* 10 (2022b): 360. https://doi.org/10.3390/microorganisms10020360.

Tiwari, Pragya, Tushar Khare, Varsha Shriram, Hanhong Bae, Vinay Kumar. "Exploring synthetic biology strategies for producing potent antimicrobial phytochemicals." *Biotechnology Advances* 48, (2021a):107729.

Tiwari, Pragya, Yashdeep Srivastava, Mangalam Bajpai, Abhishek Sharma. "Bioactive metabolites from natural sources: Prospects and significance in drug discovery and research." *Bioingene PSJ* 1 (2021b):1-14.

Tiwari, P., Y. Srivastava, H. Bae. "Trends of pharmaceutical design of Endophytes as anti-infectives." *Current Topics in Medicinal Chemistry*, 21 (2021c): 1572-1586.

Tiwari, Pragya, Subir Kumar Bose, and Hanhong Bae. "Plant Growth Promoting Soil Microbiomes: beneficial attributes and potential applications." In *Soil Microbiomes for Sustainable Agriculture* - Volume 2: Functional Annotation. edited by Dr. Ajar Nath Yadav, Springer Publication (2021d).

Tiwari, Pragya, Mangalam Bajpai, Lalit Kumar Singh, Ajar Nath Yadav, Hanhong Bae. "Portraying fungal mechanisms in stress tolerance: perspective for sustainable agriculture." In Springer-Nature book on *Recent Trends in Mycological Research*, (2021e).

Tshishonga K, Serepa-Dlamini MH. "Draft Genome Sequence of *Pseudarthrobacter phenanthrenivorans strain MHSD1*, a bacterial endophyte isolated from the medicinal plant *Pellaea calomelanos*." *Evolutionary Bioinformatics* (2020). doi:10.1177/1176934320913257

Ulrich, K., A. Ulrich, D. Ewald. "Diversity of endophytic bacterial communities in poplar grown under field conditions." *FEMS Microbiol Ecol.* 63, no. 2 (2008):169-80.

Upson, R., K.K. Newsham, P.D. Bridge, D.A. Pearce, D.J. Read. "Taxonomic affinities of dark septate root endophytes of *Colobanthus quitensis* and *Deschampsia antarctica*, the two native Antarctic vascular plant species." *Fungal Ecol.* 2 (2009): 184–196.

Vega, F.E., A. Simpkins, C. Aime, F. Posada, S.W. Peterson et al. "Fungal endophyte diversity in coffee plants from Colombia, Hawai, Mexico, and Puerto Rico." *Fungal Ecol* 3 (2010): 122–138.

Wade, M.J. "The co-evolutionary genetics of ecological interactions." *Nat. Rev. Genet.* 8 (2007):185–195.

Waqas, M., A.L. Khan, M. Kamran, M.Hamayun et al. "Endophytic fungi produce giberellins and indoleacetic acid and promotes host-plant growth during stress." *Molecules* 17 (2012): 10754–73.

Weyens, Nele, Daniel van der Lelie, Safiyh Taghavi, Lee Newman, Jaco Vangronsveld. "Exploiting plant–microbe partnerships to improve biomass production and remediation." *Trends Biotechnol.* 27 (2009): 591–598.

Wheatley, R.E. "The consequences of volatile organic compound mediated bacterial and fungal interactions." *Antonie Van Leeuwenhoek* 81 (2002): 357–364.

White Jr., James F, Mónica S Torres. "Is plant endophyte-mediated defensive mutualism the result of oxidative stress protection?" *Physiol. Plant.* 138 (2010): 440–446.

Wu, T., J. Xu, J. Liu, W.-H. Guo, X.-B. Li, J.-B. Xia, W.-J. Xie, Z.-G. Yao, Y-M. Zhang, and R.-Q. Wang, "Characterization and initial application of endophytic *Bacillus safensis* strain ZY16 for improving phytoremediation of oil-contaminated saline soils." *Front. Microbiol.* 10 (2019):991. doi: 10.3389/fmicb.2019.00991.

Wu, T., J. Xu, W. Xie, Z. Yao, H. Yang, C. Sun, X. Li. "*Pseudomonas aeruginosa* L10: A hydrocarbon-degrading, biosurfactant-producing, and plant-growth-promoting endophytic bacterium isolated from a Reed (*Phragmites australis*)." *Front Microbiol.* 9 (2018):1087. doi: 10.3389/fmicb.2018.01087.

Xu, D., B.Y. Zhang, X.L. Yang. "Antifungal monoterpene derivatives from the plant endophytic fungus *Pestalotiopsis foedan*." *Chem. Biodivers.* 13 (2016): 1422–1425.

Yadav, V., M. Kumar, D.K. Deep, H. Kumar, R. Sharma, T. Tripathi, N. Tuteja, A.K. Saxena, A.K. Johri, "A phosphate transporter from the root endophytic fungus *Piriformospora indica* plays a role in phosphate transport to the host plant." *J. Biol. Chem.* 285 (2010): 26532–26544.

Yan, J.F., S.J. Broughton, S.L. Yang, A.C. Gangeb. "Do endophytic fungi grow through their hosts systemically?" *Fungal Ecol.* 13 (2015):13:53–59.

Yan, L., J. Zhu, X. Zhao, J. Shi, C. Jiang, D. Shao. "Beneficial effects of endophytic fungi colonization on plants." *Appl. Microbiol. Biotechnol.* 103 (2019): 3327–3340.

Yang, Y., H. Zhao, R.A. Barrero, B. Zhang, G. Sun, I.W. Wilson, F. Xie, K.D. Walker, J.W. Parks, R. Bruce, G. Guo, L. Chen, Y. Zhang, X. Huang, Q. Tang, H. Liu, M.I. Bellgard, D. Qiu, J. Lai and A. Hoffman. "Genome sequencing and analysis of the paclitaxel-producing endophytic fungus *Penicillium aurantiogriseum* NRRL 62431." *BMC Genomics* 15, no. 69 (2014). https://doi.org/10.1186/1471-2164-15-69

Yang, Yang, Xiaobao Liu, Jimiao Cai, Yipeng Chen, Boxun Li, Zhikai Guo and Guixiu Huang. "Genomic characteristics and comparative genomics analysis of the endophytic fungus *Sarocladium brachiariae*." *BMC Genomics* 20 (2019): 782. https://doi.org/10.1186/s12864-019-6095-1

Yu, H., L. Zhang, L. Li, C. Zheng, L. Guo, W. Li, P. Sun, L. Qin. "Recent developments and future prospects of antimicrobial metabolites produced by endophytes." *Microbiol. Res.* 165 (2010): 437–449.

Zhang, Y., C. Bai, G. Ran, Z.-y. Zhang, Y.-h. Chen, G. Wu. "Characterization of endophytic bacterial strain YS45 from the citrus xylem and its biocontrol activity against *Sclerotinia* stem rot of rapeseed." *Acta Phytopathol. Sin.* 39 (2009):638–645.

Zuccaro, A., U. Lahrmann, U. Güldener, G. Langen, S. Pfiffi, D. Biedenkopf, P. Wong, B. Samans, C. Grimm, M. Basiewicz, C. Murat, F. Martin, K.H. Kogel. "Endophytic life strategies decoded by genome and transcriptome analyses of the mutualistic root symbiont *Piriformospora indica*." *PLoS Pathogen* 7 (2011): e1002290. https://doi.org/10.1371/journal.ppat.1002290

Chapter 12

Recent Trends and Prospects in Endophyte Biology and Research

Hiran Kanti Santra and Debdulal Banerjee*
Microbiology and Microbial Biotechnology Laboratory, Department of Botany and Forestry, Vidyasagar University, Midnapore, West Bengal, India

Abstract

Endophytes are the age-old microbial partners (including bacteria, actinobacteria, and fungi) of chlorophyllous tissue promoting host survival. They are of biotechnological, pharmaceutical, and agricultural necessity also. As endophytes are believed to be co-evolved with their hosts, they undergo vertical and horizontal gene sharing, therefore synthesizing compounds of interest similar to the host tissues. Endophytes have proved their efficacy as antimicrobial, antioxidant, anti-cancer components, industrially valuable enzymes, extracellular polymeric components and serve as the warrior of sustainable development by acting as a biological control agent, myco-fumigating agent, plant growth promotors, stress tolerators when applied in-vivo. They have been a point of scientific interest and harbor novel compounds for the past few decades and will keep surprising the scientific community for upcoming centuries. These ubiquitous microbial floras are detected throughout the plant kingdom-algae, pteridophyte, gymnosperm, angiosperm over a wide geographical region of the tropical rain forest, temperate forest, mangrove plants, etc. and shape the plant-plant interaction, plant-microbe signaling pathways, ecosystem functions and also reshapes our traditional concepts of environment and biotic interactions. The metagenomics approach and pyrosequencing studies

---

*Corresponding Author's Email: debu33@gmail.com; db@mail.vidyasagar.ac.in.

In: Endophytes
Editor: Pragya Tiwari
ISBN: 979-8-88697-045-6

are illuminating the non-culturable endophytic microorganisms in the phyllosphere, caulosphere, or rhizosphere tissues. Endophytes are the untapped biological sources that have all types of answers and exhibit a variety of functions, representing a wide range of bioactive chemical components. The present study contains the recent landmark concepts or proofs of endophyte biology that expand our endophytic horizon and establishes a new sector of plant-microbe biology; "endophyte biology". In near future, every concept of botany or plant ecology could be reassessed by the principles of endophyte biology opening up hilarious secrets of mother nature.

Keywords: multifaceted applications, illuminating prospects of endophyte biology, trending biotechnological aspects

1. Introduction

1.1. Endophytes in General

Endophytes are the hidden microorganisms that reside within the host body and help the plant to survive in odd situations. They remain as a silent microbial counterpart of the eukaryotic host and have co-evolved with the green plant cells, as a result, they carry the common type of necessary bioactive genes and produce a diverse range of bioactive compounds (Tiwari and Bae 2020). They may be involved in commensalism, mutualism, symbiosis, and trophobiotic relationship and mostly resides in the rhizosphere and phyllosphere portions of a plant-environment interface (Roy et al. 2020). They are proved to be both horizontally and vertically transmitted from one generation to another (Tiwari and Bae 2020). Endophytes are in discussion for the last 100 years since the term was coined by DeBary (1866) and has gained interest day by day by its extended application and detailed study of a wide range of host plants like cryptograms-phanerogams, angiosperms-gymnosperms, monocots-dicots, bryophytes-pteridophytes, also includes all types of habit and habitat; herbs, shrubs, and tree; terrestrial, aquatic, and xerophytic, etc. Endophytes are diverse in nature (Figure 1) e.g., bacteria, actinobacteria, archaebacteria, fungi, actinomycetes, etc. (Santra and Banerjee 2020a; 2020b). They are sometimes considered latent pathogens and may cause disease when the plant faces extremely weird conditions, but such reports are rare and the endophyte's supportive nature is well-established.

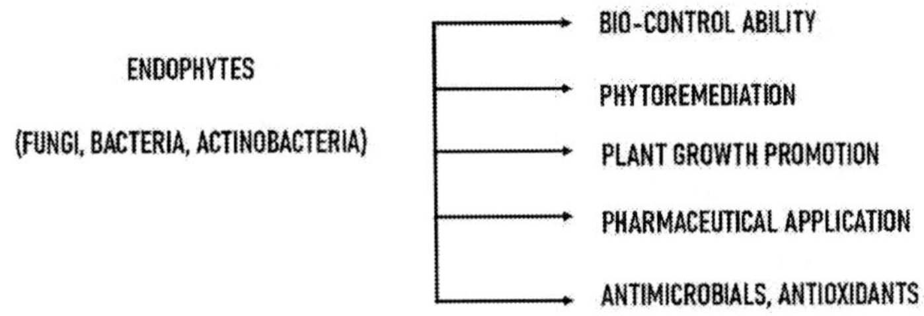

Figure 1. Different utilities of endophytic microorganisms.

1.2. Utilities of Endophyte and Endophyte Derived Products in the Modern World

These microorganisms produce a diverse range of bioactive chemicals like alcohols, alkaloids, antibiotics, butenolides, bicyclic lactones, benzopyran derivatives, butyrolactones, cytochalasin-related compounds, cyclic pentapeptides, depsipeptide, depsidoenes, diketopiperazine, ergosterol, isofurano-naphthalenone, flavonoids, long-chain hydrocarbon compounds, methoxyphenols, maleimide-bearing compounds, organic acids, polyketides, peptides, pestalols, phenol, spiro-bisnaphthalenes, sesquiterpenoids, steroids, terpenes, tannins, xanthones, etc. (Tiwari et al. 2021a; Tiwari et al. 2021b) (Figure 2) with multiple bioactivities like antibacterial (antibiotic), antifungal, antioxidative, anti-proliferative, anti-tumor, anti-cancer, antimycotic, etc. (Tiwari et al. 2022). Researchers from all over the world are gaining interest day by day on this topic and are opening new truths about plant-microbe symbiosis. In the recent past, extracellular polymeric substances obtained from fungal or bacterial endophytes and actinobacterial sources have gained interest due to their *in vivo* application in mice and rat models which have shown impressive antioxidative potency as well as the betterment of stress conditions or increase in gut microbial flora. Due to their multifunctional approaches, they have gained the interest of agriculturalists, biotechnologists, microbiologists, plant scientists, pharmaceutical sectors, industrial areas, etc. There is a detailed listing of fungal and bacterial endophytes, and actinobacteria (and their derived products) with their respective host and contributors that represent the multifunctionality of endophyte microorganisms and endophyte biology as a whole in general and specific sectors that can re-shape the modern agriculture, industrial sectors, pharmaceutical programs, build eco-friendly approaches towards plant disease management and control of xenobiotics and pollutions (Table 1).

Table 1. List of endophytic microorganisms with their respective bioactivity

S. No.	Endophyte	Host plant	Bioactivity	Explored by
1.	*Taxomyces andreanae*	*Taxus brevifolia*	Produces Taxol that has anti-cancer activity	Andrea et al. 1993
2.	*Streptomyces* NRRL 30562 *Streptomyces* NRRL 30566	Kennedia nigriscans, and *Grevillea pteridifolia*	Two antibiotics Munumbicins and kakadumycins are found	Castillo et al. 2002, 2003
3.	*Pestalotiopsis* sp., *Phoma* sp., *Xylaria* sp., *Phomopsis* sp.	*Cinnamomum iners*, *Camellia sinensis*, *Manglietia garrettii*	Produces different types of extracellular enzymes Cellulase, Mannase, Protease, Xylanase	Moy et al. 2002
4.	*Burkholderia cepacia* G4	*Lupinus luteus*	Phytoremediation of toluene	Barac et al. 2004
5.	*Methylobacterium populi* BJ001	*Populus deltoidesnigra* DN34	Bio-degradation of xenobiotic components like TNT (trinitrotoluene), RDX (Research department explosive), HMX (high melting explosive), methane	Van Aken et al. 2004
6.	*Streptomyces* sp.	*Monstera* sp.	Antimalarial and antifungal activity of Coronamycin is reported	Ezra et al. 2004
7.	*B. cepacia* Bu61	*Triticum aestivum*	Bioremediation of toluene	Taghavi et al. 2005
8.	*Pseudomonas putida* VM1450	*Populus* sp., and *Salix* sp.	Removal of 2,4-D	Germaine et al. 200
9..	*Bjerkandera* sp.	*Drimys winteri*	Cellulase, phenoloxidase	Oses et al. 2006
10.	*Epichloe festucae*, *Neotyphodium lolii*, *Monotospora* sp.	*Poa ampla*, *Cynodon dactylon*	β-1,6-Glucanase, laccase	Wang et al. 2006
11.	*Serratia proteamaculans*	*Glycine max*	Siderophore production, IAA activity, synthesis of acetoin and 2,3-butanediol synthesis, also fungi-inhibitory activity	Taghavi et al. 2009
12..	*Acinetobacter johnsonii* strain 3-1	*Beta vulgaris*	Plant growth promotion in terms of IAA production and phosphate solubilization	Yingwu et al. 2011
13.	*Azospirillum lipoferum* 4B	*Oryza sativa*, *Zea mays*, *Triticum* sp.	Nitrogen fixation and phytohormone secretion	Wisneiwski-Dye et al. 2011

S. No.	Endophyte	Host plant	Bioactivity	Explored by
14.	*Fusarium solani* SD5	-	Anti-inflammatory (erythrocyte membrane protection from hemolysis), anti-allergic (inhibits 48/80 induced mast cells degranulation), prebiotic activity (selectively stimulates *Lactobacillus* sp. growth and inhibits *Clostridium perfringens*), and antioxidative activity (scavenge DPPH radicals, reduces MDA, oxidized glutathione contents, increases SOD, GPx, and GSH levels in macrophage cells)	Mahapatra and Banerjee 2012; 2013b
15.	*Aspergillus niger* DR02, *Trichoderma atroviride* DR17 and DR19, *Alternaria* sp. DR45, *Annulohypoxylon stigyum* DR47, and *Talaromyces wortmannii* DR4	-	Produces enzymes xylanase, pectinase, β-glucosidase, CMCase, and FPase activities	Robl et al. 2013
16.	*Enterobacter* sp. strain FD17	*Zea mays*	Plant growth promotion by facilitating IAA production, solubilization of phosphate, and ACC deaminase activity	Naveed et al. 2014
17.	*Bacillus* sp., *Paenibacillus* sp., *Klebsiella* sp., *Acinetobacter* sp.	*Triticum* sp.	Plant growth promotion by IAA, P solubilization, and also bio-control of *Gaeumannomyces graminis* pathogen	Duran et al. 2014
18.	*Bacillus subtilis* var. *amyloliquifaciens*	*Oryza sativa*	Check the growth of *Rhizoctonia solani*	Nagendran et al. 2014
19.	*Fusarium* sp., *Pestalotiopsis* sp.	*Acanthus ilicifolius*, *Acrostichum aureum*,	Produces extracellular enzymes like amylase, cellulase, laccase, lipase, protease, pectinase, xylanase, phytase	Corrêa et al. 2014
20.	*Acremonium* sp., *Alternaria* sp., *Aspergillus* sp., *Dreshclera* sp., *Myrothecium* sp., *Paecilomyces* sp.	*Hedychium flavescens*, *Hedychium coronarium*	Produces extracellular enzymes amylase, cellulase, pectinase, asperginase	Uzma et al. 2016

Table 1. (Continued)

S. No.	Endophyte	Host plant	Bioactivity	Explored by
21.	*Bacillus cereus*, *Bacillus subtilis*, *Penicillium chrysogenum*, *P. crustosum*	*Teucrium polium*	IAA activity, P solubilization, ammonia production, enzyme productions (Amylase, Pectinase, Carboxy Methyl Cellulase, Cellulase, Xylanase, Gelatinase), and antimicrobial activities against *Pseudomonas aeruginosa*, *Salmonella typhi*, *Escherichia coli*, *Staphylococcus aureus*, *Bacillus subtilis*, *Candida albicans* (yeast pathogen)	Hassan 2017
22.	*Cymbidium aloifolium*	*Bipolaris* spp., *Taloromyces rotundus*, *Penicillium purpurogenum*, *Cylindrocephalum* spp.	Produces extracellular enzymes phosphatase, cellulase, amylase, protease, pectinase, lipase, and laccase.	Shubha and Srinivas 2017
23.	*Rhynchotoechum ellipticum*	*Brevibacterium* sp., *Pseudonocardia* sp., *Leifsonia* sp., *Rhodococcus* sp., *Tsukamurella* sp., *Promicromonospora* sp., and *Saccharopolyspora* sp.	Produces six antibiotics (erythromycin, ketoconazole, fluconazole, chloramphenicol, rifampicin, miconazole) and also nine phenolics compounds (catechin, kaempferol, chebulagic, chlorogenic acid, Asiatic acid, ferulic acid, arjunic acid, gallic acid, boswellic acid), an anticancerous compound like paclitaxel, plant growth-promoting activities, produces extracellular enzymes (cellulase and amylase)	Passari et al. 2017
24.	*Bacillus* sp. (14 DS-1)	*Codonopsis pilosula*	Induces macroph age activation and infiltration, TNF-α, and nitric oxide production. Suppresses cancer cell migration, arrests S phase hampers spindle orientation and positioning	Chen et al. 2018

S. No.	Endophyte	Host plant	Bioactivity	Explored by
25.	*Alternaria* sp., *Fusarium* sp.	*Costus spicatus*, *Lepisanthes tetraphylla*, *Sauropus androgynous*, *Madhuea longifolia*	Antimicrobial activity against pathogenic microorganisms *Escherichia coli*, *Salmonella typhi*, *Proteus mirabilis*, *Staphylococcus aureus*, *Corynebacterium diphtheriae*, antioxidant and anticancer activity	Palanichamy et al. 2018
26.	*Alternaria tenuissima* F1	*Angelica sinensis*	Possess *in vitro* antioxidative property confirmed hydroxy (OH) radical scavenging activity, SO (superoxide anion) scavenging effect along with reducing power ability	Wang et al. 2019
27.	*Alternaria* sp., *Didymella* sp., *Fusarium* sp., *Xylogone* sp.	*Sophora flavescens*	Indole acetic acid production, betterment of physical characters like root length, shoot length, increase in chlorophyll a and b, carotenoid contents	Turbat et al. 2020
28..	*Streptomyces polyantibioticus* T4SB028	*Thymus roseus*	*Alternaria solani*, *Valsa malicola*, and *Valsa mali*	Musa et al. 2020
29.	*Agrobacterium fabrum*, *Acinetobacter radioresistant*, *Brevibacillus brevis*, *Bacillus cereus*, *B. subtilis*, *Paenibacillus barengoltzii*, *Burkholderia cepaci*	*Pulicaria incisa*	IAA production, enzymatic activity (amylase, protease, xylanase, cellulase, chitinase, catalase etc.), anti-pathogenic against *Fusarium oxysporum*, *Alternaria alternata*, *Pythium ultimum* and enhancement of physical parameters like shoot, root length, fresh weight enhancement	Fouda et al. 2021
30.	*Selaginella stellata*, *Pilea microphylla*	*Aloe vera*, *Sida cordifolia*, *Sauropusandrogynus*, *Ocimum tenuiflorum*, *Piper nigrum*, *Selaginella stellata*, *Pilea microphylla*	Anti-microbial activity against *Staphylococcus aureus*, *Escherichia coli*, and *Mycobacterium smegmatis*, *Bacillus cereus*, *Escherichia coli*, *Klebsiella pneumoniae*, *Pseudomonas aeruginosa*, *Candida albicans*	Ashok et al. 2021

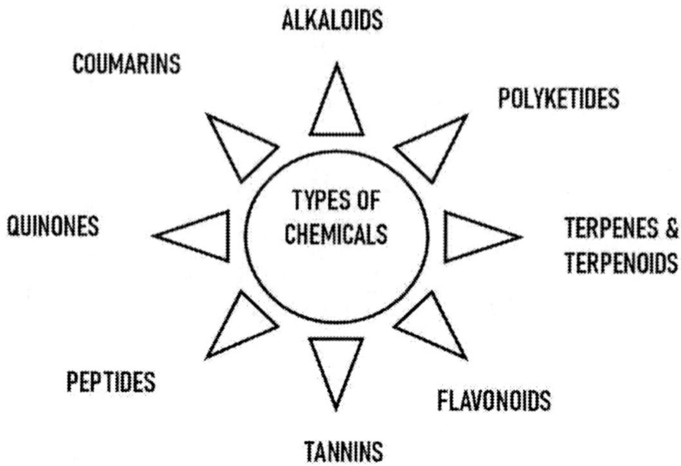

Figure 2. Diverse chemicals produced by Endophytic microorganisms.

The identification of endophytes is also a tough job because all endophyte biology research depends on the appropriate study area or collection area selection, accurate selection of disease-free plants, proper sterilization techniques of explants, and minute observations during the time of fungal endophytes hyphal emergence from dead decaying green tissues. The problem with the identification of fungal endophytes in a comparison with normal microscopic or macroscopic edible or pathogenic fungi is the sometimes-rare occurrence of fruiting bodies, reproductive structures, characteristic pigments, etc. In such situations, plate morphology or microscopic morphology cannot help, and techniques like carnation leaf agar are applied where media is supplemented with leaf extracts and sometimes media is supplemented with a mild amount of antifungal components or a lower percentage of growth media in order to induce reproductive structure development but still if there is no reproductive structure, then we have to depend on ITS sequences, tubulin genes or other conserved sequences for identification purposes and the phylogenetic tree is developed for proper identification of the genus. But the non-culturable nature of some genera has led to the popularization of metagenomics approaches, and other advanced molecular techniques like other than ITS sequencing i.e., pyrosequencing, DNA barcoding, FAME (fatty acid methyl ester analysis) analysis, MALDI-TOF (matrix-associated laser desorption/ionization-time of flight). A better understanding of endophyte microflora is performed by analyzing the overall microorganisms present on that particular plant tissue as they are extremely tissue-specific, age-dependent, genotype influenced, etc. A well-known endophyte biologist Gary Strobel did some remarkable findings with his all-square endophyte *Muscodor albus*, a member of Xylariaceae, which produces volatile antimicrobials (VOCs-volatile organic compounds) with antifungal potency and inhibits the growth of huge numbers of plant pathogenic fungus that cause huge crop loss each year. He earned some remarkable patents also by studying the diverse bioactivities of *M. albus*. Later different strains of *Muscodor* were isolated and studied by a variety of researchers and scientists from all over the globe. Endophytic components and culture-derived exopolysaccharides are reported to reduce salt stress and alleviate drought stress in rice seedlings. Endophytes are sometimes considered as "biological weapons" to combat parasitic infection, burrowing nematodes, fungal contaminations, harmful insects, leaf-cutting ants, and other biotic perils in case of economically valuable cash crops. Actually, extremophiles with their extremozymes are the best candidate for phytoremediation or biological agent-based cleaning of the environment but nowadays endophytes are also becoming the most prominent agent for this purpose (Santra and Banerjee 2021a). Endophytes are also prominent agents of phytoremediation that can extract the toxic components, heavy metals, and other pollutants from soil, air, and water and break them into simple or less toxic components. One such example includes the host plant *Festuca arundinacea* and *F. pratensis* which when infected with endophytes (*Neotyphodium* sp.) show higher root and shoot biomass than the un-inoculated one in the case of cadmium phytoremediation (Soleimani et al. 2010a). Endophytes act as the phytoremediation agent in two ways either directly by enhancing plant growth vigor, vitality, and ability to withstand stress or indirectly by degrading toxins or accelerating the process of degradation, decomposition, or accumulation. Endophytic *Penicillium funiculosum* promotes the bioremediation of copper in agricultural landfills (Khan and Lee 2013).

These microorganisms are vital components of the global ecosystem and are predominantly found in tropical rainforest areas like Brazilian forests, Amazonian rain forests, Arunachal Pradesh, and allied areas (In India) due to the extra humid-warm weather of these portions that support high growth of internal microorganisms. Though a large fungal genus represents endophytic microorganisms but is actually two particular groups of endophytes that dominate the whole endophyte biology discussion are- class-I-clavicipitaceous and class-II-non-clavicipitaceous. This classification has been proposed based on serious factors like plant and fungal evolutionary lineage, plant-fungal taxonomic relationship, plant-endophyte ecology, and also hosts' nature and response methods towards the infection. Other than these the factors like range of hosts for proper infection, transmission or inoculation of endophytes from one host to another, tissue specificities, and level of symbiosis are also taken into consideration (Figure 3) and further two more effective classes are obtained like class III and class IV (Rodriguez et al. 2009). This chapter focuses on the different types of diverse applications of endophytic microorganisms in popular angles that can uplift the modern pharmaceutical and agricultural system and also help ameliorate, protect and prevent disease occurrences in case of both plants and animals.

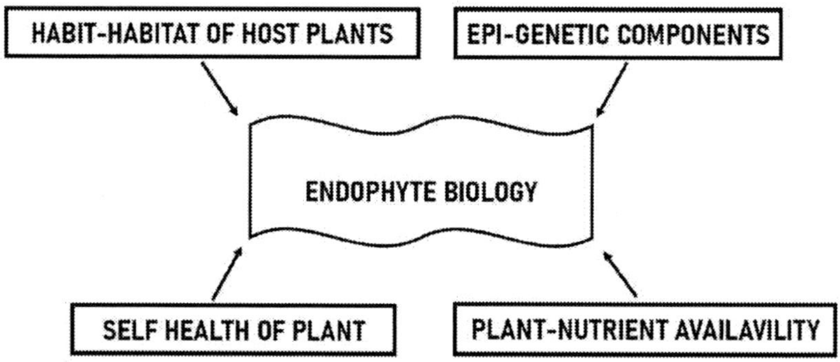

Figure 3. Factors affecting the endophyte colonization in host plants.

2. Endophytic Microorganisms as the Weapon of "Green Immunization" or "Plant Vaccination"

Fungal endophytes can check the growth of plant pathogenic fungi and protect the host plant from a variety of diseases. Actually, endophytes are sometimes considered localized entities but most of the time they are considered systemic microorganisms. Therefore, they share some common genes with the eukaryotic host plants and can produce bioactive components that support the plant in fighting against dreadful pathogens (Tiwari et al. 2022). Endophyte biologists are nowadays trying to develop techniques that can solve pathogen-related crop loss in multifaceted ways (Figure 4). Their target is to incorporate endophytic microorganisms into the host plant and as a result, the plants will develop resistance and all the defense-related genes (salicylic acid, jasmonic acid, and ethylene) will be up-regulated and the plant will develop hard immunity.

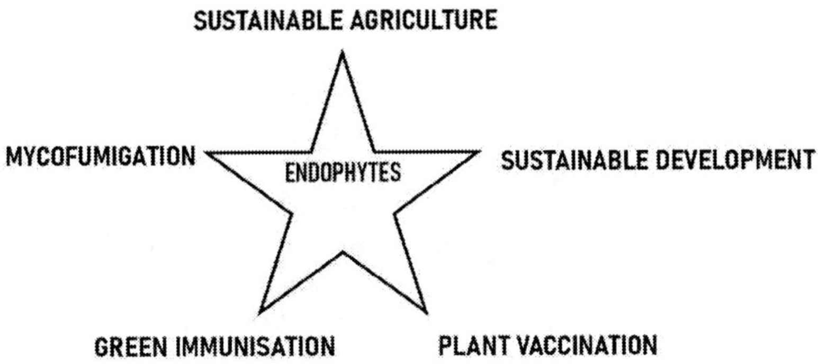

Figure 4. Role of endophyte in agricultural management.

This can be explained in a nutshell as "GREEN IMMUNIZATION" or "PLANT VACCINATION". Therefore, in one way the infection-related problem will be solved, the economic loss will be checked, and also the use of unwanted xenobiotics like chemical pesticides, insecticides, and fungicides will be minimized. This will not only reduce chemical toxicity but also will open up a new horizon for sustainable agricultural practices. Also, those chemicals or xenobiotics induced disease occurrence, cancer or tumor development, oxidative stress generation and aging-related problems, insulin resistance, low immunity, and thousands of unexplained or medically explained symptoms or syndrome will be cured or to some extent can be avoided. Here some examples are included that emphasize the fact that endophytes are not only effective in *in-vitro* conditions with proper fermentation manipulation or culture broth extraction but also can directly confront pathogenic microorganisms and protect the host plant from crop loss or growth retardation. The endophyte biologists and associated agriculturalists, botanists, and microbiologists are working in this field for the past 25-30 years and still it is a popular trend in this arena of research and scientists are showing their positive bends in this prosperous aspect. One such reports include that the endophyte of banana can minimize the mycelial growth of *Fusarium oxysporum F. sp. Cubense* is known as the causal agent of Panama-wilt disease and also restricts the growth of burrowing nematode *Radopholus similis* (Pocasangre 2000). Therefore, the endophytes open up approaches of biocontrol or sustainable development; sustainable agricultural practices. Another example includes the case of the "tall fescue" plant as the removal of fungal endophyte (*Acremonium coenophialum* Morgan-Jones and Grams) from the host tissues of tall fescue causes the higher susceptibility of the nematode pest *Helicotylenchus dihystera*, and *Paratrichodorous minor*. So, the harmful chemicals used for the control of these types of pests can be replaced by this type of endophyte microorganisms. Stewart (1993) and his scientific team did a remarkable experiment on a serious pest *Meloidogyne naasi* (causal agent of gall formation on perennial ryegrass) where the endophyte *Acremonium* spp. the infected plant shows the reduced occurrence of the nematode infection e.g., reduced number of galls on root portions, decreased number of female individuals, and also the absence of juvenile ones. It could be explained that the infection of the endophyte caused the activation of a defense-related gene that provides protection against the nematode infection. Amatuzzia et al. (2018) worked on the European pepper moth and revealed that the strawberry plant-derived endophytes are diverse in the genus (*Alternaria* sp., *Aspergillus* sp., *Paecilomyces* sp., *Trichoderma* sp., etc.) and are effective against the pathogenic moth larvae *Duponchelia fovealis* (third instar larvae). The qualitative and quantitative occurrence of the diverse endophytic genera also influences the disease occurrence in host plants; this has been proved by Wang et al. (2019) while working on *Pinus* species (nine different species of *Pinus*) that the infection chance of Sirex wood-wasp (*Sirex noctilio*) is negatively related to the number of endophytes present on the host tissue. The dominant fungal endophyte genera were *Ascocoryne* sp., *Aspergillus* sp., *Capronia* sp., *Chaetomium* sp., *Coprinopsis,* sp., and *Pinus tabuliformis* was found to be the lowest susceptible to wood-wasp as it harbors the highest endophyte diversity. Cotton leafworm (*Spodoptera littoralis*) is a serious pest to cotton that can harm the cotton cultivation; economic output and that can be controlled by fungal endophyte members of *Pelargonium graveolens*, *Melia azedarach*, *Chenopodium album*, *Malva parviflora*. Saad et al. (2019) worked on the endophytes of the above said host plants that inhibited the third instar larvae *Spodoptera littoralis* which can provide a long-term solution to the nematode infection. Therefore, endophytes are acting as the biological weapons for these infectious agents. Another dreadful nematode pest is *Radopilus similis* (root burrowing nematode) which causes serious situations in a banana plantation and causes fatal devastation to the overall economic output from this field. Niere et al. (1999) isolated *Fusarium oxysporum* from the banana host plant and *in vivo* application on the banana plant cause the reduction in disease occurrence i.e., reduced number of root burrowing nematodes confirmed through pot experiments. *Aphis gossypii* is a serious aphid of agricultural lands that can drastically deteriorate the cash crop populations in a remarkable situation. The use of chemical components damages the ecosystem equilibrium. Gonzalez-Mas et al. (2019) studied the anti-aphid activity of *Beauveria bassiana in vivo* conditions in the case of the melon plant (*Cucumis melo*var cantalupensis). Citrus black spot pathogen (*Phyllosticta citricarpa*) is a dangerous causal agent of citrus yield deterioration that has been controlled to some extent by endophyte of citrus i.e., *Phyllosticta capitalensis* and *P. paracapitalensis* (Tran et al. 2019). Bio-control ability of *Theobroma cacao*-derived endophytes (*Lasiodiplodia theobromae, Colletotrichum gloeosporioides, Phanerochaete chrysosporium, Pestalotiopsis microspora*) towards *Moniliophthora* spp. (Arnold et al. 2003) also support the same conclusion that endophytes can positive response as a weapon for green immunization. Biocontrol ability of endophytes of *Acer platanoides* (Norway Maple) towards *Rhizoctonia solani* (potato tuber pathogen) is a new feather to the success story of endophyte as this pathogen cause a huge loss each and every season; every year and the isolates-*Trichoderma atroviride*, *Epicoccum nigrum,* etc. are able enough to restrict its growth (Lahlali and Hijri 2010). Fungal endophyte isolates of Srilankan rice varieties *Cylindrocladium* sp., *Absidia* sp., *Cladosporium* sp., and *Penicillium* sp. were found to be antagonistic towards rice blast pathogen *Magnaporthe grisea*, thus providing a green solution towards this pathogen related loss of rice (Atugala and Deshappriya 2015). Gibberella ear rot is a worldwide severe disease that leads to wheat destruction and breakage in agricultural earnings. Abdallah et al. (2018) studied endophytic *Epicoccum nigrum* and *Sordaria fimicola*which can significantly reduce mycotoxins (DON-deoxynivalenol and 15-acetyldeoxynivalenol-15-ADON) associated with the causal organism- *Fusarium graminearum*, thus provides an eco-friendly approach in disease management. Another wheat pathogen is a causal agent of tan spot disease *Pyrenophora tritici-repentis* can be minimized by the endophytic *Trichoderma hamatum* confirmed through *in vivo* experiments (reduction in spore germination percentage) (Larran et al. 2016). Another species of *Trichoderma-T. polysporum*is found to be effective against ginseng pathogens (*Alternaria panax*, *Botrytis cinerea,* etc.) and offers an effective alternative way to control ginseng crop loss due to harmful pathogens (Park et al. 2015). Pepper being a globally utilized vegetable crop is largely affected by root rot pathogens like *Phytophthora capsici*, *P. nicotianae* var *parasitica*, *Fusarium solani*, *F. oxysporum,* etc., and these pathogens could be to some extent controlled by the endophytes of *Cornus florida* e.g., *Dothideomycetes* sp., *Hypoxylon howeanum*, *Hypoxylon* sp., *Nigrospora sphaerica,* etc. (Mmbaga et al. 2018). Therefore, these microorganisms could further broad-spectrum utilized for disease prevention. In near future with the accelerating development in plant-endophyte research it is quite hopefully expected that endophytes' indirect or endophyte-derived products will solve problems of plant disease and in 2021 purposing of endophyte biology in solving agricultural sector-related problems is the most common and trending call.

3. *Muscodor albus*: The All-Rounder Fungal Endophyte

Volatile organic compounds are basically a unique mixture of chemical components that are volatile at normal room temperature and can perform various biological functions like antibacterial, antifungal, etc. Numerous endophytes are there that can produce bioactive volatiles e.g., *Gliocladium* sp. *Nodulisporium* sp., *Oidium* sp., *Botrytis* sp., *Cladosporium* sp., *Oxyporus latemarginatus*, *Myrothecium inundatum*, *Penicillum* sp., *Candida intermedia*, *Phomopsis* sp., *Phoma* sp., *Xylaria* sp., *Fusarium oxysporum,* etc. but *Muscodor* is an all-rounder in this field and is highly praised among endophyte biologists. *Muscodor albus* is such an endophyte that has a unique mixture of VOCs and perform various functions. Several species of *Muscodor* like *M. strobelii*, *M. darjeelingensis*, *M. tigerii*, *M. suthepensis*, *M. heveae*, *M. ghoomensis*, *M. indica*, *M. camphora*, *M. roseus*, *M. vitigenus*, *M. fengyangensis*, *M. sutura*, *M. cinnamomi*, *M. equiseti*, *M. musae*, *M. kashayum* have been isolated from a variety of angiosperm and gymnosperm host plants like *Guazuma ulmifolia*, *Myristica fragrans*, *Terminalia prostrata*, *Cinnamomum zeylanicum*, *Ginkgo biloba*, *Eucryphia cordifolia*, *Callicarpa acuminata*, *Ananas ananassoides*, *Cinnamomum bejolghata*, *Pseudotaxus chienii*, *Actinidia chinensis*, *Abiesbeshanzuensis*, *Musa acuminata*, *Prestonia trifidi*, *Equisetum debile*, *Hevea brasiliensis*, *Aegle marmelos*, *Cinnamomum zeylanicum*, *C. camphora,* etc. (Suwannarach et al. 2010, 2012, 2013, 2015; Meshram et al. 2013, 2014, 2015 and 2017). The isolates produce a unique mixture of volatile compounds like Thujopsene, chamigrene, isocaryophyllene, butanoic acid, 2-methyl, 1-butanol, 3-methyl-acetate, naphthalene, tetrohydrofuran, 2-methyl furan, 2-butanone, aciphyllene, azulene, germacrene B, acetic acid, methyl ester, 1-Butanol, benzeneethanol, acetate, Pyrrolidine, 2-Heptanoic acid, 4-cyclopropyl-, Bicyclo[3.1.1] heptane, 6-methyl-2- propanoic acid, 2-butenoic acid, ethyl ester, 1,2,4-trimethylbenzene, 2,3-nonadiene, Naphthalene, azulene, 3-methylbutan-1-ol, 3-methylbutyl acetateetc. that are antifungal against a wide range of fungal pathogens like *Aphanomyces cochliodiodes*, *Aspergillus fumigatus*, *Fusarium culmorum*, and *Glomerella cingulate*, *Botrytis cinerea*, *Aspergillus clavatus*, *Colletotrichum fragariae*, *Didymella bryoniae*, *Magnaporthe oryzae*, *Pythium ultimum*, *Rhizoctonia solani*, Sclerotium *rolfsii*, *Verticillium dahlia*, *Penicillium digitatum*, *Botrytis cinerea*, *Colletotrichum lagenarium*, *Ceratocystis ulmi*, *Cercospora beticola*, *Geotrichum candidum*, *Mycosphaerella fijiensis*, *Phytophthora cinnamomi*, *Phytophthora palmivora*, *Pythium ultimum*, *Rhizoctonia solani*, *Sclerotinia sclerotiorum*, *Verticillium dahli*, etc. (Stinson et al. 2003; Banerjee et al. 2010b; Saxena et al. 2014, 2015; Siri-udom et al. 2016). *Muscodor* has been widely used as a post-harvest decay management organism, a suitable component for transportation of fruits and vegetables as it can slow down the decay rate of such organisms, and provide protection from decaying pathogenic microorganisms. The volatiles are effective in combating parasitic infections so, it will be a great opportunity to use these fungi as a biological control agent. Myco-fumigation is known to be the best tool to minimize the infections and *Muscodor*-derived VOCs are efficient in treating codling moth (*Cydia pomonella*) infection (storage pathogen), verticillium wilt pathogen in case of egg-plant-brinjal. These types of *in vitro* and *in vivo* efficiency of these potent fungi open up a new horizon for sustainable agricultural practices.

4. Sustainable Agricultural Approaches

Our whole world depends on agricultural outputs but fungal, bacterial, and nematode infections cause a massive loss in this sector, and in order to cope with this loss, pesticides, insecticides, and chemical components are used. But these chemical xenobiotics are very much harmful to nature and natural components, toxifies the food chain-food web, food pyramids, etc. so, to overcome this unpleasant situation researchers are trying to find out new and novel solutions. Biocontrol agents are necessary for this respect that can combat these problems and endophytic microorganisms are the best option in this respect. Endophytic microorganisms from different plant sources act on the same or different plants and can enhance disease resistance, promote stress tolerance, enhance tissue longevity and also provide the long-term "deep ecological solution' to these problems. Thus, in one way the environment will be protected from xenobiotics toxicity and endophytes will solve the infection-related crop loss issue. In the 21st century, the crying need is sustainable development that will stabilize the environmental parameters and will support the growth and development of mankind. Not only disease prevention but also growth promotion in terms of phosphate solubilization or nitrogen fixation. The plant-endophyte nexus is seen as a pillar of sustainable agricultural practices and can promote plant health and survival in a variety of safe ways (Figure 5). Sustainability has been a trending topic nowadays and scientists are trying their best in all possible ways to cope with the challenging problems and to restore environmental balance or nature's equilibrium. In near future, it could easily be expected that endophyte-mediated plant disease resistance will be the first and foremost prime option for agricultural experts and chemical components will be the least used in agricultural lands, and Mother Earth's sustainability will be maintained. Not only as biocontrol agents but also endophytes are potent ones as plant growth regulators. Reports focus on the fact that endophytes promote the physical growth of plants (like leaf chlorophyll content, dry weight or wet weight of tissue samples, flowering time, fruiting time, early maturation of plantlets, first seed germination, plant height, etc.), betterment of chemical constituents (soluble sugar content, proline content, malonaldehyde contents, activation of defense-related genes, antioxidant enzymes-peroxidase, catalase, superoxide dismutase, etc.) of plant cells and tissues. They promote phosphate solubilization, indole acetic acid production, siderophore production, ammonia production, ACC deaminase activity, supply necessary vitamins from soil to plant, facilitate osmotic adjustment, regulate stomatal movement, modify root morphology, accelerate mineral absorption, alter nitrogen fixation and metabolisms, etc. (Compant et al. 2005a, 2005b). Endophytes provide the base to fight against biotic and abiotic factors and enhance the tolerance level of the plant. They induce disease resistance through ISR (induced systemic resistance) like *Bacillus cereus* (Gram-positive bacteria) is known to activate ISR against the pathogenic infection of *Botrytis cinerea* by accumulating pathogenesis-related proteins and activating MAPK signaling pathways, switching on WRKY53 gene expressions (as a result of jasmonic acid and ethylene signaling). Another method of *Botrytis cinerea* infection minimization is the utilization of trichodiene produced from *Trichoderma arundinaceum* (Mengistu 2020). Endophytes can shape the forest vegetations also and promote forest regeneration.

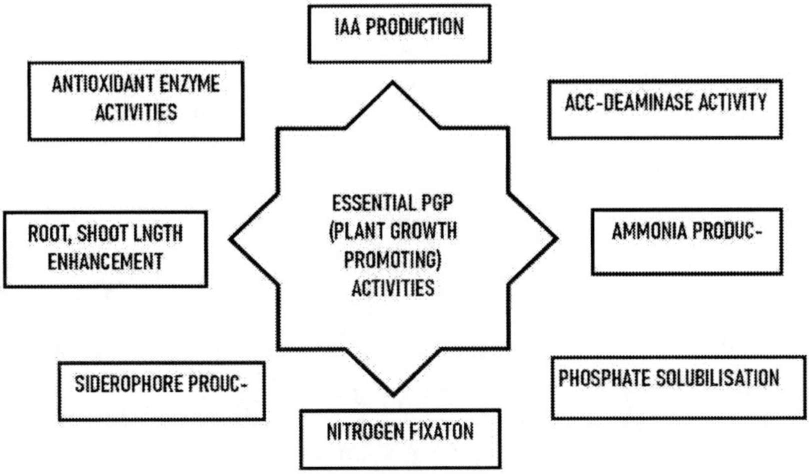

Figure 5. Multidisciplinary plant growth-promoting activities of endophytic microorganisms.

5. Pharmaceutical Applications

Endophyte-derived compounds are of immense pharmacological importance ranging from antibacterial, antioxidant, antitumor, antiproliferative, and antiallergic, to anti-inflammatory actions, antimalarial, antidiabetic, antiviral, neuroprotective, anti-leishmanial, and cytotoxic activity (Toghueo 2019; Falade et al. 2021). Figure 6. Represents the multifaceted pharmacological functions of endophytic microorganisms. Endophytes represent a deep pool of extracellular enzymes that can serve as a new tool for biotechnologists and pharmaceutical authorities. Figure 7. Represents the different types of extracellular enzymes reported so far from endophytic microbiota. As leishmaniases and inflammatory diseases are very tough to handle like anti-microbial resistance or the development of MDR (multi-drug resistant) pathogens endophyte-derived components can replace the popular common chemical components without any drawbacks, toxicity, or side effects and emphasizes the modern drug discovery from natural resources. Endophytes harbor unique chemicals with potent bioactivity. In Figure 8. Some bioactive chemicals of endophyte origin are reported. Especially exopolysaccharides from fungal or actinobacterial origin can play a major role in this aspect. Other than these industrially valuable enzymes are also derived from endophyte culture extract with digestive, hydrolyzing anticancer properties. Nowadays the trend has been shifted to endophytes for the discovery of life-saving molecules and soon or later scientists will surely declare life-saving components from biological sources like endophytes. Some recent publications related to the pharmaceutical application of endophytic microorganisms are included here.

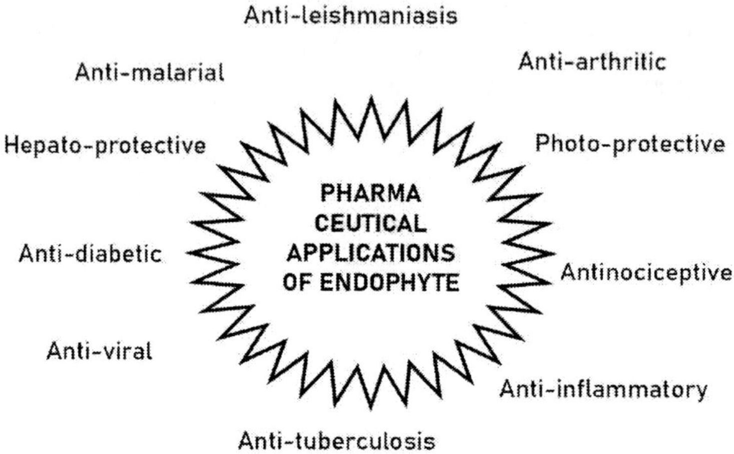

Figure 6. Multifaceted pharmaceutical applications of endophytic microorganisms.

Kuo et al. (2021) isolated fungal endophyte *Penicillium* sp., *Colletotrichum* sp., *Diaporthe* sp., *Daldinia* sp., *Alternaria* sp., *Didymella* sp. from *Zanthoxylum simulans* with antibiotics potential against Gram-positive and Gram-negative bacteria, yeast cells, anti-inflammatory activity. Ngashangva et al. (2021) isolated *Paenibacillus peoriae* IBSD35 from a medicinal plant *Millettia pachycarpa* and obtained antimicrobial peptide (N-terminal-methionine) from this isolate with antibacterial activity against *Staphylococcus aureus* ATCC 25923, *Escherichia coli* ATCC 25922, *Candida albicans* ATCC 10231. The structure of the bioactive peptide was confirmed by genomics and proteomics studies. The use of high-resolution structural or biotechnological tools in elucidating endophyte-derived components has been a new recent trend in modern biology.

Mbekou et al. (2021) isolated fungal endophyte *Aspergillus* sp. N454, *Aspergillus* sp. N13, *Curvularia* sp. N101, *Aspergillus* sp. N18 from *Terminalia mantaly*, *T.* catappa, *Canaga odorata* with antibacterial activity against *E. coli*, *Haemophilus influenzae*, antioxidative activity by DPPH, FRAP assay. The mode of action of endophyte culture extract has been revealed by the authors that bacterial lysis has occurred, membrane permeability increased, salt tolerance reduced, and bacterial catalase activity disturbed.

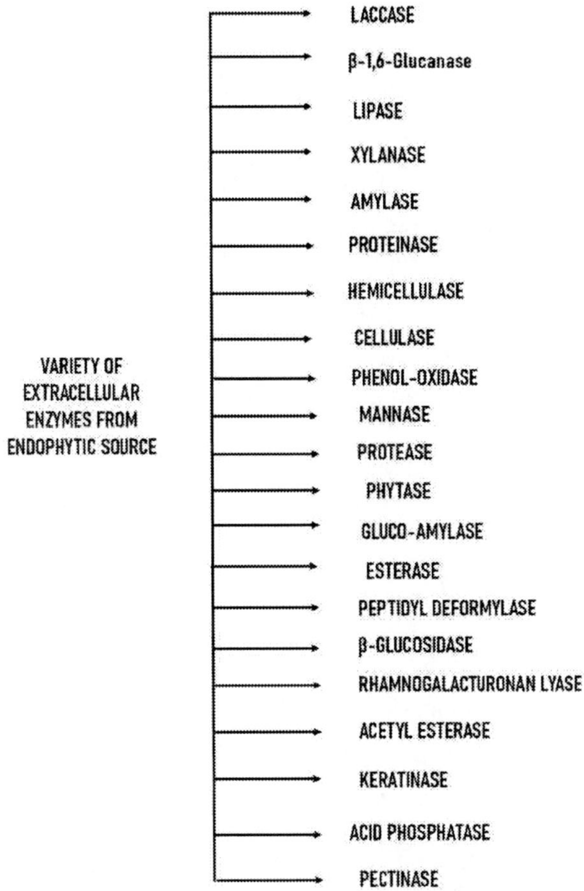

Figure 7. Various types of extracellular enzymes of endophyte origin.

Ababutain (2021) evaluated the antibacterial activity of fungal endophytes (*Aspergillus oryzae*, *A. niger*, *Penicillium chrysogenum*) against Gram-positive and Gram-negative bacteria. Dhakshinamoorthy et al. (2021) isolated camptothecin from fungal endophyte *Diaporthe caatingaensis* (isolated from *Buchanania axillaris*) with anticancer activity, antibacterial activity and opens up new positive aspects for the medicinal industry. Mazumder et al. (2021) isolated five fungal endophytes (*Aspergillus niger*, *A. fumigatus*, *Rhizoctonia solani*, *Rhizopus oryzae*, *Syncephalastrum racemosum*) from the ethnopharmacologically important plant *Elaeocarpus floribundus* Blume and the isolated derived bioactive components (tensyuic acid, hexylitaconic acid, chlorogenic acid, nigragillin, asnipyrone B, asperenone, fumaric acid, fusarubin) are reported to exhibit anti-inflammatory (*in vitro* cyclo-oxygenase and 5-lipoxygenase activity), antinociceptive and anti-arthritic activity.

Calado et al. (2021) isolated bioactive components (echinulin and neoechinulin A) from endophytic fungi (Eurotiomycetes, Sordariomycetes, Dothidiomycetes; *Penicillium* sp., *Exilicaulis* sp., *Aspergillus chevalieri*), from a coastal seaweed *Halopteris scoparia*, which represents anti-oxidative, extra-cellular enzymatic (hyaluronidase, collagenase, elastase, tyrosinase) activities, anti-inflammatory, photoprotective (protection from ultra-violet rays), anti-bacterial (against pathogenic microorganisms- *Cutibacterium acnes*, *Staphylococcus epidermidis*, *Malassezia furfur*). Many areas are there where endophytes have excelled in their applications. Some of such promising areas are incorporated herein in this book chapter.

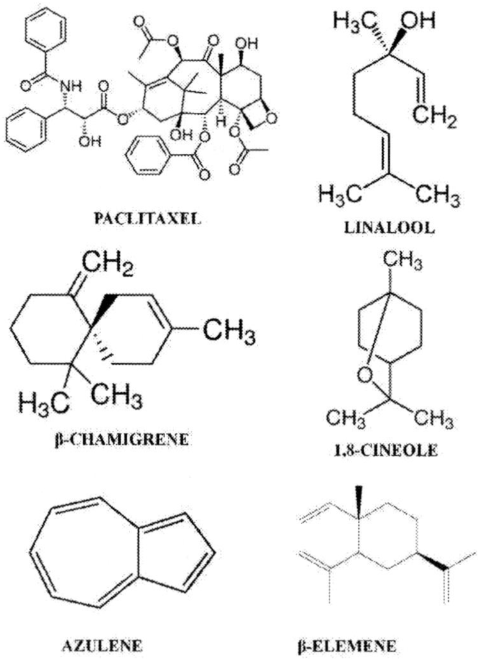

Figure 8. Different chemical components of endophyte origin.

Conclusion

Science is a never-ending topic full of secrets and prizes at its every step when deals with nature and its natural components. There is no doubt that endophyte is just a new domain on-to that list. It is really a trending topic in the present scenario. Eminent scientists from all over the globe are working on it. Reputed publishers like Springer Nature group, Elsevier, Taylor and Francis, Plos One. Frontiers in microbiology, CRC Press, Nova Publishers are investing their interest in inviting papers, organizing reviews, and editing books on different aspects of endophyte biology research just to put some clear light on this sector of science. Popular web search engines or online libraries like repositories like Web of Science, Scopus, Pubmed (NCBI), research gate, and Academia. Educators gather a lot of endophyte-related databases and promote upward interest in this topic. Even the recent threat to our mankind the COVID-19 disease could also be minimized or checked by the endophyte-derived components that type of research is also carried out in order to draw the interest of the scientific world. Also, they offer therapeutic management for the NTDs (neglected tropical diseases) and endophyte is reported to be a major part of attaining SDGs (Sustainable development goals) of the UN (United States). In order to achieve much accuracy and high output in this field, modern techniques like bioinformatics-based approaches are utilized; metabolomics, next-generation sequencing, metagenomic multi-locus sequencing, etc. along with modern genomics approaches like CRISPR-Cas9 (Clustered regulatory interspaced short palindromic repeats- Cas9), ZFN-Zinc finger nuclease, (TALEN) Transcription activator-like effector nuclease, etc. So, it could be assumed that in the near future endophyte and endophyte-derived molecules will hold the most potent aspects in the field of science and technology and act as a savior of mankind.

Acknowledgments

The authors extend their gratitude to their respective research organizations.

Conflict of Interests

No conflict of interest was declared.

References

Ababutain, Ibtisam Mohammed, Sahar Khamees Aldosary, Amal Abdulaziz Aljuraifani, Azzah Ibrahim Alghamdi, Amira Hassan Alabdalall, Eida Marshid Al-Khaldi, Sumayh A. Aldakeel, Noor B. Almandil, Sayed AbdulAzeez, and J. Francis Borgio. "Identification and Antibacterial Characterization of Endophytic Fungi from Artemisia sieberi." *International journal of microbiology* 2021 (2021).

Amatuzzia R. F., Cardosob N., Poltronieria A. S., Poitevina C. G., Dalzotoa P., Zawadeneaka M. A., Pimentela I. C. (2018) "Potential of endophytic fungi as biocontrol agents of *Duponchelia fovealis* (Zeller) (Lepidoptera:Crambidae)." *Braz J Biol* 78, no. 3 (2018):429–435.

Arnold A. E., Mejía L. C., Kyllo D., Rojas E. I., Maynard Z., Robbins N. "Fungal endophytes limit pathogen damage in a tropical tree." *Proc Natl Acad Sci* 100, no. 26 (2003):15649–15654.

Ashok, Alka, Usha Nandhini, Ayswarya Sreenivasan, Manigundan Kaari, Revathy Kalyanasundaram, and Radhakrishnan Manikkam. "AntiMycobacterial activity of endophytic actinobacteria from selected medicinal plants." *Biomedical and Biotechnology Research Journal (BBRJ)* 4, no. 3 (2020): 193.

Atugala D. M., Deshappriya N. "Effect of endophytic fungi on plant growth and blast disease incidence of two traditional rice varieties." *J Natn Sci* 43, no. 2(2015):173–187.

Banerjee D., Strobel G. A., Geary B., Sears J., Ezra D., Liarzi O., Coombs J. "*Muscodor albus* strain GBA, an endophytic fungus of *Ginkgo biloba* from United States of America, produces volatile antimicrobials." *Mycology* 1, no. 3(2010b):179–186.

Barac, Tanja, Safiyh Taghavi, Brigitte Borremans, Ann Provoost, Licy Oeyen, Jan V. Colpaert, Jaco Vangronsveld, and Daniel van der Lelie. "Engineered endophytic bacteria improve phytoremediation of water-soluble, volatile, organic pollutants." *Nature biotechnology* 22, no. 5 (2004): 583-588.

Calado, Maria da Luz, Joana Silva, Celso Alves, Patrícia Susano, Débora Santos, Joana Alves, Alice Martins, Helena Gaspar, Rui Pedrosa, and Maria Jorge Campos. "Marine endophytic fungi associated with *Halopteris scoparia* (Linnaeus) Sauvageau as producers of bioactive secondary metabolites with potential dermocosmetic application." *PloS one* 16, no. 5 (2021): e0250954.

Castillo, U. F., Strobel, G. A., Ford, E. J., Hess, W. M., Porter, H., Jensen, J. B., Albert, H., Robison, R., Condron, M. A. M., Teplow, D. B., Stevens, D., & Yaver, D. (2002)."Munumbicins, wide-spectrum antibiotics produced by Streptomyces NRRL 30562, endophytic on *Kennedia nigriscansaa*. The GenBank accession number for the sequence determined in this work is AY127079." *Microbiology* 148, no. 9 (2002): 2675-2685.

Castillo, Uvidelio, James K Harper, Gary A Strobel, Joseph Sears, Kara Alesi, Eugene Ford, Janine Lin, Michelle Hunter, Michelle Maranta, Haiyan Ge, Debbie Yaver, James B Jensen, Heidi Porter, Richard Robison, D Millar, Wilford M Hess, Margret Condron, David Teplow. "Kakadumycins, novel antibiotics from *Streptomyces sp.* NRRL 30566, an endophyte of *Grevillea pteridifolia*." *FEMS Microbiology Letters* 224, no. 2 (2003): 183-190.

Chen M., Li Y., Liu Z., Qu Y., Zhang H., Li D., Zhou J., Xie S., Liu M. "Exopolysaccharides from a *Codonopsis pilosula* endophyte activate macrophages and inhibit cancer cell proliferation and migration." *Thorac Cancer* 9, no. 5 (2018):630–639.

Compant S., Duffy B., Nowak J., Cl C. & Barka E. A. Use of plant growth-promoting bacteria for biocontrol of plant diseases: principles, mechanisms of action, and future prospects. *Appl Environ Microbiol* 71 (2005a): 4951–4959.

Compant S., Reiter B., Sessitsch A., Nowak J., Clement C. and Barka' E. A. Endophytic colonization of *Vitis vinifera* L. by a plant growth-promoting bacterium, *Burkholderia sp. strain PsJN. Appl Environ Microbiol* 71 (2005b): 1685–1693.

Compant, Stéphane, Birgit Reiter, Angela Sessitsch, Jerzy Nowak, Christophe Clément, and Essaïd Ait Barka. "Endophytic colonization of *Vitis vinifera* L. by plant growth-promoting bacterium Burkholderia sp. strain PsJN." *Applied and environmental microbiology* 71, no. 4 (2005): 1685-1693.

Compant, Stéphane, Brion Duffy, Jerzy Nowak, Christophe Clément, and Essaïd Ait Barka. "Use of plant growth-promoting bacteria for biocontrol of plant diseases: principles, mechanisms of action, and future prospects." *Applied and environmental microbiology* 71, no. 9 (2005a): 4951-4959.

Corrêa, Rúbia Carvalho Gomes, Sandro Augusto Rhoden, Thatiane Rodrigues Mota, João Lúcio Azevedo, João Alencar Pamphile, Cristina Giatti Marques de Souza, Maria de Lourdes Teixeira de Moraes Polizeli, Adelar Bracht, and Rosane Marina Peralta. "Endophytic fungi: expanding the arsenal of industrial enzyme producers." *Journal of Industrial Microbiology and Biotechnology* 41, no. 10 (2014): 1467-1478.

Dhakshinamoorthy, Madhankumar, Kannan Kilavan Packiam, Ponnusamy Senthil Kumar, and Tamilselvi Saravanakumar. "Endophytic fungus *Diaporthe caatingaensis* MT192326 from *Buchanania axillaris*: An indicator to produce biocontrol agents in plant protection." *Environmental Research* 197 (2021): 111147.

De Bary A. (1866) Morphologie und Physiologie Pilze, Flechten, und myxomyceten. Hofmeister's *Handbook of Physiological Botany*, vol. 2. Leipzig.

Durán, Paola, Jacquelinne J. Acuña, Milko A. Jorquera, Rosario Azcón, Cecilia Paredes, Zed Rengel, and María de la Luz Mora. "Endophytic bacteria from selenium-supplemented wheat plants could be useful for plant-growth promotion, biofortification and *Gaeumannomyces graminis* biocontrol in wheat production." *Biology and fertility of soils* 50, no. 6 (2014): 983-990.

Ezra, D., Castillo, U. F., Strobel, G. A., Hess, W. M., Porter, H., Jensen, J. B., Condron, M. A. M., Teplow, D. B., Sears, J., Maranta, M., Hunter, M., Weber, B., & Yaver, D. (2004) "Coronamycins, peptide antibiotics produced by a verticillate *Streptomyces* sp. (MSU-2110) endophytic on *Monstera* sp." *Microbiology* 150, no. 4 (2004): 785-793.

Fouda, Amr, Ahmed M. Eid, Albaraa Elsaied, Ehab F. El-Belely, Mohammed G. Barghoth, Ehab Azab, Adil A. Gobouri, and Saad El-Din Hassan. "Plant growth-promoting endophytic bacterial community inhabiting the leaves of *Pulicaria incisa (Lam.) DC* inherent to arid regions." *Plants* 10, no. 1 (2021): 76.

Germaine, Kieran J., Xuemei Liu, Guiomar Garcia Cabellos, Jill P. Hogan, David Ryan, and David N. Dowling. "Bacterial endophyte-enhanced phytoremediation of the organochlorine herbicide 2, 4-dichlorophenoxyacetic acid." *FEMS microbiology ecology* 57, no. 2 (2006): 302-310.

González-Mas, N., M. Cuenca-Medina, F. Gutiérrez-Sánchez, and E. Quesada-Moraga. "Bottom-up effects of endophytic *Beauveria bassiana* on multitrophic interactions between the cotton aphid, *Aphis gossypii*, and its natural enemies in melon." *Journal of Pest Science* 92, no. 3 (2019): 1271-1281.

Hassan, Saad El-Din. "Plant growth-promoting activities for bacterial and fungal endophytes isolated from medicinal plant of *Teucrium polium* L." *Journal of advanced research* 8, no. 6 (2017): 687-695.

Khan, Abdul Latif, and In-Jung Lee. "Endophytic *Penicillium funiculosum LHL06* secretes gibberellin that reprograms *Glycine max* L. growth during copper stress." *BMC plant biology* 13, no. 1 (2013): 1-14.

Kuo, J., Chang, CF. and Chi, WC. "Isolation of endophytic fungi with antimicrobial activity from medicinal plant *Zanthoxylum simulans* Hance." *Folia Microbiol* 66 (2021): 385–397.

Lahlali R, Hijri M. "Screening, identification and evaluation of potential biocontrol fungal endophytes against *Rhizoctonia solani AG3* on potato plants." *FEMS Microbiol Lett* 311 (2010):152–159.

Larran S, Simón MR, Moreno MV, Santamarina Siurana MP, Perelló A. "Endophytes from wheat as biocontrol agents against tan spot disease." *Biol Control* 92 (2016):17–23.

Mahapatra, Subhadip, and Debdulal Banerjee. "Fungal exopolysaccharide: production, composition and applications." *Microbiology insights* 6 (2013): MBI-S10957.

Mahapatra, Subhadip, and Debdulal Banerjee. "Structural elucidation and bioactivity of a novel exopolysaccharide from endophytic *Fusarium solani SD5*." *Carbohydrate polymers* 90, no. 1 (2012): 683-689.

Mazumder, Kishor, Yasmeen Nazim Ruma, Rasheda Akter, Asma Aktar, Mir Monir Hossain, Zinnat Shahina, Santosh Mazumdar, and Philip G. Kerr. "Identification of bioactive metabolites and evaluation of *in vitro* anti-inflammatory and in vivo antinociceptive and antiarthritic activities of endophyte fungi isolated from *Elaeocarpus floribundus blume*." *Journal of Ethnopharmacology* 273 (2021): 113975.

Mbekou, Michele Ines Kanko, Darline Dize, Victorine Lorette Yimgang, Fred Djague, Rufin Marie Kouipou Toghueo, Norbert Sewald, Bruno Njakou Lenta, and Fabrice Fekam Boyom. "Antibacterial and Mode of Action of extracts from endophytic fungi derived from *Terminalia mantaly*, *Terminalia catappa*, and *Cananga odorata*." *BioMed research international* 2021 (2021).

Mengistu, Anteneh Ademe. "Endophytes: Colonization, behaviour, and their role in defence mechanism." *International Journal of Microbiology* 2020 (2020).

Mengistu, Anteneh Ademe. "Endophytes: Colonization, behaviour, and their role in defense mechanism." *International Journal of Microbiology* 2020 (2020).

Meshram V., Gupta M., Saxena S. "*Muscodor ghoomensis* and *Muscodor indica*: new endophytic species based on morphological features, molecular and volatile organic analysis from Northeast India." *Sydowia* 67 (2015):133–146.

Meshram V., Kapoor N., Chopra G., Saxena S. "*Muscodor camphora*, a new record from *Cinnamomum camphora*." *Mycosphere* 8, no. 4 (2017):568–582.

Meshram V., Kapoor N., Saxena S. "*Muscodor kashayum sp. nov.* – a new volatile antimicrobial producing endophytic fungus." *Mycology* 4, no. 4 (2013):196–204.

Meshram V., Saxena S., Kapoor N. "*Muscodor strobelii*, a new endophytic species from South India." *Mycotaxon* 128 (2014):93–104.

Mmbaga M. T., Gurung S, Maheshwari A. "Screening of plant endophytes as biological control agents against root rot pathogens of pepper (*Capsicum annum* L.)." *J Plant Pathol Microbiol* 9 (2018):3.

Moy, Melinda, Huaijun Michael Li, Ray Sullivan, James F. White Jr, and Faith C. Belanger. "Endophytic fungal β-1, 6-glucanase expression in the infected host grass." *Plant Physiology* 130, no. 3 (2002): 1298-1308.

Musa, Zulpiya, Jinbiao Ma, Dilfuza Egamberdieva, Osama Abdalla Abdelshafy Mohamad, Gulsumay Abaydulla, Yonghong Liu, Wen-Jun Li, and Li Li. "Diversity and antimicrobial potential of cultivable endophytic actinobacteria associated with the medicinal plant thymus roseus." *Frontiers in microbiology* 11 (2020): 191.

Musa, Zulpiya, Jinbiao Ma, Dilfuza Egamberdieva, Osama Abdalla Abdelshafy Mohamad, Gulsumay Abaydulla, Yonghong Liu, Wen-Jun Li, and Li Li. "Diversity and antimicrobial potential of cultivable endophytic actinobacteria associated with the medicinal plant thymus roseus." *Frontiers in microbiology* 11 (2020): 191.

Nagendran, K., G. Karthikeyan, P. Mohammed Faisal, P. Kalaiselvi, M. Raveendran, K. Prabakar, and T. Raguchander. "Exploiting endophytic bacteria for the management of sheath blight disease in rice." *Biological agriculture & horticulture* 30, no. 1 (2014): 8-23.

Naveed M., Mitter B., Yousaf S., Pastar M., Afzal M., Sessitsch A. The endophyte *Enterobacter sp. FD17*: a maize growth enhancer selected based on rigorous testing of plant beneficial traits and colonization characteristics. *Biology and fertility of soils* 50, no. 2 (2014):249–262.

Naveed, Muhammad, Birgit Mitter, Sohail Yousaf, Milica Pastar, Muhammad Afzal, and Angela Sessitsch. "The endophyte *Enterobacter sp. FD17*: a maize growth enhancer selected based on rigorous testing of plant beneficial traits and colonization characteristics." *Biology and fertility of soils* 50, no. 2 (2014): 249-262.

Niere, B. I., P. R. Speijer, C. S. Gold, and R. A. Sikora. "Fungal endophytes from bananas for the biocontrol of *Radopholus similis*." *Proceedings of Mobilizing IPM for sustainable banana production in Africa* (1999).

Ngashangva, Ng, Pulok Mukherjee, K. Chandradev Sharma, M. C. Kalita, and Sarangthem Indira. "Analysis of antimicrobial peptide Metabolome of bacterial endophyte isolated from traditionally used medicinal plant *Millettia pachycarpa Benth*." *Frontiers in microbiology* 12 (2021): 1192.

Oses, Rómulo, Sofía Valenzuela, Juanita Freer, Jaime Baeza, and Jaime Rodríguez. "Evaluation of fungal endophytes for lignocellulolytic enzyme production and wood biodegradation." *International Biodeterioration & Biodegradation* 57, no. 2 (2006): 129-135.

Palanichamy, Prabha, Govindan Krishnamoorthy, Suganya Kannan, and Murugan Marudhamuthu. "Bioactive potential of secondary metabolites derived from medicinal plant endophytes." *Egyptian Journal of Basic and Applied Sciences* 5, no. 4 (2018): 303-312.
Palanichamy, Prabha, Govindan Krishnamoorthy, Suganya Kannan, and Murugan Marudhamuthu. "Bioactive potential of secondary metabolites derived from medicinal plant endophytes." *Egyptian Journal of Basic and Applied Sciences* 5, no. 4 (2018): 303-312.
Park Y. H., Kim Y., Mishra R. C., Bae H. Fungal endophytes inhabiting mountain-cultivated ginseng (*Panax ginseng Meyer*): Diversity and biocontrol activity against ginseng pathogens. *Sci Rep* 7 (2015):16221

Passari, Ajit Kumar, Vineet Kumar Mishra, Garima Singh, Pratibha Singh, Brijesh Kumar, Vijai Kumar Gupta, Rupak Kumar Sarma, Ratul Saikia, and Bhim Pratap Singh. "Insights into the functionality of endophytic actinobacteria with a focus on their biosynthetic potential and secondary metabolites production." *Scientific reports* 7, no. 1 (2017): 1-17.

Pocasangre L. E. "Biological enhancement of banana tissue culture plantlets with endophytic fungi for the control of the burrowing nematode *Radopholus similis* and the Panama disease (*Fusarium oxysporum f. sp. cubense*)." Ph.D. thesis, University of Bonn, Germany, (2000), p 107.

Robl, Diogo, Priscila da Silva Delabona, Patrícia dos Santos Costa, Deise Juliana da Silva Lima, Sarita Candida Rabelo, Ida Chapaval Pimentel, Fernanda Büchli, Fabio Marcio Squina, Gabriel Padilla, and Jose Geraldo da Cruz Pradella. "Xylanase production by endophytic *Aspergillus niger* using pentose-rich hydrothermal liquor from sugarcane bagasse." *Biocatalysis and Biotransformation* 33, no. 3 (2015): 175-187.

Rodriguez R. J., White J. F., Arnold A. E., Redman R. S. "Fungal endophytes: diversity and functional roles." *New Pathol* 182 (2009):314–330.

Rodriguez, R. J., J. F. White Jr, A. E. Arnold, and A. RS and Redman. "Fungal endophytes: diversity and functional roles." *New phytologist* 182, no. 2 (2009): 314-330.

Roy, Sudipta, Hiran Kanti Santra, and Debdulal Banerjee. "Diversity and biotechnological potential of culturable rhizospheric Actinomicrobiota." In *Advances in Plant Microbiome and Sustainable Agriculture*, pp. 159-187. Springer, Singapore, 2020.

Saad, Mona M. G., Rehab Y. Ghareeb, and Ali A. Saeed. "The potential of endophytic fungi as bio-control agents against the cotton leafworm, Spodoptera littoralis (Boisd.) (Lepidoptera: Noctuidae)." *Egyptian Journal of Biological Pest Control* 29, no. 1 (2019): 1-7.

Santoyo, Gustavo, Gabriel Moreno-Hagelsieb, Ma del Carmen Orozco-Mosqueda, and Bernard R. Glick. "Plant growth-promoting bacterial endophytes." *Microbiological research* 183 (2016): 92-99.

Santra, Hiran Kanti, and Debdulal Banerjee. "Extremophilic Microbes and their Extremozymes for Industry and Allied Sectors." In *Microbiomes of Extreme Environments*, pp. 1-42. CRC Press, 2021a.

Santra, Hiran Kanti, and Debdulal Banerjee. "Fungal endophytes: A source for biological control agents." In *Agriculturally Important Fungi for Sustainable Agriculture*, pp. 181-216. Springer, Cham, 2020b.

Santra, Hiran Kanti, and Debdulal Banerjee. "Microbial Exopolysaccharides: Structure and Therapeutic Properties." In *Microbial Polymers*, pp. 375-420. Springer, Singapore, 2021b.

Santra, Hiran Kanti, and Debdulal Banerjee. "Natural products as fungicide and their role in crop protection." *Natural Bioactive Products in Sustainable Agriculture* (2020a): 131.

Saxena S., Meshram V., Kapoor N. "*Muscodor darjeelingensis*, a new endophytic fungus of *Cinnamomum camphora* collected from northeastern Himalayas." *Sydowia* 66, no. 1 (2014):55–67.

Saxena S., Meshram V., Kapoor N. "*Muscodor tigerii sp. nov.*-volatile antibiotic producing endophytic fungus from the Northeastern Himalayas." *Ann Microbiol* 65, no. 1 (2015):47–57.

Shi, Yingwu, Kai Lou, and Chun Li. "Growth promotion effects of the endophyte *Acinetobacter johnsonii strain 3-1* on sugar beet." *Symbiosis* 54, no. 3 (2011): 159-166.

Shubha, J., and C. Srinivas. "Diversity and extracellular enzymes of endophytic fungi associated with *Cymbidium aloifolium* L." *African Journal of Biotechnology* 16, no. 48 (2017): 2248-2258.

Siri-udom S., Suwannarach N., Lumyong S. "Existence of *Muscodor vitigenus*, *M. equiseti* and *M. heveae sp. nov.* in leaves of the rubber tree (*Hevea brasiliensis Müll. Arg.*), and their biocontrol potential." *Ann Microbiol* 66 (2016):437–448.

Soleimani, Mohsen, Majid Afyuni, Mohammad A. Hajabbasi, Farshid Nourbakhsh, Mohammad R. Sabzalian, and Jan H. Christensen. "Phytoremediation of an aged petroleum contaminated soil using endophyte infected and non-infected grasses." *Chemosphere* 81, no. 9 (2010): 1084-1090.

Stierle, Andrea, Gary Strobel, and Donald Stierle. "Taxol and taxane production by *Taxomyces andreanae*, an endophytic fungus of Pacific yew." *Science* 260, no. 5105 (1993): 214-216.

Stinson A. M., Zidack N. K., Strobel G. A., Jacobsen B. J. "Mycofumigation with *Muscodor albus* and *Muscodor roseus* for control of seedling diseases of sugar beet and *Verticillium* Wilt of eggplant." *Plant Disease* 87 (2003):1349–1354.

Strobel G. A., Daisy B. "Bioprospecting for microbial endophytes and their natural products." *Microbiol Mol Biol Rev* 67 (2003):491–502.

Strobel G. A., Kluck K., Hess W. M., Sears J., Ezra D., Vargas P. N. "*Muscodor albus* E-6, an endophyte of *Guazuma ulmifolia* making volatile antibiotics: isolation, characterization and experimental establishment in the host plant." *Microbiology* 153 (2007):2613–2620.

Suwannarach N., Bussaban B., Hyde K. D., Lumyong S. "*Muscodor cinnamomi*, a new endophytic species from *Cinnamomum bejolghota*." *Mycotaxon* 15, no. 114 (2010):15–23.

Suwannarach N., Bussaban B., Nuangmek W., Pithakpol W., Jirawattanakul B., Matsui K., Lumyong S." *J Sci Food Agric* 96, no. 1 (2015):339–345.

Suwannarach N., Kumla J., Bussaban B., Hyde K. D., Matsui K., Lumyong S. "Molecular and morphological evidence support four new species in the genus *Muscodor* from northern Thailand." *Annals of Microbiology* 63 (2013):1341–1351.

Suwannarach N., Kumla J., Bussaban B., Lumyong S. Biocontrol of Rhizoctonia solani AG-2, the causal agent of damping-off by *Muscodor cinnamomi CMU-Cib 461. World J Microbiol Biotechnol* 28 (2012):3171–3177.

Taghavi, Safiyh, Craig Garafola, Sébastien Monchy, Lee Newman, Adam Hoffman, Nele Weyens, Tanja Barac, Jaco Vangronsveld, and Daniel van der Lelie. "Genome survey and characterization of endophytic bacteria exhibiting a beneficial effect on growth and development of poplar trees." *Applied and environmental microbiology* 75, no. 3 (2009): 748-757.

Taghavi, Safiyh, Tanja Barac, Bill Greenberg, Brigitte Borremans, Jaco Vangronsveld, and Daniel van der Lelie. "Horizontal gene transfer to endogenous endophytic bacteria from poplar improves phytoremediation of toluene." *Applied and environmental microbiology* 71, no. 12 (2005): 8500-8505.

Tiwari, Pragya, Hanhong Bae. "Horizontal gene transfer and Endophytes: An implication for the acquisition of novel traits." *Plants* 9(2020): 305.

Tiwari, Pragya, Yashdeep Srivastava, and Hanhong Bae. "Trends of pharmaceutical design of Endophytes as anti-infective." *Current Topics in Medicinal Chemistry* 21, no. 17 (2021a): 1572-1586.

Tiwari, Pragya, Yashdeep Srivastava, Y, Mangalam Bajpai, M, Abhishek Sharma. "Bioactive metabolites from natural sources: Prospects and significance in drug discovery and research." *Bioingene PSJ*, no. 1, (2021b): 1-14.

Tiwari, Pragya, Hanhong Bae. "Endophytic fungi: Key insights, emerging prospects, and challenges in natural product drug discovery." *Microorganisms* 10 (2022): 360. https://doi.org/10.3390/microorganisms10020360.

Toghueo, Rufin Marie Kouipou. "Anti-leishmanial and anti-inflammatory agents from endophytes: a review." *Natural products and bioprospecting* 9, no. 5 (2019): 311-328.

Tran, Nga T., Andrew K. Miles, Ralf G. Dietzgen, and André Drenth. "*Phyllosticta capitalensis* and *P. paracapitalensis* are endophytic fungi that show potential to inhibit pathogenic P. citricarpa on citrus." *Australasian Plant Pathology* 48, no. 3 (2019): 281-296.

Turbat, A., Rakk, D., Vigneshwari, A., Kocsubé, S., Thu, H., Szepesi, Á., Bakacsy, L., D. Škrbić, B., Jigjiddorj, E. A., Vágvölgyi, C., & Szekeres, A. (2020). "Characterization of the plant growth-promoting activities of endophytic fungi isolated from *Sophora flavescens*." *Microorganisms* 8, no. 5 (2020): 683.

Uzma, Fazilath, Narasimha Murthy Konappa, and Srinivas Chowdappa. "Diversity and extracellular enzyme activities of fungal endophytes isolated from medicinal plants of Western Ghats, Karnataka." *Egyptian journal of basic and applied sciences* 3, no. 4 (2016): 335-342.

Van Aken, Benoit, Caroline M. Peres, Sharon Lafferty Doty, Jong Moon Yoon, and Jerald L. Schnoor. "Methylobacterium populi sp. nov., a novel aerobic, pink-pigmented, facultatively methylotrophic, methane-utilizing bacterium isolated from poplar trees (Populus deltoides× nigra DN34)." *International journal of systematic and evolutionary microbiology* 54, no. 4 (2004): 1191-1196.

Wang Y., Li Y., Li S., Li Q., Fan W., Kiatoukosin L., Chen J. "Extracellular polysaccharides of endophytic fungus *Alternaria tenuissima* F1 from *Angelica sinensis*: Production conditions, purification, and antioxidant properties." *Int J Biol Macromol* 133 (2019):172–183.

Wang, J. W., J. H. Wu, W. Y. Huang, and R. X. Tan. "Laccase production by Monotospora sp., an endophytic fungus in Cynodon dactylon." *Bioresource Technology* 97, no. 5 (2006): 786-789.

Wang, Lixiang, Chunchun Li, Juan Shi, Chengcheng Li, Jiale Li, Lili Ren, and Youqing Luo. "Incidental fungi in host trees disrupt the development of *Sirex noctilio* (Hymenoptera: Siricidae) symbiotic fungus and larvae." *Journal of economic entomology* 113, no. 2 (2020): 832-838.

Wang, Shan-Shan, Jia-Meng Liu, Jing Sun, Yu-Feng Sun, Jia-Ni Liu, Ning Jia, Bei Fan, and Xiao-Feng Dai. "Diversity of culture-independent bacteria and antimicrobial activity of culturable endophytic bacteria isolated from different *Dendrobium* stems." *Scientific reports* 9, no. 1 (2019): 1-12.

Wisniewski-Dyé, F., Borziak, K., Khalsa-Moyers, G., Alexandre, G., Sukharnikov, L. O., Wuichet, K., Hurst, G. B., McDonald, W. H., Robertson, J. S., Barbe, V., Calteau, A., Rouy, Z., Mangenot, S., Prigent-Combaret, C., Normand, P., Boyer, M., Siguier, P., Dessaux, Y., Elmerich, C., Zhulin, I. B. "A*zospirillum* genomes reveal transition of bacteria from aquatic to terrestrial environments." *PLoS genetics* 7, no. 12 (2011): e1002430.

Yingwu S., Kai L, Chun L. "Growth promotion effects of the endophyte *Acinetobacter johnsonii strain 3-1* on sugar beet." *Symbiosis* 54 (2011):159–166.

Falade, Ayodeji O., Kayode E. Adewole, and Temitope C. Ekundayo. "Therapeutic potentials of endophytes for healthcare sustainability." *Egyptian Journal of Basic and Applied Sciences* 8, no. 1 (2021): 117-135.

About the Editor

Dr. Pragya Tiwari is working as Research Professor in the Department of Biotechnology, Yeungnam University, Republic of Korea. She possesses research/teaching experience of more than 8 years in national/international institutions in the field of plant sciences, with research experience in areas of phytomolecules from medicinal plants, plant-associated endophytes, plant-microbe interactions and computational biology in drug designing. She has been a recipient of various research fellowships from CSIR and ICAR and gained scientific recognitions for her work in inter-disciplinary areas of plant sciences. She has been an active participant/presenter in more than 30 symposiums/conferences on international/national platforms and holds life membership of scientific and professional societies including Korean Society of Biotechnology and Bioengineering and The Society of Biological Chemist (SBC), India. Her research work has gained broad interest through highly cited publications, book chapters and invited lectures. Presently, she is working as Research Professor in the Department of Biotechnology, Yeungnam University, Republic of Korea. Her area of research comprises of plant-endophyte interactions in *Panax ginseng* and the significance of bioactive metabolite (ginsenoside) in promoting plant growth and as bio-stimulants in sustainable agriculture.

List of Contributors

A. Sankarnarayanan

Sri Sathya Sai University for Human Excellence, Navanihal, Okali Post, Kamalapur, Kalaburagi, Karnataka – 585313, India. Email-drsankarkamal@gmail.com

Abhay Tiwari

Centre for Rural Development and Technology, Indian Institute of Technology Delhi, Hauz Khas, New Delhi, 110016. Email-tiwariabhay.19@gmail.com

Abhishek Sharma

Department of Biotechnology and Bioengineering, Institute of Advanced Research, Koba Institutional Area, Gandhinagar – 382007, India. Email-abhi19ind@gmail.com

Anjani Devi Chintagunta

Department of Biotechnology, Vignan's Foundation for Science, Technology & Research, Vadlamudi, Andhra Pradesh, India. Email-sumapriya.ch@gmail.com

Ankit Singh

Centre for Rural Development and Technology, Indian Institute of Technology Delhi, Hauz Khas, New Delhi, 110016. Email-singhankit1991@gmail.com

Anu Dalal

Department of Chemistry, Indian Institute of Technology Delhi, Hauz Khas, New Delhi, 110016. Email- anudalal8@gmail.com

Anuj Rana
Department of Microbiology, CCS Haryana Agriculture University, Hisar, Haryana-125004. Email- anuj23rana@gmail.com

Anupam Maurya
Pharmacopoeia Commission for Indian Medicine and Homoeopathy (PCIM&H), (Ministry of AYUSH, Govt. of India), PLIM Campus, Kamla Nehru Nagar, Kavi Nagar Post, Ghaziabad-201002, Uttar Pradesh, India. Email- manupamk@gmail.com

Ashutosh Gautam
Spices Board, Regional Office, Srinagar, 190 008, (J&K UT), India. Email- ashutoshgautam.sb@gov.in

Avneesh Kumar
Department of Botany, Akal University, Talwandi Sabo, Bathinda, Punjab-151302. Email- avnishenator@gmail.com

Avnish Kumar
Department of Biotechnology, School of Life Sciences, Khandari Campus, Dr. Bhimrao Ambedkar University, Agra-282002, Uttar Pradesh, India. Email- akbiotech81@gmail.com

Bhawana Mishra
Department of Environmental Sciences, Central University of Himachal Pradesh, Dharamshala, Himachal Pradesh, India. Email- mishrabhawna990@gmail.com

Debdulal Banerjee
Microbiology and Microbial Biotechnology Laboratory, Department of Botany and Forestry, Vidyasagar University, Midnapore-721102, West Bengal, India. Email- db@mail.vidyasagar.ac.in

Gourav Choudhir
Centre for Rural Development and Technology, Indian Institute of Technology Delhi, Hauz Khas, New Delhi, 110016. Email- gouravbotany@gmail.com

Hiran Kanti Santra
Microbiology and Microbial Biotechnology Laboratory, Department of Botany and Forestry, Vidyasagar University, Midnapore-721102, West Bengal, India. Email- hiran.kanti@gmail.com

H. I. Atagana
Department of Environmental Sciences Institute for Science and Technology Education, University of South Africa, Pretoria, South Africa 0003. Email- eanyasro@unisa.ac.za

Janki S. Patel
Department of Biological Sciences, P. D. Patel Institute of Applied Sciences, Charotar University of Science and Technology, Changa- 388421, Gujarat, India. Email- jankipatel.bio@charusat.ac.in

Jen-Tsung Chen
Department of Life Sciences, National University of Kaohsiung, Kaohsiung 811, Taiwan. Email- jentsung@nuk.edu.tw

Jyotirmoyee Das
Global Product Compliances, GPC Regulatory India Private Limited, Civil Lines, Nagpur, India. Email- jyotirmoyee18das.jd@gmail.com

K. Pallavi
Vignan Pharmacy College, Vadlamudi, Guntur, Andhra Pradesh, India. Email- pallavi@vignanpharmacycollege.in

Mangalam Bajpai
Global Product Compliances, GPC Regulatory India Private Limited, Civil Lines, Nagpur, India. Email- mangalambajpai084@gmail.com

Muhammad Adil
Pharmacology and Toxicology Section, University of Veterinary and Animal Sciences, Lahore, Jhang Campus, Pakistan. Email address: muhammad.adil@uvas.edu.pk

Muktesh Chandra
SL Education Institute, Moradabad, MJP Rohilkhand University, Bareilly, UP, India. Email- mchandra33@gmail.com

Nagaraj Basavegowda
Department of Biotechnology, Yeungnam University, Gyeongsan, Gyeongbuk 38541, Republic of Korea. Email- drbn5275@gmail.com

N. S. Sampath Kumar
Department of Biotechnology, Vignan's Foundation for Science, Technology & Research, Vadlamudi, Andhra Pradesh, India. Email- nssk84@gmail.com

N. Satya Vijaya Kumar
Centre of Excellence (CoExAMMPC), Vignan's Foundation for Science, Technology & Research, Vadlamudi, Andhra Pradesh, India. Email- satyavijayakumar@gmail.com

Pragya Tiwari
Department of Biotechnology, Yeungnam University, Gyeongsan, Gyeongbuk 38541, Republic of Korea. Email- pragyatiwari@ynu.ac.kr

Pravin Kumar
Department of Computer Science and Engineering, Akal University, Talwandi Sabo, Bathinda, Punjab-151302. Email- pravinkum10@gmail.com

R. O. Anyasi
Department of Environmental Sciences Institute for Science and Technology Education, University of South Africa, Pretoria, South Africa 0003. Email- ghnlwl@yahoo.com

Siddhi Darji
C. G. Bhakta Institute of Biotechnology, Uka Tarsadia University, Gopal-Vidyanagar, Maliba Campus, Surat-394350, India. Email- siddhiadarji@gmail.com

Simranjit Kaur
Department of Botany, Akal University, Talwandi Sabo, Bathinda, Punjab-151302. Email- simranbotany@gmail.com

Subir Kumar Bose
Department of Biotechnology and Microbiology, Meerut Institute of Engineering and Technology, N.H. 58, Delhi-Roorkee Highway, Baghpat Road Bypass Crossing, Meerut-250005, Uttar Pradesh, India. Email- bbausubir01@gmail.com

V. Gopi
Microbiology Section, Pharmacopoeia Commission for Indian Medicine and Homoeopathy (PCIM&H), Ministry of Ayush, Ghaziabad, India. Email- gopimicro2010@gmail.com

Vyoma Mistry
C. G. Bhakta Institute of Biotechnology, Uka Tarsadia University, Gopal-Vidyanagar, Maliba Campus, Surat-394350, India. Email- vyoma_mistry@yahoo.in

Index

P

R

S

T

Y